Understanding Computational Bayesian Statistics

WILEY SERIES IN COMPUTATIONAL STATISTICS

Wiley Series in Computational Statistics is comprised of practical guides and cutting edge research books on new developments in computational statistics. It features quality authors with a strong applications focus. The texts in the series provide detailed coverage of statistical concepts, methods and case studies in areas at the interface of statistics, computing, and numerics.

With sound motivation and a wealth of practical examples, the books show in concrete terms how to select and to use appropriate ranges of statistical computing techniques in particular fields of study. Readers are assumed to have a basic understanding of introductory terminology.

The series concentrates on applications of computational methods in statistics to fields of bioinformatics, genomics, epidemiology, business, engineering, finance and applied statistics.

A complete list of titles in this series appears at the end of the volume.

Understanding Computational Bayesian Statistics

William M. Bolstad
University of Waikato
Department of Statistics
Hamilton, New Zealand

A John Wiley & Sons, Inc., Publication

Published by John Wiley & Sons, Inc., Hoboken, New Jersey.
Published simultaneously in Canada.

For general information on our other products and services or for technical support, please contact our Customer Care Department within the United States at (800) 762-2974, outside the United States at (317) 572-3993 or fax (317) 572-4002.

Wiley also publishes its books in a variety of electronic formats. Some content that appears in print may not be available in electronic format. For information about Wiley products, visit our web site at www.wiley.com.

Library of Congress Cataloging-in-Publication Data:

Bolstad, William M., 1943–
Understanding Computational Bayesian statistics / William M. Bolstad.
p. cm.
Includes bibliographical references and index.
ISBN 978-0-470-04609-8 (cloth)
1. Bayesian statistical decision theory—Data processing. I. Title.
QA279.5.B649 2010
519.5'42--dc22 2009025219

Printed in the United States of America.

10 9 8 7 6 5 4 3 2 1

This book is dedicated to

Sylvie,
Ben, Rachel,
Mary, and Elizabeth

Contents

Preface

In theory, Bayesian statistics is very simple. The posterior is proportional to the prior times likelihood. This gives the shape of the posterior, but it is not a density so it cannot be used for inference. The exact scale factor needed to make this a density can be found only in a few special cases. For other cases, the scale factor requires a numerical integration, which may be difficult when there are multiple parameters. So in practice, Bayesian statistics is more difficult, and this has held back its use for applied statistical problems.

Computational Bayesian statistics has changed all this. It is based on the big idea that statistical inferences can be based on a random sample drawn from the posterior. The algorithms that are used allow us to draw samples from the exact posterior even when we only know its shape and we do not know the scale factor needed to make it an exact density. These algorithms include direct methods where a random sample drawn from an easily sampled distribution is reshaped by only accepting some candidate values into the final sample. More sophisticated algorithms are based on setting up a Markov chain that has the posterior as its long-run distribution. When the chain is allowed to run a sufficiently long time, a draw from the chain can be considered a random draw from the target (posterior) distribution. These algorithms are particularly well suited for complicated models with many parameters. This is revolutionizing applied statistics. Now applied statistics based on these computational Bayesian methods can be easily accomplished in practice.

Features of the text

This text grew out of a course I developed at Waikato University. My goal for that course and this text is to bring these exciting developments to upper-level undergraduate and first-year graduate students of statistics. This text introduces this big idea to students in such a way that they can develop a strategy for making statistical inferences in this way. This requires an understanding of the pitfalls that can arise when using this approach, what can be done to avoid them, and how to recognize them if they are occurring. The practitioner has many choices to make in using this approach. Poor choices will lead to incorrect inferences. Sensible choices will lead to satisfactory inferences in an efficient manner.

This text follows a step-by-step development. In Chapter 1 we learn about the similarities and differences between the Bayesian and the likelihood approaches to statistics. This is important because when a flat prior is used, the posterior has the same shape as the likelihood function, yet they have different methods for inferences. The Bayesian approach allows us to interpret the posterior as a probability density and it is this interpretation that leads to the advantages of this approach. In Chapter 2 we examine direct approaches to drawing a random sample from the posterior even when we only know its shape by reshaping a random sample drawn from another easily sampled density by only accepting some of the candidates into the final sample. These methods are satisfactory for models with only a few parameters provided the candidate density has heavier tails than the target. For models with many parameters direct methods become very inefficient. In these models, direct methods still may have a role as a small step in a larger Markov chain Monte carlo algorithm. In Chapter 3 we show how statistical inferences can be made from a random sample from the posterior in a completely analogous way to the corresponding inferences taken from a numerically calculated posterior. In Chapter 4 we study the distributions from the one-dimensional exponential family. When the observations come from a member of this family, and the prior is from the conjugate family, then the posterior will be another member of the conjugate family. It can easily be found by simple updating rules. We also look at the normal distribution with unknown mean and variance, which is a member of two-dimensional exponential family, and the multivariate normal and normal regression models. These exponential family cases are the only cases where the formula for the posterior can be found analytically. Before the development of computing, Bayesian statistics could only be done in practice in these few cases. We will use these as steps in a larger model. In Chapter 5 we introduce Markov chains. An understanding of Markov chains and their long-run behavior is needed before we study the more advanced algorithms in the book. Things that can happen in a Markov chain can also happen in a Markov chain Monte Carlo (MCMC) model. This chapter finishes with the Metropolis algorithm. This algorithm allows us take a Markov chain and find a new Markov chain from it that will have the target (posterior) as its long-run distribution. In Chapter 6 we introduce the Metropolis-Hastings algorithm and show that how it performs depends on whether we use a random-walk or independent candidate density. We show how, in a multivariate case, we can either draw all the parameters at once, or blockwise, and that the Gibbs sampler is a special case of blockwise Metropolis-Hastings. In Chapter 7 we investigate how the mixing properties of the chain depend on the choice of the candidate density. We show how to find a heavy-tailed candidate density starting from the maximum likelihood estimator and matched curvature covariance matrix. We show that this will lead to a very efficient MCMC process. We investigate several methods for deciding on burn-in time and thinning required to get an approximately random sample from the posterior density from the MCMC output as the basis for inference. In Chapter 8 we apply this to the logistic regression model. This is a generalized linear model, and we find the maximum likelihood estimator and matched curvature covariance matrix using iteratively reweighted least squares. In the cases where we have a normal prior, we can find the approximate normal posterior by the simple updating rules we studied in Chapter 4. We use the *Student's t* equivalent as the heavy-tailed independent candidate density for the Metropolis-Hastings algorithm. After burn-in, a draw from the Markov chain will be random draw from the exact posterior, not the *normal* approximation. We discuss how to determine priors for this model. We also investigate strategies to remove variables from the model to get a better prediction model. In Chapter 9, we apply these same ideas to the Poisson regression model. The Proportional hazards model turns out to have the same likelihood as a Poisson, so these ideas apply here as well. In Chapter 10 we investigate the Gibbs sampling algorithm. We demonstrate it on the $normal(\mu, \sigma^2)$ model where both parameters are unknown for both the independent prior case and the joint conjugate prior case. We see the Gibbs sampler is particularly well suited when we have a hierarchical model. In that case, we can draw a directed acyclic graph showing the dependency structure of the parameters. The conditional distribution of each block of parameters given all other blocks has a particularly easy form. In Chapter 11, we discus methods for speeding up convergence in Gibbs sampling. We also direct the reader to more advanced topics that are beyond the scope of the text.

Software

I have developed Minitab macros that perform the computational methods shown in the text. My colleague, Dr. James Curran has written corresponding R-Functions. These may be downloaded from the following website.

http://www.stats.waikato.ac.nz/publications/bolstad/UnderstandingComputationalBayesianStatistics/

Acknowledgments

I would like to acknowledge the help I have had from many people. First, my students over the past three years, whose enthusiasm with the early drafts encouraged me to continue writing. My colleague, Dr. James Curran, who wrote the R-functions and wrote Appendix B on how to implement them, has made a major contribution to this book. I want to thank Dr. Gerry Devlin, the Clinical director of Cardiology at the Waikato Hospital for letting me use the data from the Waikato District Health

Board Cardiac Survival Study, Dr. Neil Swanson and Gaelle Dutu for discussing this dataset with me, and my student Yihong Zhang, who assisted me on this study. I also want to thank Dr. David Fergusson from the Dept. of Psychological Medicine at the Christchurch School of Medicine and Health Sciences for letting me use the circumcision data from the longitudinal study of a birth cohort. I want to thank my colleagues at the University of Waikato, Dr Murray Jorgensen, Dr. Judi McWhirter, Dr. Lyn Hunt, Dr. Kevin Broughan, and my former student Dr. Jason Catchpole, who all proofread parts of the manuscript. I appreciated their helpful comments, and any errors that remain are solely my responsibility. I would like to thank Cathy Akritas at Minitab for her help in improving my Minitab macros. I would like to thank Steve Quigley, Jackie Palmieri, Melissa Yanuzzi, and the team at John Wiley & Sons for their support, as well as Amy Hendrickson of TEXnology, Inc. for help with LaTex.

Finally, last but not least, I wish to thank my wife Sylvie for her constant love and support.

WILLIAM M. "BILL" BOLSTAD

Hamilton, New Zealand

1

Introduction to Bayesian Statistics

In the last few years the use of Bayesian methods in the practice of applied statistics has greatly increased. In this book we will show how the development of computational Bayesian statistics is the key to this major change in statistics. For most of the twentieth century, frequentist statistical methods dominated the practice of applied statistics. This is despite the fact that statisticians have long known that the Bayesian approach to statistics offered clear cut advantages over the frequentist approach. We will see that Bayesian solutions are easy in theory, but were difficult in practice. It is easy to find a formula giving the shape of the posterior. It is often more difficult to find the formula of the exact posterior density. Computational Bayesian statistics changed all this. These methods use algorithms to draw samples from the incompletely known posterior and use these random samples as the basis for inference. In Section 1.1 we will look briefly at the the ideas of the frequentist approach to statistics. In Section 1.2 we will introduce the ideas of Bayesian statistics. In Section 1.3 we show the similarities and differences between the likelihood approach to inference and Bayesian inference. We will see that the different interpretations of the parameters and probabilities lead to the advantages of Bayesian statistics.

1.1 THE FREQUENTIST APPROACH TO STATISTICS

In frequentist statistics, the parameter is considered a fixed but unknown value. The sample space is the set of all possible observation values. Probability is interpreted as long-run relative frequency over all values in the sample space given the unknown parameter. The performance of any statistical procedure is determined by averaging over the sample space. This can be done prior to the experiment and does not depend on the data.

There were two main sources of frequentist ideas. R. A. Fisher developed a theory of statistical inference based on the likelihood function. It has the same formula as the joint density of the sample, however, the observations are held fixed at the values that occurred and the parameter(s) are allowed to vary over all possible values. He reduced the complexity of the data through the use of sufficient statistics which contain all the relevant information about the parameter(s). He developed the theory of maximum likelihood estimators (MLE) and found their asymptotic distributions. He measured the efficiency of an estimator using the Fisher information, which gives the amount of information available in a single observation. His theory dealt with nuisance parameters by conditioning on an ancillary statistic when one is available. Other topics associated with him include analysis of variance, randomization, significance tests, permutation tests, and fiducial intervals. Fisher himself was a scientist as well as a statistician, making great contributions to genetics as well as to the design of experiments and statistical inference. As a scientist, his views on inference are in tune with science. Occam's razor requires that the simplest explanation (chance) must be ruled out before an alternative explanation is sought. Significance testing where implausibility of the chance model is required before accepting the alternative closely matches this view.

Jerzy Neyman and Egon Pearson developed decision theory, and embedded statistical inference in it. Their theory was essentially deductive, unlike Fisher's. They would determine criteria, and try to find the optimum solution in the allowed class. If necessary, they would restrict the class until they could find a solution. For instance, in estimation, they would decide on a criterion such as minimizing squared error. Finding that no uniformly minimum squared error estimator exists, they would then restrict the allowed class of estimators to unbiased ones, and find uniformly minimum variance unbiased estimators (UMVUE). Wald extended these ideas by defining a loss function, and then defining the risk as the expected value of the loss function averaged over the sample space. He then defined as inadmissible any decision rule that is dominated by another for all values of the parameter. Any rule that is not inadmissible is admissible. Unexpectedly, since he was using frequentist criteria, he found that the class of admissible rules is the class of Bayesian rules. Other topics in this school include confidence intervals, uniformly most powerful tests of hypothesis, uniformly most powerful unbiased tests, and James-Stein estimation.

The disputes Fisher had with the Neyman are legendary (Savage, 1976). Fisher strongly opposed the submerging of inference into decision theory and Neyman's denial that inference uses inductive logic. His specific criticisms about the Neyman-Pearson methods include:

- Unbiased estimators are not invariant under one-to-one reparameterizations.
- Unbiased estimators are not compatible with the likelihood principle.
- Unbiased estimates are not efficient. He scathingly criticized this waste of information as equivalent to throwing away observations.

Nevertheless, what currently passes for frequentist parametric statistics includes a collection of techniques, concepts, and methods from each of these two schools, despite the disagreements between the founders. Perhaps this is because, for the very important cases of the normal distribution and the binomial distribution, the MLE and the UMVUE coincided. Efron (1986) suggested that the emotionally loaded terms (unbiased, most powerful, admissible, etc.) contributed by Neyman, Pearson, and Wald reinforced the view that inference should be based on likelihood and this reinforced the frequentist dominance. Frequentist methods work well in the situations for which they were developed, namely for exponential families where there are minimal sufficient statistics. Nevertheless, they have fundamental drawbacks including:

- Frequentist statistics have problems dealing with nuisance parameters, unless an ancillary statistic exists.
- Frequentist statistics gives prior measures of precision, calculated by sample space averaging. These may have no relevance in the post-data setting.

Inference based on the likelihood function using Fisher's ideas is essentially constructive. That means algorithms can be found to construct the solutions. Efron (1986) refers to the MLE as the "original jackknife" because it is a tool that can easily be adapted to many situations. The maximum likelihood estimator is invariant under a one-to-one reparameterization. Maximum likelihood estimators are compatible with the likelihood principle. Frequentist inference based on the likelihood function has some similarities with Bayesian inference as well as some differences. These similarities and differences will be explored in Section 3.3.

1.2 THE BAYESIAN APPROACH TO STATISTICS

Bayesian statistics is based on the theorem first discovered by Reverend Thomas Bayes and published after his death in the paper *An Essay Towards Solving a Problem in the Doctrine of Chances* by his friend Richard Price in *Philosophical Transactions of the Royal Society*. Bayes' theorem is a very clever restatement of the conditional probability formula. It gives a method for updating the probabilities of unobserved events, given that another related event has occurred. This means that we have a prior probability for the unobserved event, and we update this to get its posterior probability, given the occurrence of the related event. In Bayesian statistics, Bayes' theorem is used as the basis for inference about the unknown parameters of a statistical distribution. Key ideas forming the basis of this approach include:

- Since we are uncertain about the true values of the parameters, in Bayesian statistics we will consider them to be random variables. This contrasts with the frequentist idea that the parameters are fixed but unknown constants. Bayes' theorem is an updating algorithm, so we must have a prior probability distribution that measures how plausible we consider each possible parameter value before looking at the data. Our prior distribution must be subjective, because somebody else can have his/her own prior belief about the unknown values of the parameters.

- Any probability statement about the parameters must be interpreted as "degree of belief."

- We will use the rules of probability directly to make inferences about the parameters, given the observed data. Bayes' theorem gives our posterior distribution, which measures how plausible we consider each possible value after observing the data.

- Bayes' theorem combines the two sources of information about the unknown parameter value: the prior density and the observed data. The prior density gives our relative belief weights of every possible parameter value before we observe the data. The *likelihood function* gives the relative weights to every possible parameter value that comes from the observed data. Bayes' theorem combines these into the posterior density, which gives our relative belief weights of the parameter value after observing the data.

Bayes' theorem is the only consistent way to modify our belief about the parameters given the data that actually occurred. A Bayesian inference depends only on the data that occurred, not on the data that could have occurred but did not. Thus, Bayesian inference is consistent with the *likelihood principle*, which states that if two outcomes have proportional likelihoods, then the inferences based on the two outcomes should be identical. For a discussion of the likelihood principle see Bernardo and Smith (1994) or Pawitan (2001). In the next section we compare Bayesian inference with likelihood inference, a frequentist method of inference that is based solely on the likelihood function. As its name implies, it also satisfies the likelihood principle.

A huge advantage of Bayesian statistics is that the posterior is always found by a single method: Bayes' theorem. Bayes' theorem combines the information about the parameters from our prior density with the information about the parameters from the observed data contained in the likelihood function into the posterior density. It summarizes our knowledge about the parameter given the data we observed.

Finding the posterior: easy in theory, hard in practice

Bayes' theorem is usually expressed very simply in the unscaled form, *posterior proprotional to prior times likelihood*:

$$g(\theta_1, \ldots, \theta_p | y_1, \ldots, y_n) \propto g(\theta_1, \ldots, \theta_p) \times f(y_1, \ldots, y_n | \theta_1, \ldots, \theta_p)\,. \qquad (1.1)$$

This formula does not give the posterior density $g(\theta_1, \ldots, \theta_p | y_1, \ldots, y_n)$ exactly, but it does give its shape. In other words, we can find where the modes are, and relative heights at any two locations. However, it cannot be used to find probabilities or to find moments since it is not a density. We can't use it for inferences. The actual posterior density is found by scaling it so it integrates to 1:

$$g(\theta_1, \ldots, \theta_p | y_1, \ldots, y_n) = \frac{g(\theta_1, \ldots, \theta_p) \times f(y_1, \ldots, y_n | \theta_1, \ldots, \theta_p)}{K} \qquad (1.2)$$

where the divisor needed to make this a density is

$$K = \int \cdots \int g(\theta_1, \ldots, \theta_p) \times f(y_1, \ldots, y_n | \theta_1, \ldots, \theta_p)\, d\theta_1 \ldots d\theta_p\,. \tag{1.3}$$

A closed form for the p-dimensional integral only exists for some particular cases.[1] For other cases the integration has to be done numerically. This may be very difficult, particularly when p, the number of parameters, is large. When this is true, we say there is a high dimensional parameter space.

Finding the posterior using Bayes' theorem is easy in theory. That is, we can easily find the unscaled posterior by Equation 1.1. This gives us all the information about the shape of the posterior. The exact posterior is found by scaling this to make it a density and is given in Equation 1.2. However, in practice, the integral given in Equation 1.3 can be very difficult to evaluate, even numerically. This is particularly difficult when the parameter space is high dimensional. Thus we cannot always find the divisor needed to get the exact posterior density. In general, the incompletely known posterior given by Equation 1.1 is all we have.

In Bayesian statistics we do not have to assume that the observations come from an easily analyzed distribution such as the normal. Also, we can use any shape prior density. The posterior would be found the same way. Only the details of evaluating the integral would be different.

Example 1 *A random sample $y_1, \ldots, y_n$ is drawn from a distribution having an unknown mean and known standard deviation. Usually, it is assumed the draws come from a normal (μ, σ^2) distribution. However, the statistician may think that for this data, the observation distribution is not normal, but from another distribution that is also symmetric and has heavier tails. For example, the statistician might decide to use the Laplace(a, b) distribution with the same mean μ and variance σ^2. The mean and variance of the Laplace(a, b) distribution are given by a and $2b^2$, respectively. The observation density is given by*

$$f(y|\mu, \sigma) = \frac{1}{\sqrt{2}\sigma} e^{-\frac{|y-\mu|}{\sigma/\sqrt{2}}}\,.$$

The posterior distribution is found by the same formula

$$g(\mu|y_1, \ldots, y_n) = \frac{g(\mu) \times f(y_1, \ldots, y_n|\mu)}{\int g(\mu) \times f(y_1, \ldots, y_n|\mu) d\mu}\,.$$

The only difference would be the details of the integration. In most cases, it would have to be done numerically.

[1] Where the observation distribution comes from an exponential family, and the prior comes from the family that is conjugate to the observation distribution.

1.3 COMPARING LIKELIHOOD AND BAYESIAN APPROACHES TO STATISTICS

In this section we graphically illustrate the similarities and differences between the likelihood and Bayesian approaches to inference; specifically, how a parameter is estimated using each of the approaches. We will see that:

1. The likelihood and the posterior density are found in a similar manner, by cutting a surface with the same vertical hyperplane. However, the surfaces used in the two approaches have different interpretations and in most cases they will have different shapes.

2. Even when the surfaces are the same (when flat priors are used) the estimators are chosen to satisfy different criteria.

3. The two approaches have different ways of dealing with nuisance parameters.

The observation(s) come from the observation density $f(y|\theta)$ where θ is the fixed parameter value. It gives the probability density over all possible observation values for the given value of the parameter. The parameter space, Θ, is the set of all possible parameter values. The parameter space ordinarily has the same dimension as the total number of parameters, p. The sample space, S, is the set of all possible values of the observation(s). The dimension of the sample space is the number of observations n. Many of the commonly used observation distributions come from the one-dimensional exponential family of distributions. When we are in the one-dimensional exponential family, the sample space may be reduced to a single dimension due to the single sufficient statistic.

The Inference Universe

We define the inference universe of the problem to be the Cartesian product of the parameter space and the sample space. It is the $p + n$ dimensional space where the first p dimensions are the parameter space, and the remaining n dimensions are the sample space. We do not ever observe the parameter, so the position in those coordinates is always unknown. However, we do observe the sample, so we know the last n coordinates.

We will let the dimensions be $p = 1$ and $n = 1$ for illustrative purposes. This is the case when we have a single parameter and a single observation (or we have a random sample of observations from a one-dimensional exponential family). The inference universe has two dimensions. The vertical dimension is the parameter space and is unobservable. The horizontal dimension is the sample space and is observable. We wish to make inference about where we are in the vertical dimension given that we know where we are in the horizontal dimension.

Let $f(y|\theta)$ be the observation density. For each value of the parameter θ, it gives the probability density of the observation y for that parameter value. Actually, this formula is a function of both the value of the observation and the parameter value. It

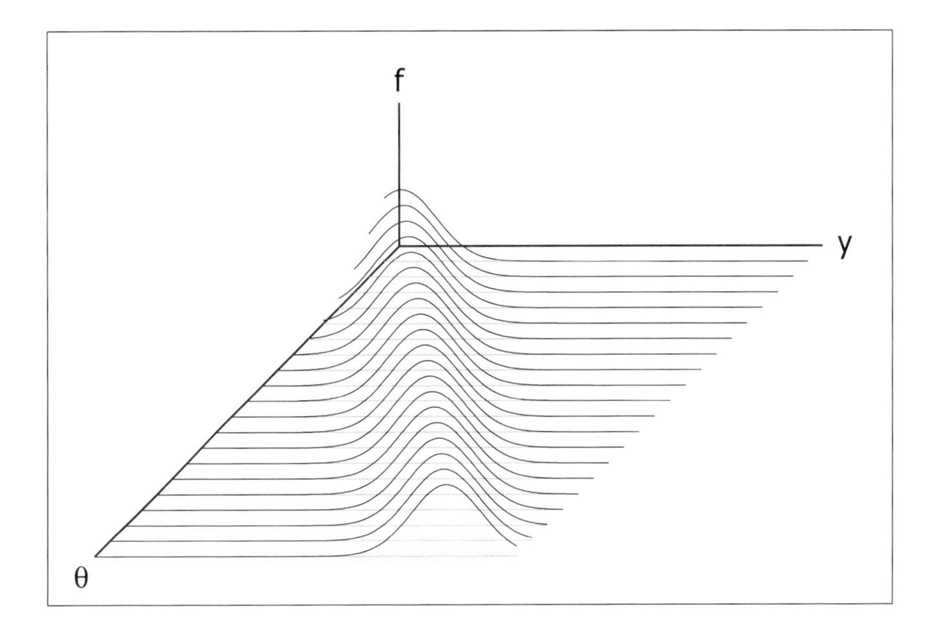

Figure 1.1 The observation density surface in 3D perspective.

is defined for all points in the inference universe, thus it forms a surface defined on the inference universe. It forms a probability density in the observation dimension for each particular value in the parameter dimension. However in general, it is not a probability distribution in the parameter dimension. Figure 1.1 shows the observation density surface in 3D perspective.

The likelihood function, first defined by R. A. Fisher (1922), has the same functional form as the observation density, only y is held at the observed value, and θ is allowed to vary over all possible values. Thus, it is a function of the parameter θ. It is found by cutting the observation density surface with a vertical plane parallel to the θ axis through the observed value. This is shown in Figure 1.2. Likelihood inference is based entirely on the likelihood function.

Maximum Likelihood Estimation

We are trying to choose an estimator (function of the observations) to represent the unknown value of the parameter. In likelihood inference, the likelihood function cannot be considered to be a probability density in general. Because of this, Fisher (1922) decided that the best way to estimate the parameter is to choose the parameter value that has the highest value of the likelihood function, i.e., its mode. This is the parameter value that gives the observed data the highest probability. He named this the *maximum likelihood estimator* (MLE). The mode will be invariant under any

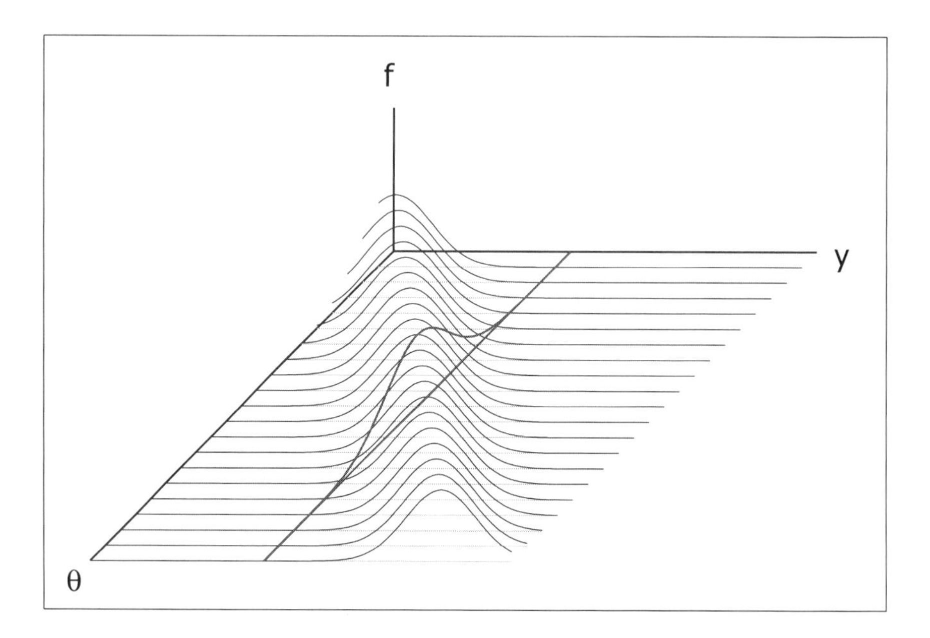

Figure 1.2 Observation density surface with the likelihood function shown in 3D perspective.

one-to-one transformation of the parameter space. Hence, the MLE will be invariant under any reparameterization of the problem.[2]

Bayesian Estimation

Bayesian estimation requires that we have a probability distribution defined on the parameter space before we look at the data. It is called the *prior* density because it gives our belief weights for each of the possible parameter values before we see the data. This requires that we allow a different interpretation of probability on the parameter space than on the sample space. It is measuring our belief, and thus is subjective. The probability on the sample space has the usual long-run relative frequency interpretation. The prior density of the parameter is shown with the observation density surface in Figure 1.3. The joint density of the parameter and the observation is found by multiplying each value of the observation density surface by the corresponding height of the prior density. This is shown in Figure 1.4. Bayesians call joint density of the parameters and the observation "the full Bayesian model." It is clear that the full Bayesian model surface will not be the same shape as the sampling surface unless we use a flat prior that gives all possible parameter values equal weight. To find the posterior density of the parameter given the observed value we cut the

[2]Fisher was well aware of Bayes' theorem, and wanted his method to work on the same type of problems. He viewed Bayes' use of flat prior to be very arbitrary, and realized that the Bayesian estimator would not be invariant under the reparameterization.

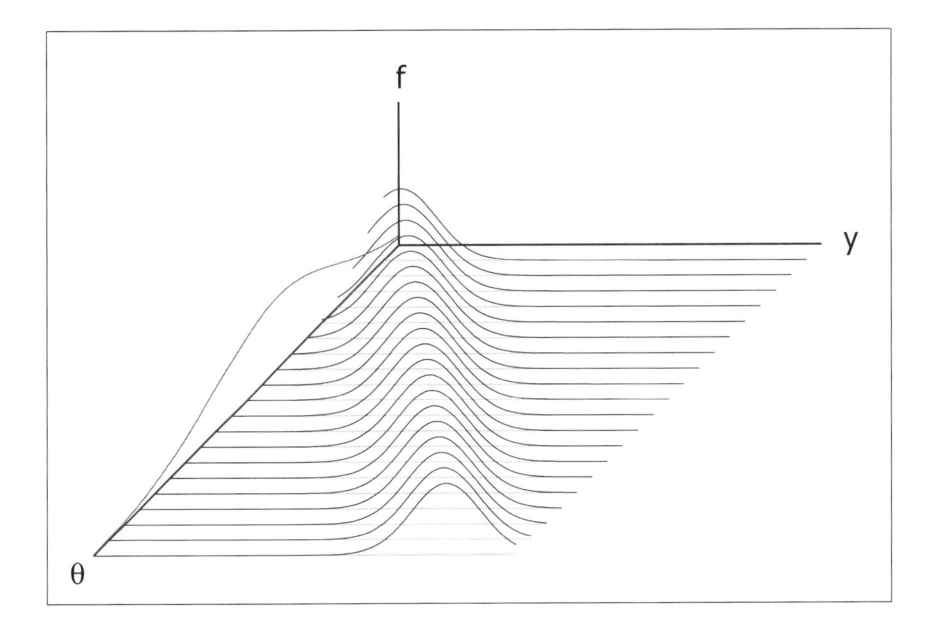

Figure 1.3 Prior density and observation density surface in 3D perspective.

joint density of the parameter and the observation with a vertical plane parallel to the parameter axis through the observed value of y. Thus, the likelihood and the posterior density are found by cutting different surfaces with the same hyperplane. The posterior density is shown in 3D perspective in Figure 1.5. The posterior density is the complete inference in the Bayesian approach. It summarizes the belief we can have about all possible parameter values, given the observed data. The posterior will always be a probability density, conditional on the observed data. Because of this, we can use the mean of the posterior distribution as the estimate of the parameter. The mean of a distribution is the value that minimizes the mean-squared deviation. Hence, the Bayesian posterior mean is the estimator that minimizes the mean-squared deviation of the posterior distribution.[3]

The Likelihood Function Can Be a Posterior

If we decide to use a flat prior density that gives equal weight to all values of the parameter, the joint density on the inference universe will be the same as the observation density surface. This is shown in Figure 1.6. Note that this prior density will be improper (the integral over the whole range will be infinite) unless the parameter values have finite lower and upper bounds. When the prior is improper, we do not have a joint probability density for the full Bayesian model. However,

[3]In decision theory, this means the posterior mean is the optimal estimator when we are using a *squared error* loss function. We can find the optimal Bayesian estimator for any particular loss function. For example, the posterior median is the optimal estimator when we are using an *absolute value* loss function.

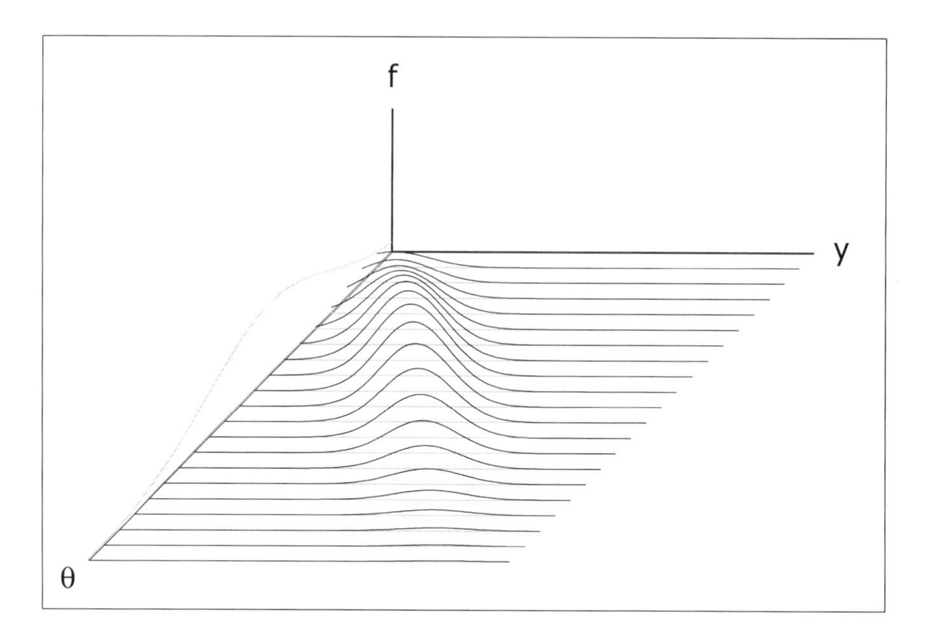

Figure 1.4 The joint density of parameter and observation in 3D perspective. The prior density of the parameter is shown on the margin.

the normed likelihood (likelihood function divided by its integral over the whole range of parameter values) will usually be a probability density. Thus, the likelihood function will have the same shape as the posterior density in this case. The Bayesian posterior mean estimator would be the mean value (balance point) of the likelihood function. This would not generally be the same value as the maximum likelihood estimator, unless the likelihood function is symmetric and unimodal such as in the *normal* likelihood. Figure 1.7 illustrates the difference between these estimators on a nonsymmetric likelihood function that could also be considered a Bayesian posterior density with a flat prior density. The maximum likelihood estimator is the mode of this curve, while the Bayesian posterior estimate is its mean, the balance point. This shows the two estimators are based on different ideas, even when the likelihood function and the posterior density have the same shape.[4]

Note; we are not advocating always using flat priors. We only want to illustrate that when we do, the posterior will be the same shape as the likelihood. Hence, the likelihood can be thought of as an unscaled posterior when we have used flat priors. When the integral of the flat prior over its whole range is infinite, the flat prior will be improper. Despite this, the resulting posterior which is the same shape as the likelihood will usually be proper. For many models, such as the regression-type models that we will discuss in Chapters 8 and 9, it is ok to use improper flat priors.

[4]Jaynes and Bretthorst (2000) show that the maximum likelihood estimator implies that we are using a 0:1 loss function, 0 at the true value of θ, and 1 at every other value. This means getting it exactly right is everything, and getting it close is of no value.

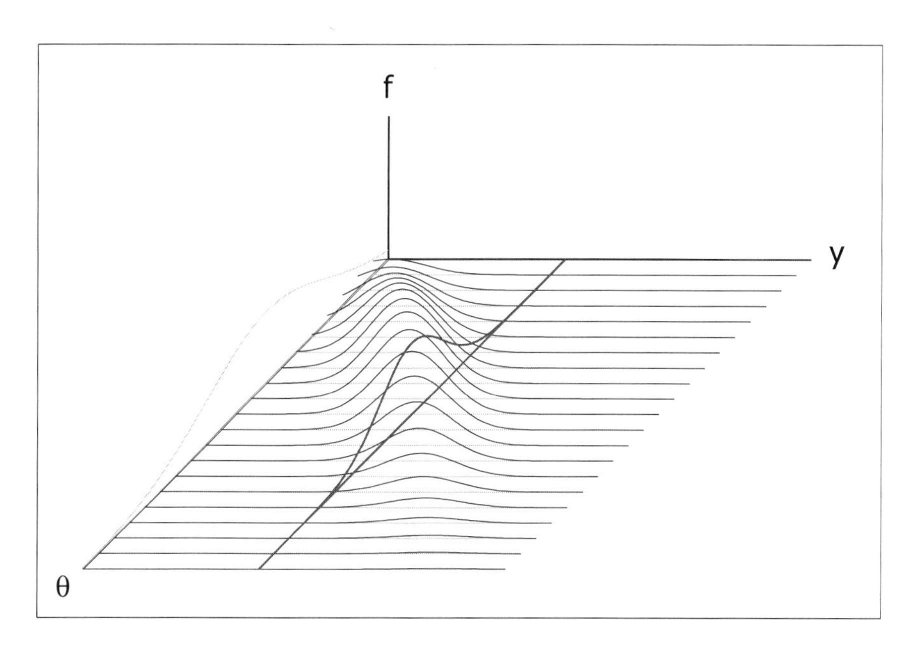

Figure 1.5 Posterior density of the parameter in the inference universe. The prior density is shown in the margin.

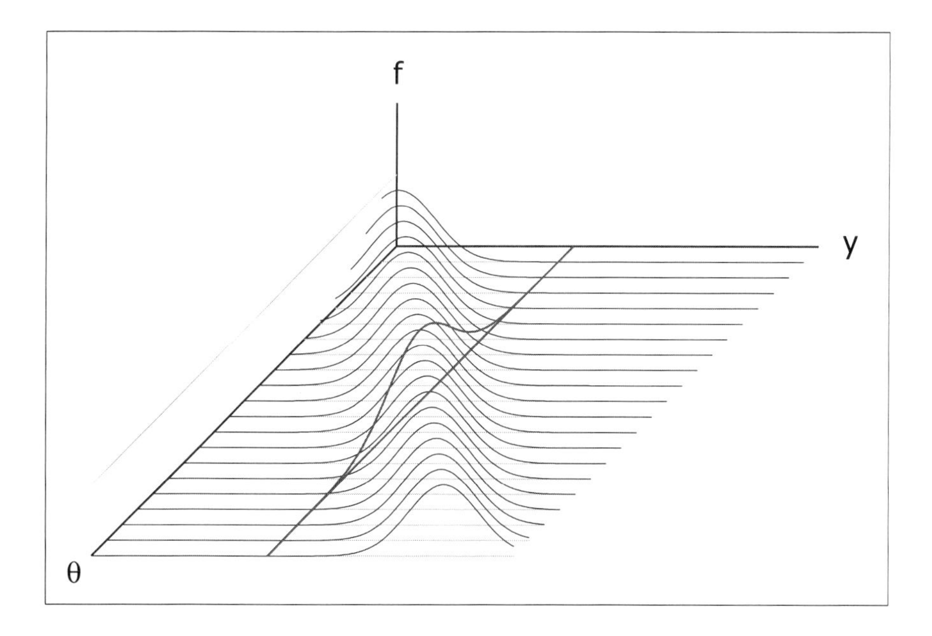

Figure 1.6 Posterior density in the inference universe using a flat prior. It has the same shape as the likelihood function.

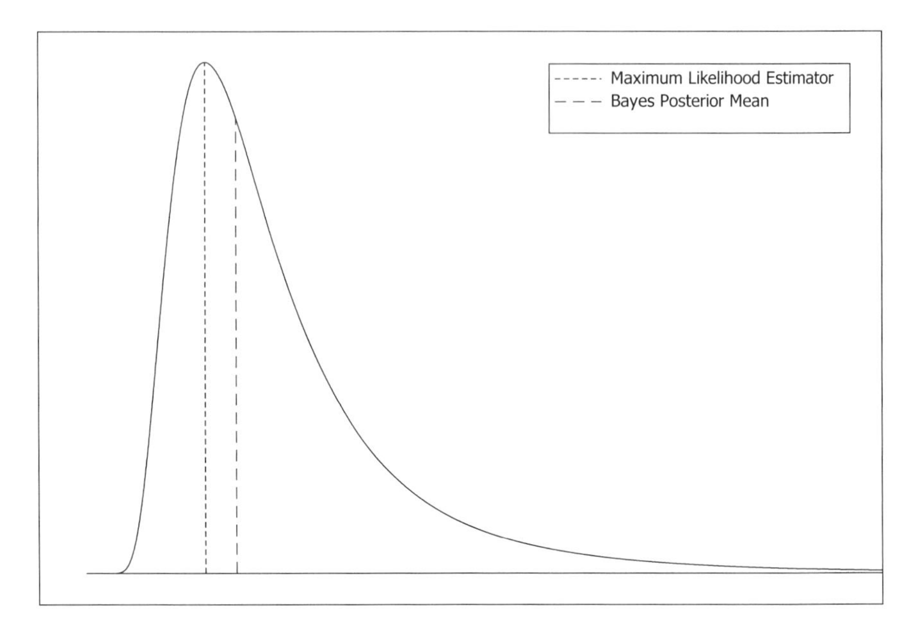

Figure 1.7 Maximum likelihood estimator and Bayesian posterior estimator for a non-symmetric likelihood (posterior with flat prior).

However, there are situations such as when we have a hierarchical normal model where improper priors should not be used for variance components. This will be discussed more fully in Chapter 10.

Multiple Parameters

When we have $p \geq 2$ the same ideas hold. However, we cannot project the surface defined on the inference universe down to a two-dimensional graph. With multiple parameters, Figures 1.1, 1.2, 1.3, 1.4, 1.5, and 1.6 can be considered to be schematic diagrams that represent the ideas rather than exact representations.

We will use the two-parameter case to show what happens when there are multiple parameters. The inference universe has at least four dimensions, so we cannot graph the surface on it. The likelihood function is still found by cutting through the surface with a hyperplane parallel to the parameter space passing through the observed values. The likelihood function will be defined on the the two parameter dimensions as the observations are fixed at the observed values and do not vary. We show the bivariate likelihood function in 3D perspective in Figure 1.8. In this example, we have the likelihood function where θ_1 is the mean and θ_2 is the variance for a random sample from a normal distribution. We will also use this same curve to illustrate the Bayesian posterior since it would be the joint posterior if we use independent flat priors for the two parameters.

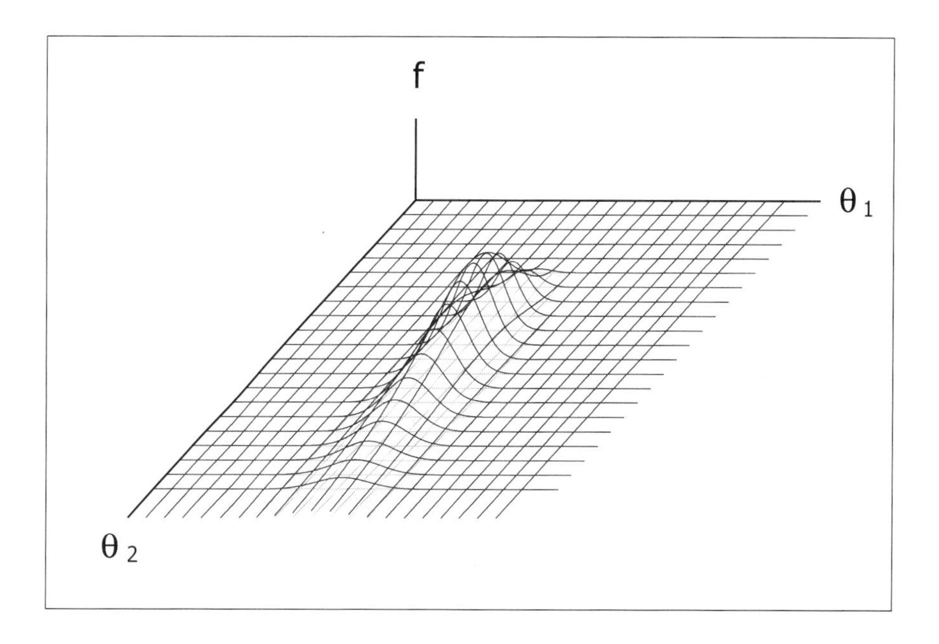

Figure 1.8 Joint Likelihood Function. Note: this can also be considered the joint posterior density when independent flat priors are used for θ_1 and θ_2.

Inference in the Presence of Nuisance Parameters

Sometimes, only one of the parameters is of interest to us. We don't want to estimate the other parameters and call them "nuisance" parameters. All we want to do is make sure the nuisance parameters don't interfere with our inference on the parameter of interest. Because using the Bayesian approach the joint posterior density is a probability density, and using the likelihood approach the joint likelihood function is not a probability density, the two approaches have different ways of dealing with the nuisance parameters. This is true even if we use independent flat priors so that the posterior density and likelihood function have the same shape.

Likelihood Inference in the Presence of Nuisance Parameters

Suppose that θ_1 is the parameter of interest, and θ_2 is a nuisance parameter. If there is an ancillary[5] sufficient statistic, conditioning on it will give a likelihood that only depends on θ_1, the parameter of interest, and inference can be based on that conditional likelihood. This can only be true in certain exponential families, so is of limited general use when nuisance parameters are present. Instead, likelihood

[5]Function of the data that is independent of the parameter of interest. Fisher developed ancillary statistics as a way to make inferences when nuisance parameters are present. However, it only works in the exponential family of densities so it cannot be used in the general case. See Cox and Hinkley (1974).

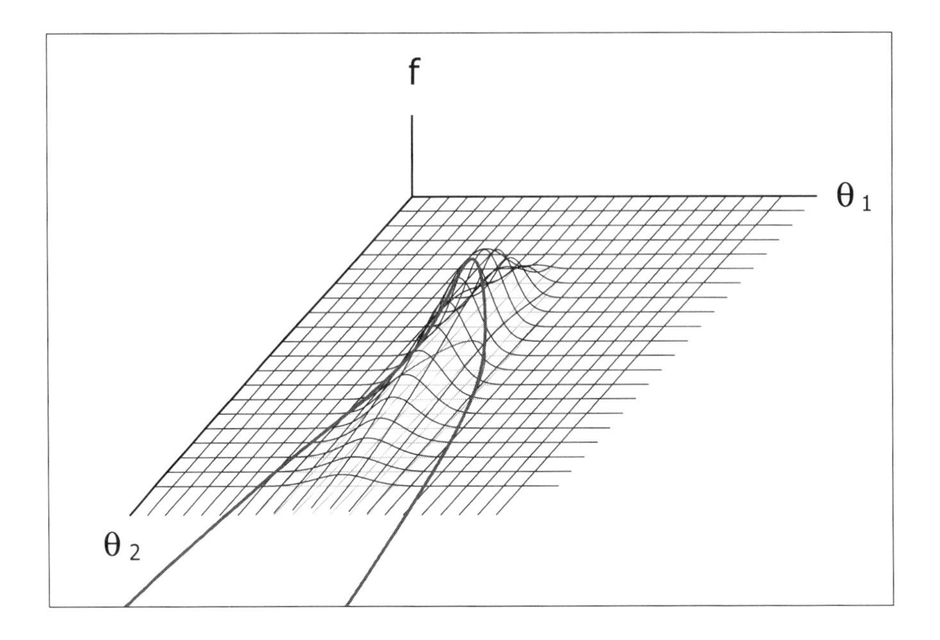

Figure 1.9 Profile likelihood of θ_1 in 3D.

inference on θ_1 is often based on the *profile likelihood* function given by

$$L_p(\theta_1; data) = \sup_{\theta_2|\theta_1} L(\theta_1, \theta_2; data)$$

where $L(\theta_1, \theta_2; data)$ is the joint likelihood function. Essentially, the nuisance parameter has been eliminated by plugging $\hat{\theta}_2|\theta_1$, the conditional maximum likelihood value of θ_2 given θ_1, into the joint likelihood. Hence

$$L_p(\theta_1; data) = L(\theta_1, \hat{\theta}_2|\theta_1; data)\,.$$

The profile likelihood function of θ_1 is shown in three-dimensional space in Figure 1.9. The two-dimensional profile likelihood function is found by projecting it back to the $f \times \theta_1$ plane and is shown in Figure 1.10. (It is like the "shadow" the curve $L(\theta_1, \hat{\theta}_2|\theta_1, data)$ would project on the $f \times \theta_1$ plane from a light source infinitely far away in the θ_2 direction.) The profile likelihood function may lose some information about θ_1 compared to the joint likelihood function. Note that the maximum profile likelihood value of θ_1 will be the same as its maximum likelihood value. However, confidence intervals based on profile likelihood may not be the same as those based on the joint likelihood.

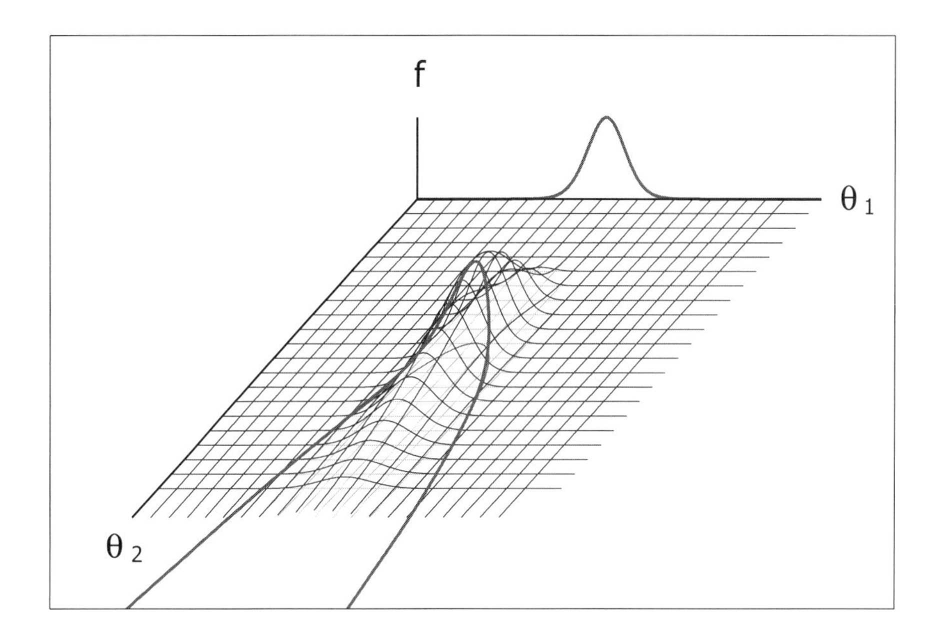

Figure 1.10 Profile likelihood of θ_1 projected onto $f \times \theta_1$ plane.

Bayesian Inference in the Presence of Nuisance Parameters

Bayesian statistics has a single way of dealing with nuisance parameters. Because the joint posterior is a probability density in all dimensions, we can find the marginal densities by integration. Inference about the parameter of interest θ_1 is based on the marginal posterior $g(\theta_1|data)$, which is found by integrating the nuisance parameter θ_2 out of the joint posterior, a process referred to as *marginalization*:

$$g(\theta_1|data) = \int g(\theta_1, \theta_2|data)\, d\theta_2 \,.$$

Note: we are using independent flat priors for both θ_1 and θ_2, so the joint posterior is the same shape as the joint likelihood in this example. The marginal posterior density of θ_1 is shown on the $f \times \theta_1$ plane in Figure 1.11. It is found by integrating θ_2 out of the joint posterior density. (This is like sweeping the probability in the joint posterior in a direction parallel to the θ_2 axis into a vertical pile on the $f \times \theta_1$ plane.) The marginal posterior has all the information about θ_1 that was in the joint posterior.

The Bayesian posterior estimator for θ_1 found from the marginal posterior will be the same as that found from the joint posterior when we are using the posterior mean as our estimator. For this example, the Bayesian posterior density of θ_1 found by marginalizing θ_2 out of the joint posterior density, and the profile likelihood function of θ_1 turn out to have the same shape. This will not always be the case. For instance, suppose we wanted to do inference on θ_2, and regarded θ_1 as the nuisance parameter.

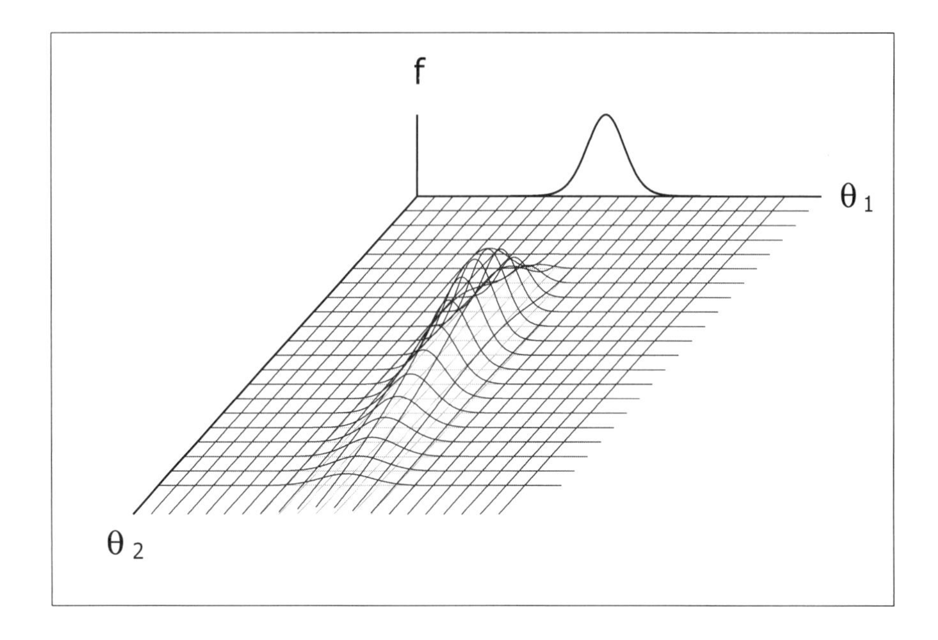

Figure 1.11 Marginal posterior of θ_1.

We have used independent flat priors for both parameters, so the joint posterior has the same shape as the joint likelihood. The profile likelihood of θ_2 is shown in 3D perspective in Figure 1.12 and projected onto the $f \times \theta_2$ plane in Figure 1.13. The marginal posterior of θ_2 is shown in Figure 1.14.

Figure 1.15 shows both the profile likelihood function and the marginal posterior density in 2D for θ_2 for this case. Clearly they have different shapes despite coming from the same two-dimensional function.

Conclusion

We have shown that both the likelihood and Bayesian approach arise from surfaces defined on the inference universe, the observation density surface and the joint probability density respectively. The sampling surface is a probability density only in the observation dimensions, while the joint probability density is a probability density in the parameter dimensions as well (when proper priors are used). Cutting these two surfaces with a vertical hyperplane that goes through the observed value of the data yields the likelihood function and the posterior density that are used for likelihood inference and Bayesian inference, respectively.

In likelihood inference, the likelihood function is not considered a probability density, while in Bayesian inference the posterior always is. The main differences between these two approaches stem from this interpretation difference; certain ideas arise naturally when dealing with a probability density. There is no reason to use the

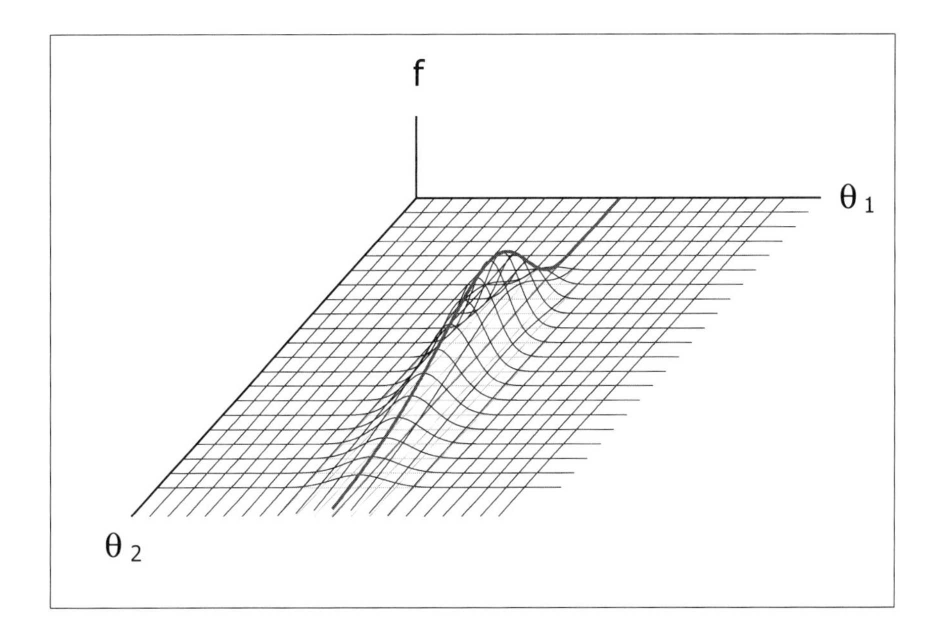

Figure 1.12 Profile likelihood function of θ_2 in 3D.

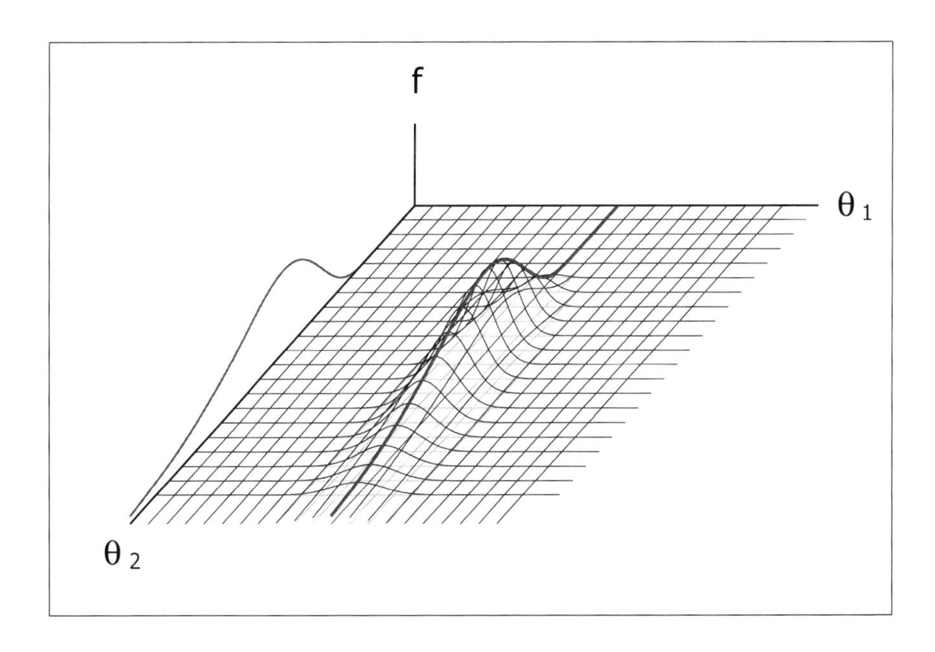

Figure 1.13 Profile likelihood of θ_2 projected onto $f \times \theta_2$ plane.

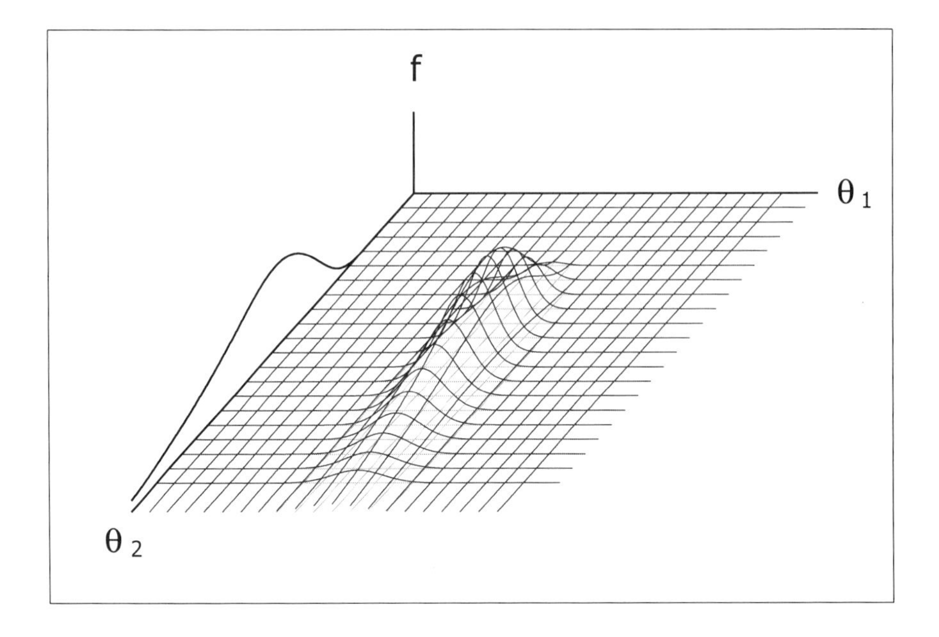

Figure 1.14 Marginal posterior of θ_2.

first moment of the likelihood function without the probability interpretation. Instead, the maximum likelihood estimator is the value that gives the highest value on the likelihood function. When a flat prior is used, the posterior density has the same shape as the likelihood function. Under the Bayesian approach it has a probability interpretation, so the posterior mean will be the estimator since it minimizes the mean squared deviation.

When there are nuisance parameters, there is no reason why they could not be integrated out of the joint likelihood function, and the inference be based on the marginal likelihood. However, without the probability interpretation on the joint likelihood, there is no compelling reason to do so. Instead, likelihood inference is commonly based on the profile likelihood function, where the maximum conditional likelihood values of the nuisance parameters given the parameters of interest are plugged into the joint likelihood. This plug-in approach does not allow for all the uncertainty about the nuisance parameters. It treats them as if it were known to have their conditional maximum likelihood values, rather than treating them like unknown parameters. This may lead to confidence intervals that are too short to have the claimed coverage probability. Under the Bayesian approach the joint posterior density is clearly a probability density. Hence Bayesian inference about the parameter of interest will be based on the marginal posterior where the nuisance parameters have been integrated out. The Bayesian approach has allowed for all the uncertainty about the nuisance parameters.

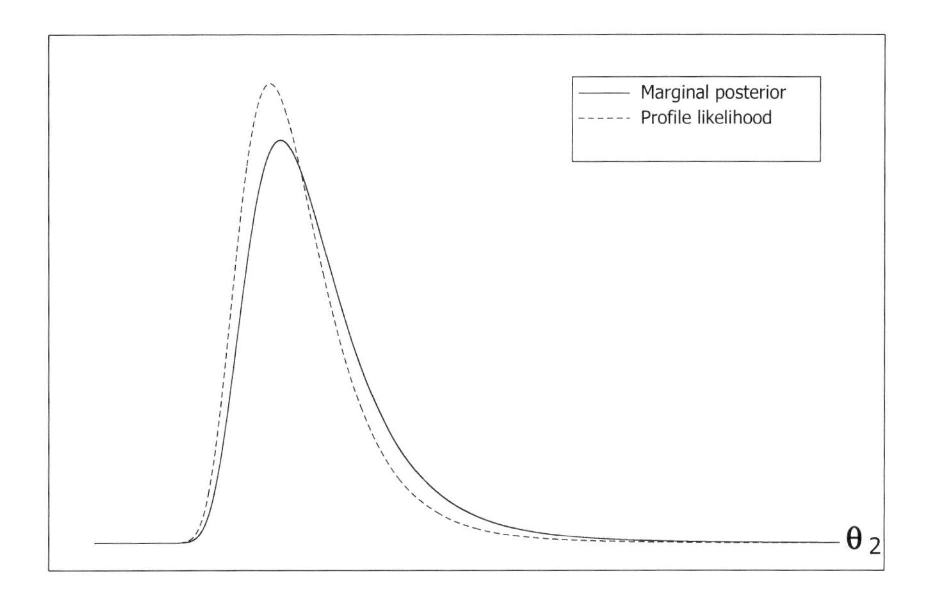

Figure 1.15 Profile likelihood and marginal posterior density of θ_2.

1.4 COMPUTATIONAL BAYESIAN STATISTICS

In this section we introduce the main ideas of computational Bayesian statistics. We show how basing our inferences on a random sample from the posterior distribution has overcome the main impediments to using Bayesian methods. The first impediment is that the exact posterior cannot be found analytically except for a few special cases. The second is that finding the numerical posterior requires a difficult numerical integration, particularly when there is a large number of parameters.

Finding the posterior using Bayes' theorem is easy in theory. That is, we can easily find the unscaled posterior by Equation 1.1. This gives us all the information about the shape of the posterior. The exact posterior is found by scaling this to make it a density and is given in Equation 1.2. However, in practice, the integral given in Equation 1.3 can be very difficult to evaluate, even numerically. This is particularly difficult when the parameter space is high dimensional. Thus we cannot always find the divisor needed to get the exact posterior density. In general, the incompletely known posterior given by Equation 1.1 is all we have.

Computational Bayesian statistics is based on developing algorithms that we can use to draw samples from the true posterior, even when we only know the unscaled version. There are two types of algorithms we can use to draw a sample from the true posterior, even when we only know it in the unscaled form. The first type are direct methods, where we draw a random sample from an easily sampled density, and reshape this sample by only accepting some of the values into the final sample, in such a way that the accepted values constitute a random sample from the posterior. These methods quickly become inefficient as the number of parameters increase.

The second type is where we set up a Markov chain that has the posterior as its long-run distribution, and letting the chain run long enough so a random draw from the Markov chain is a random draw from the posterior. These are known as Markov chain Monte Carlo (MCMC) methods. The Metropolis-Hastings algorithm and the Gibbs sampling algorithm are the two main Markov chain Monte Carlo methods. The Markov chain Monte Carlo samples will not be independent. There will be serial dependence due to the Markov property. Different chains have different mixing properties. That means they move around the parameter space at different rates. We show how to determine how much we must thin the sample to obtain a sample that well approximates a random sample from the posterior to be used for inference.

Inference from the Posterior Sample

The overall goal of Bayesian inference is knowing the posterior. The fundamental idea behind nearly all statistical methods is that as the sample size increases, the distribution of a random sample from a population approaches the distribution of the population. Thus, the distribution of the random sample from the posterior will approach the true posterior distribution. Other inferences such as point and interval estimates of the parameters can be constructed from the posterior sample. For example, if we had a random sample from the posterior, any parameter could be estimated by the corresponding statistic calculated from that random sample. We could achieve any required level of accuracy for our estimates by making sure our random sample from the posterior is large enough. Existing exploratory data analysis (EDA) techniques can be used on the sample from the posterior to explore the relationships between parameters in the posterior.

1.5 PURPOSE AND ORGANIZATION OF THIS BOOK

The development and implementation of computational methods for drawing random samples from the incompletely known posterior has revolutionized Bayesian statistics. Computational Bayesian statistics breaks free from the limited class of models where the posterior can be found analytically. Statisticians can use observation models, and choose prior distributions that are more realistic, and calculate estimates of the parameters from the Monte Carlo samples from the posterior. Computational Bayesian methods can easily deal with complicated models that have many parameters. This makes the advantages that the Bayesian approach offers accessible to a much wider class of models.

This book aims to introduce the ideas of computational Bayesian statistics to advanced undergraduate and first-year graduate students. Students should enter the course with some knowledge in Bayesian statistics at the level of Bolstad (2007). This book builds on that background. It aims to give the reader a big-picture overview of the methods of computational Bayesian statistics, and to demonstrate them for some common statistical applications.

In Chapter 2, we look at methods which allow us to draw a random sample directly from an incompletely known posterior distribution. We do this by drawing a random sample from an easily sampled distribution, and only accepting some draws into the final sample. This reshapes the accepted sample so it becomes a random sample from the incompletely known posterior. These methods work very well for a single parameter. However, they can be seen to become very inefficient as the number of parameters increases. The main use of these methods is as a small step that samples a single parameter as part of a Gibbs sampler.

Chapter 3 compares Bayesian inferences drawn from a numerical posterior with Bayesian inferences from the posterior random sample.

Chapter 4 reviews Bayesian statistics using conjugate priors. These are the classical models that have analytic solutions to the posterior. In computational Bayesian statics, these are useful tools for drawing an individual parameter as steps of the Markov chain Monte Carlo algorithms.

In Chapter 5, we introduce Markov chains, a type of process that evolves as time passes. They move through a set of possible values called *states* according to a probabilistic law. The set of all possible values is called the *state space* of the chain. Markov chains are a particular type of stochastic process where the probability of the next state given the past history of the process up to and including the current state only depends on the current state, that is, it is memoryless. The future state depends only on the current state and the past states can be forgotten. We study the relationship between the probabilistic law of the process and the long-run distribution of the chain. Then we see how to solve the inverse problem. In other words, we find a probabilistic law that will give a desired long-run distribution.

In Chapter 6, we introduce the Metropolis-Hastings algorithm as the fundamental method for finding a Markov chain that has the long-run distribution that is the same shape as the posterior. This achieves the fundamental goal in computational Bayesian statistics. We can easily find the shape of the posterior by the proportional form of Bayes' theorem. The Metropolis-Hastings algorithm gives us a way to find a Markov chain that has that shape long-run distribution. Then by starting the chain, and letting it run long enough, a draw from the chain is equivalent to a draw from the posterior. Drawing samples from the posterior distribution this way is known as Markov chain Monte Carlo sampling. We have lots of choices in setting up the algorithm. It can be implemented using either random-walk or independent candidate densities. We can either draw a multivariate candidate for all parameters at once, or draw the candidates blockwise, each conditional on the parameters in the other blocks. We will see that the Gibbs sampling algorithm is a special case of the Metropolis-Hastings algorithm. These algorithms replace very difficult numerical calculations with the easier process of drawing random variables using the computer. Sometimes, particularly for high-dimensional cases, they are the only feasible method.

In Chapter 7 we develop a method for finding a Markov chain that has good mixing properties. We will use the Metropolis-Hastings algorithm with heavy-tailed independent candidate density. We then discuss the problem of statistical inference on the sample from the Markov chain. We want to base our inferences on an approximately random sample from the posterior. This requires that we determine the burn-in time and the amount of thinning we need to do so the thinned sample will be approximately random.

In Chapter 8 we show how to do computational Bayesian inference on the logistic regression model. Here we have independent observations from the *binomial* distribution where each observation has its own probability of success. We want to relate the probability of success for an observation to known values of the predictor variables taken for that observation. Probability is always between 0 and 1, and a linear function of predictor variables will take on all possible values from $-\infty$ to ∞. Thus we need a link function of the probability of success so that it covers the same range as the linear function of the predictors. The logarithm of the odds ratio

$$\log_e\left(\frac{\pi}{1-\pi}\right)$$

gives values between $-\infty$ to ∞ so it is a satisfactory link function. It is called the *logit* link function. The logistic regression model is a generalized linear model, so we can find the vector of maximum likelihood estimates, along with their matched curvature covariance matrix, by using iteratively reweighted least squares. We approximate the likelihood by a multivariate normal having the maximum likelihood vector as its mean vector, and the matched curvature covariance. We can find the approximate normal posterior by the simple normal updating rules for normal linear regression. We develop a heavy-tailed candidate density from the approximate normal posterior that we use in the Metropolis-Hastings algorithm to find a sample from the exact posterior. The computational Bayesian approach has a significant advantage over the likelihood approach. The computational Bayesian approach gets a random sample from the true posterior, so credible intervals will have the correct coverage probabilities. The covariance matrix found by the likelihood approach does not actually relate to the spread of the likelihood, but rather to its curvature so coverage probabilities of confidence intervals may not be correct.

In Chapter 9 we develop the Poisson regression model, and the proportional hazards model. We follow the same approach we used for the logistic regression model. We find the maximum likelihood vector and matched curvature covariance matrix. Then we find the normal approximation to the posterior, and modify it to have heavy tails so it can be used as the candidate density. The Metropolis-Hastings algorithm is used to draw a sample from the exact posterior. We find that when we have censored survival data, and we relate the linear predictor, the censoring variable has the same shape as the Poisson, so we can use the same algorithm for the proportional hazards model. Again we will find that the computational Bayesian approach has the same advantages over the likelihood approach since the sample is from the true posterior.

In Chapter 10, we show how the Gibbs sampling algorithm cycles through each block of parameters, drawing from the conditional distribution of that block given all the parameters in other blocks at their most recent value. In general these conditional distributions are complicated. However, they will be quite simple when the parameters have a hierarchical structure. That means we can draw a graph where each node stands for a block of parameters or block of data. We connect the nodes with arrows showing the direction of dependence. When we look at the resulting graph, we will find there are no connected loops. All nodes that lead into a specific node are called its parent nodes. All nodes that lead out of a specific node are called its child nodes. The other parent nodes of a child node are called coparent nodes. For this model, the conditional distribution of a specified node, given all the other nodes, will be proportional to its distribution given its parent nodes (the prior) times the joint distribution of all its child nodes given it and the coparent nodes of the child nodes (the likelihood). They will be particularly simple if the likelihood distributions are from the exponential family and the priors are from the conjugate family.

The biggest advantage of Markov chain Monte Carlo methods is that they allow the applied statistician to use more realistic models because he/she is not constrained by analytic or numerical tractability. Models that are based on the underlying situation can be used instead of models based on mathematical convenience. This allows the statistician to focus on the statistical aspects of the model without worrying about calculability.

Main Points

- Bayesian statistics does inference using the rules of probability directly.

- Bayesian statistics is based on a single tool, Bayes' theorem, which finds the posterior density of the parameters, given the data. It combines both the prior information we have given in the *prior* $g(\theta_1, \ldots, \theta_p)$ and the information about the parameters contained in the observed data given in the *likelihood* $f(y_1, \ldots, y_n|\theta_1, \ldots, \theta_p)$.

- It is easy to find the unscaled posterior by *posterior* proportional to *prior* times *likelihood*

  $$g(\theta_1, \ldots, \theta_p|y_1, \ldots, y_n) \propto g(\theta_1, \ldots, \theta_p) \times f(y_1, \ldots, y_n|\theta_1, \ldots, \theta_p)\,.$$

  The unscaled posterior has all the shape information. However, it is not the exact posterior density. It must be divided by its integral to make it exact.

- Evaluating the integral may be very difficult, particularly if there are lots of parameters. It is hard to find the exact posterior except in a few special cases.

- The *Likelihood principle* states that if two experiments have proportional likelihoods, then they should lead to the same inference.

- The *Likelihood* approach to statistics does inference solely using the likelihood function, which is found by cutting the sampling surface with a vertical hyperplane through the observed value of the data. It is not considered to be a probability density.

- The *maximum likelihood estimate* is the mode of the likelihood function.

- The complete inference in Bayesian statistics is the posterior density. It is found by cutting the joint density of parameters and observations with the same vertical hyperplane through the observed values of the data.

- The usual Bayesian estimate is the mean of the posterior, as this minimizes mean-squared error of the posterior. Note: this will be different from the MLE even when flat priors are used and the posterior is proportional to the likelihood!

- The two approaches have different ways of dealing with nuisance parameters. The likelihood approach often uses the *profile likelihood* where the maximum conditional likelihood value of the nuisance parameter is plugged into the joint likelihood. The Bayesian approach is to integrate the nuisance parameter out of the joint posterior.

- Computational Bayesian statistics is based on drawing a Monte Carlo random sample from the unscaled posterior. This replaces very difficult numerical calculations with the easier process of drawing random variables. Sometimes, particularly for high dimensional cases, this is the only feasible way to find the posterior.

- The distribution of a random sample from the posterior approaches the exact posterior distribution. Estimates of parameters can be calculated from statistics calculated from the random sample.

2

Monte Carlo Sampling from the Posterior

In Bayesian statistics, we have two sources of information about the parameter θ: our prior belief and the observed data. The prior distribution summarizes our belief about the parameter before we look at the data. The prior density $g(\theta)$ gives the relative belief weights we have for all possible values of the parameter θ before we look at the data. In Bayesian statistics, all the information about the parameter θ that is in the observed data y is contained in the likelihood function $f(y|\theta)$. However, the parameter is considered a random variable, so the likelihood function is written as a conditional distribution. The likelihood function gives the relative support weight each possible value of the parameter θ has from the observed data.

Bayes' theorem gives us a unified approach that combines the two sources into a single relative belief weight distribution after we have observed the data. The final belief weight distribution is known as the posterior distribution and it takes into account both the prior belief and the support from the data. Bayes' Theorem is usually expressed very simply in the unscaled form *posterior* proportional to *prior* times *likelihood*. In equation form this is

$$g(\theta|y) \propto g(\theta) \times f(y|\theta)\,. \tag{2.1}$$

This formula does not give the posterior density $g(\theta|y)$ exactly, but it does give its shape. In other words, we can find where the modes are, and relative values at any two locations. However, it does not give the scale factor needed to make it a density. This means we cannot calculate probabilities or moments from it. Thus it is not possible to do any inference about the parameter θ from the unscaled posterior. The actual posterior density is found by scaling it so it integrates to one.

$$g(\theta|y) = \frac{g(\theta) \times f(y|\theta)}{\int g(\theta) \times f(y|\theta) d\theta}\,. \tag{2.2}$$

The posterior distribution found using Bayes' theorem summarizes our knowledge of the parameters given the data that was actually observed. It combines the information from our prior distribution with that from the observed data. A closed form for the integral in the denominator only exists for some particular cases.[1] We will investigate some of these cases in Chapter 4. For other cases the posterior density has to be approximated numerically. This requires integrating

$$\int f(y|\theta) g(\theta) d\theta$$

numerically, which may be very difficult, particularly when the parameter θ is high dimensional, and in such cases, numerical approximations are often used.

Computational Bayesian Statistics

The computational approach to Bayesian statistics allows the posterior to be approached from a completely different direction. Instead of using the computer to calculate the posterior numerically, we use the computer to draw a Monte Carlo sample from the posterior. Fortunately, all we need to know is the shape of the posterior density, which is given by the prior times the likelihood. We do not need to know the scale factor necessary to make it the exact posterior density. These methods replace the very difficult numerical integration with the much easier process of drawing random samples. A Monte Carlo random sample from the posterior will approximate the true posterior when the sample size is large enough. We will base our inferences on the Monte Carlo random sample from the posterior, not from the numerically calculated posterior. Sometimes this approach to Bayesian inference is the only feasible method, particularly when the parameter space is high dimensional.

The great advantage of computational Bayesian methods is that they allow the applied statistician to use more realistic models because he or she is not constrained by analytic or numerical tractability. Models that are based on the underlying situation can be used instead of models based on mathematical convenience. This allows the statistician to focus on the statistical aspects of the model without worrying about calculability. Computational Bayesian methods, based on a Monte Carlo random sample drawn from the posterior, have other advantages as well, even when there are alternatives available. These include:

1. They are not approximations. Estimates found from the Monte Carlo random sample from the posterior can achieve any required accuracy by setting the sample size large enough.

[1] Where the observation distribution comes from an exponential family, and the prior comes from the family that is conjugate to the observation distribution.

2. Existing exploratory data analysis (EDA) techniques can be used to explore the posterior. This essentially is the overall goal of Bayesian inference.

3. They allow sensitivity analysis to be made on the model in a simple fashion.

In Section 2.1 we will look at acceptance-rejection-sampling (ARS). This is a Monte Carlo method for generating a sample directly from an unscaled posterior by reshaping a sample from an easily sampled distribution. In Section 2.2 we will look at sampling-importance-resampling. Here we will draw a sample from an easily sampled distribution, then resample from this sample using sampling weights to obtain a sample from the target distribution. In Section 2.3 we will look at adaptive rejection sampling, a method for drawing a sample from a log concave distribution. In Section 2.4 we look at the reasons why these direct methods for sampling from the posterior become inefficient when there are many parameters. Most of the values in the initial sample are not accepted into the final sample.

2.1 ACCEPTANCE-REJECTION-SAMPLING

Acceptance-rejection-sampling (ARS) is a method for drawing random samples directly from a target distribution even though we only know the unscaled target. In Bayesian statistics, the posterior distribution is the target. All we know is the unscaled target given by the prior times the likelihood. This gives the shape of the target, but not the scale factor needed to make it an exact density. The acceptance-rejection-sampling algorithm allows us to draw a random sample from a target distribution when we only know its shape. We do not need to know the scale factor necessary to make it a density. This means we can draw a random sample from the posterior, when we only know the unscaled posterior given by the proportional form of Bayes' theorem. The acceptance-rejection algorithm works by reshaping a random sample drawn from an easily sampled candidate distribution (sometimes called the starting distribution) into a random sample from the target by only accepting some of the candidate values into the final sample. Let the unscaled posterior $g(\theta) \times f(y|\theta)$ be the unscaled target, and let the easily sampled candidate density be $g_0(\theta)$. It is very important that the candidate density *dominates* the unscaled target. That means that we can find a number M such that

$$M \times g_0(\theta) \quad \geq \quad g(\theta)f(y|\theta)$$

for all θ. acceptance-rejection-sampling works by only accepting some of the candidates into the final sample. We can think of this as cutting out part of the scaled up candidate density to give it the shape of the target density. We must start with the scaled up candidate density being greater than the target for all values of θ. We can cut part of the candidate density out, but we can't put any part back in. The acceptance-rejection-sampling algorithm proceeds as follows:

1. Draw a random value of θ from the candidate density $g_0(\theta)$.

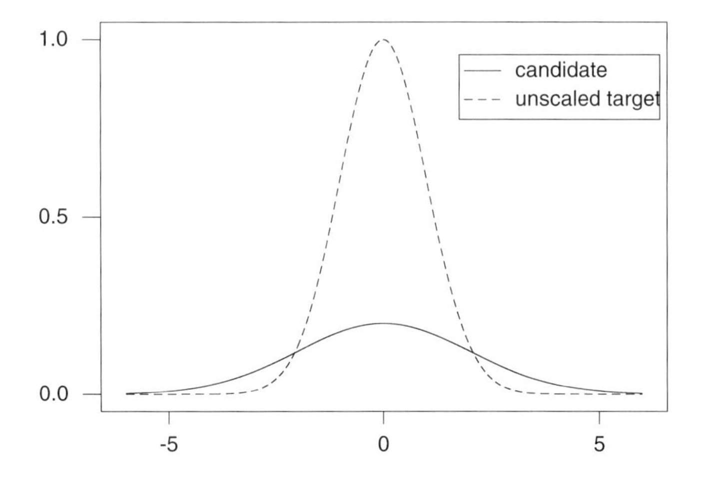

Figure 2.1 Unscaled target and starting distributions.

2. Calculate the weight for that θ value as the ratio of the target to the scaled up candidate density
$$w(\theta) = \frac{g(\theta)f(y|\theta)}{Mg_0(\theta)}$$
for that value. The weight will always be be between 0 and 1.

3. Draw (independently) u from a uniform (0,1) distribution.

4. If $u < w(\theta)$ accept θ .

5. Otherwise reject θ and return to step 1.

An accepted θ is a random draw from the exact posterior distribution $g(\theta|y)$, even though we only knew the unscaled posterior distribution $f(y|\theta)g(\theta)$. We can draw a random sample from the posterior by repeating the steps.

Example 2 *Suppose the unscaled target has the formula*
$$g(\theta) \times f(y|\theta) = e^{-\frac{\theta^2}{2}} .$$
(Note: We recognize this is the shape of a normal$(0, 1)$ *distribution. However, we will only use the formula giving the shape, and not our knowledge of what the posterior having that shape actually is.) Let us use the normal*$(0, 2^2)$ *candidate density. Figure 2.1 shows the unscaled target and the candidate density. We want the smallest value of* M *such that for all* θ,
$$M \times g_0(\theta) \quad \geq \quad g(\theta) \times f(y|\theta) .$$
Hence
$$M \quad \geq \quad \frac{e^{-\frac{\theta^2}{2}}}{\frac{1}{\sqrt{2\pi}2}e^{-\frac{\theta^2}{2\times 2^2}}}$$

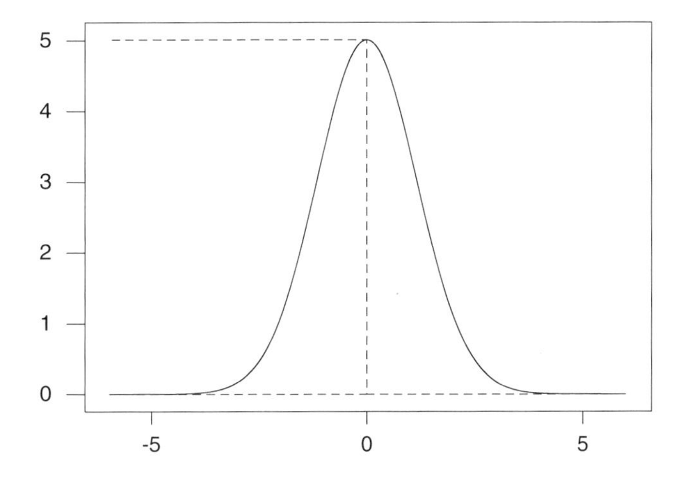

Figure 2.2 The ratio $h(\theta)$ and the maximum M.

so M will be the maximum value of the function

$$h(\theta) = \frac{e^{-\frac{\theta^2}{2}}}{\frac{1}{\sqrt{2\pi}2} e^{-\frac{\theta^2}{2\times 2^2}}} \,.$$

We can find M exactly by using calculus. First set the derivative $h'(\theta) = 0$ and solve for θ^2. In this case the solution is $\theta = 0$. Then we substitute that value in the equation, which gives the maximum value of $h(\theta)$. In this case $M = 5.01326$. (Note: an approximate value of M can be found by calculating $h(\theta)$ at a closely spaced grid of values over the range in which the maximum will occur, and letting the computer find the maximum. This will give a very good approximation when $h(\theta)$ is smooth, and your grid of values are close enough together.) Figure 2.2 shows the function $h(\theta)$ with the maximum value M. We scale up the candidate density by multiplying by M. Figure 2.3 shows the scaled up candidate density dominates the unscaled target. Suppose that the first value drawn from the candidate density $g_0(\theta)$ is $\theta = 2.60$. The value of the unscaled target will be

$$\begin{aligned} g(2.60) \times f(y|2.60) &= e^{-\frac{2.60^2}{2}} \\ &= .0340475 \end{aligned}$$

and the value of the candidate density will be

$$\begin{aligned} g_0(2.60) &= \frac{1}{\sqrt{2\pi}\,2} e^{-\frac{2.60^2}{2\times 2^2}} \\ &= .0856843\,. \end{aligned}$$

[2]This finds maximums and minimums for a continuous differentiable function. When there are multiple solutions to the equation, we must evaluate them all to find which one is the maximum.

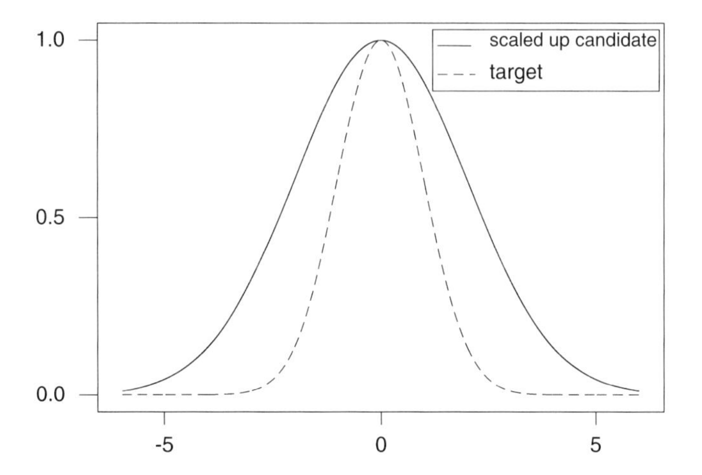

Figure 2.3 Scaled up candidate density and unscaled target.

Hence the weight function at the value $\theta = 2.60$ will be

$$
\begin{aligned}
w(2.60) &= \frac{.0340475}{5.01326 \times .0856843} \\
&= .0792617\,.
\end{aligned}
$$

The next step is to draw a uniform$(0, 1)$ random variable. Suppose we draw $u = .435198$. Since $u > w(2.6)$, we do not accept the value 2.60. Suppose the second candidate value drawn is $\theta = -.94$. The value of the unscaled target is

$$
\begin{aligned}
g(-.94) \times f(y|-.94) &= e^{-\frac{-.94^2}{2}} \\
&= .642878\,,
\end{aligned}
$$

the value of the candidate density will be

$$
\begin{aligned}
g_0(-.94) &= \frac{1}{\sqrt{2\pi 2}} e^{-\frac{9.4^2}{2\times 2^2}} \\
&= .178613\,,
\end{aligned}
$$

and the value of the weight function at the value $\theta = -.94$ will be

$$
\begin{aligned}
w(-.94) &= \frac{.642878}{5.01326 \times .178613} \\
&= .717953\,.
\end{aligned}
$$

Suppose the next uniform$(0, 1)$ value drawn is $u = .577230$. We have $u < w(-.94)$, so $-.94$ is accepted as a draw from the target.

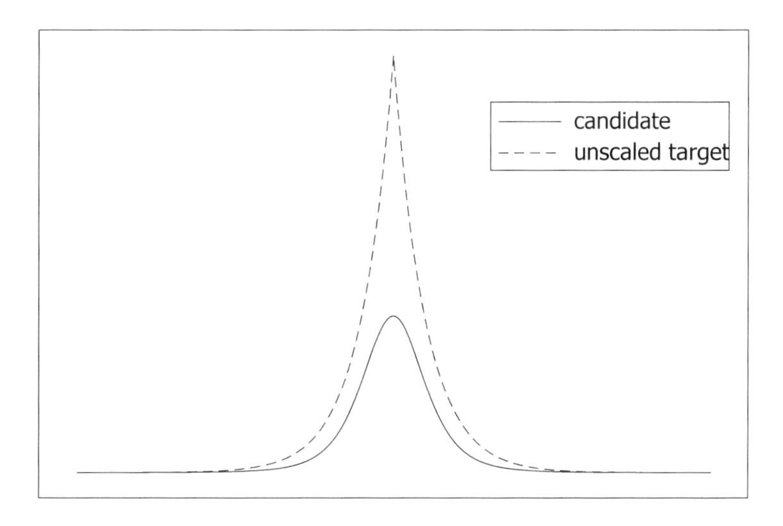

Figure 2.4 The unscaled target and proposed candidate distributions.

Some Important Points About the Candidate Distribution

acceptance-rejection-sampling does not work if $M = \sup w(\theta)$ is not finite. This may happen if the candidate distribution has lighter tails than the target. In Figure 2.4 we show the graph of an unscaled target and a possible candidate density. The candidate density will not be suitable unless it has heavier tails than the unscaled target. We cannot tell from the graph which one has heavier tails, the candidate or the unscaled target. Both are going to zero, and whether the proposed candidate density has heavier or lighter tails is lost. To resolve this, we take logarithms of both the candidate density and the target and graph them in Figure 2.5. We see that the relative tail weights can easily be determined by looking at the graphs of the logarithms of the unscaled target and the candidate density. Clearly the proposed candidate density has heavier tails, so it will be suitable for acceptance-rejection-sampling. Heavy-tailed candidate distributions such as *Student's t* with low degrees of freedom are recommended. acceptance-rejection-sampling will be more efficient (higher proportion of θ accepted into final sample) when the candidate distribution has shape that is similar to the shape of the posterior. In fact, if the candidate distribution is exactly the same shape as the posterior, every value will be accepted.

Smith and Gelfand (1992) show that the sampling analogue to Bayes theorem is the updating of a sample from the prior distribution to a sample from the posterior distribution through the likelihood. This means that if we let our candidate density be the same as the prior, the acceptance weights for the acceptance-rejection-sampling will be proportional to the likelihood. Bayes' theorem goes from the prior to the posterior by multiplying by the likelihood. Thus the acceptance-rejection-sampling algorithm using the prior as the candidate density is the sampling equivalent to Bayes theorem.

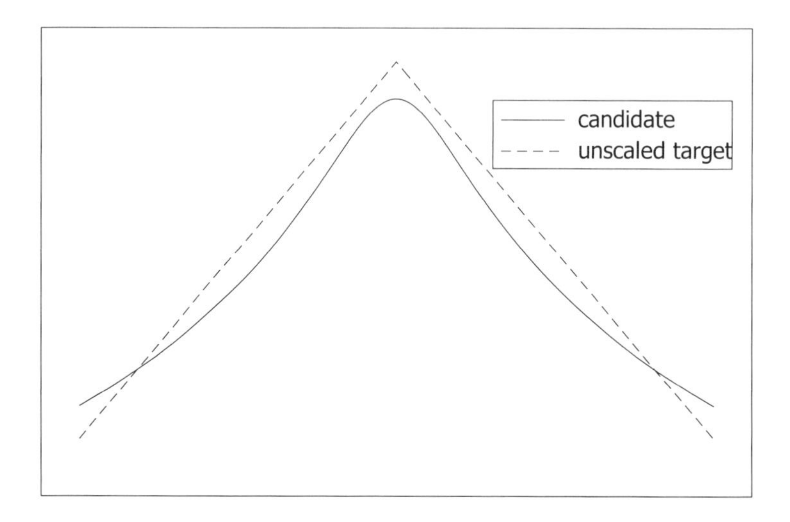

Figure 2.5 Logarithms of unscaled target and candidate distributions.

Steps for Drawing a Random Sample from the Posterior Using Acceptance-Rejection-Sampling

The preliminary steps are to first calculate the unscaled posterior and the candidate density over a fine mesh of θ values over the main range of the posterior and graph them. Take logarithms of the unscaled posterior and candidate density and graph them to make sure the candidate density dominates the unscaled posterior in the tails. Use the computer to find the maximum value of unscaled posterior divided by the candidate density

$$M = \max_{\theta} \left(\frac{g(\theta) \times f(y_1, \ldots, y_n|\theta)}{g_0(\theta)} \right)$$

at our fine mesh of values. (We could use calculus to find the exact θ value that will give the maximum, and then plug it in to find M. However, for practical purposes it is sufficient to find the maximum value calculated by computer, as long as the θ values are close enough together.) The seven steps fill in Table 2.1 column by column and are easily implemented on a computer.

1. Draw a random sample of size N from the candidate distribution $g_0(\theta)$. This will be $\theta_1, \ldots, \theta_N$. (First column)

2. Calculate the value of the unscaled target density at each value in the sample. (Second column)

3. Calculate the candidate density at each value in the sample, then multiply them by M to get the scaled up candidate density. (Third column)

4. For each value in the sample, the weight is the unscaled target density divided by the scaled up starting density. These weights will always be less than or

Table 2.1 Table for performing acceptance-rejection-sampling

θ	Target (unscaled)	Candidate (scaled)	Weight	u	d	Ind
θ_1	$g(\theta_1)f(y_1,\ldots,y_n\|\theta_1)$	$M \times g_0(\theta_1)$	w_1	u_1	d_1	Ind_i
θ_2	$g(\theta_2)f(y_1,\ldots,y_n\|\theta_2)$	$M \times g_0(\theta_2)$	w_2	u_2	d_2	Ind_2
$\vdots$	$\vdots$	$\vdots$	$\vdots$	$\vdots$	$\vdots$	$\vdots$
θ_N	$g(\theta_N)f(y_1,\ldots,y_n\|\theta_N)$	$M \times g_0(\theta_N)$	w_N	u_N	d_N	Ind_N

equal to one.

$$w_i = \frac{g(\theta_i) \times f(y_1,\ldots,y_n|\theta_i)}{M \times g_0(\theta_i)}$$

(Fourth column)

5. Draw a random sample of size $u_1,\ldots,u_N$ from a *Uniform*$(0,1)$ distribution. (Fifth column) Note: If $u_i < w_i$ we will accept θ_i into the final sample, otherwise we will reject θ_i. The following steps enable us to do this.

6. Let $d_i = w(i) - u(i)$. (Sixth column)

7. Then code negative values to 0, and positive values to 1 into an indicator variable Ind_i. (Seventh column)

Finally, take all the θ values that have indicator value $Ind_i = 1$ into our accepted sample. These are the values θ_i where $u_i \leq w_i$. In Minitab we use the **Unstack** command. These accepted values will be a random sample of size n from the posterior distribution $g(\theta|y_1,\ldots,y_n)$. Generally the final sample size n will be less than the initial sample size N, except in the case where the initial candidate distribution is proportional to the unscaled target and all candidates will be accepted. To get as many candidates accepted as possible, the candidate density should have a shape as similar as possible to the target, yet still dominate it.

2.2 SAMPLING-IMPORTANCE-RESAMPLING

Sampling-importance-resampling (SIR) is a two stage method for sampling from the posterior distribution $g(\theta|y_1,\ldots,y_n)$ when all we know is its unscaled version $g(\theta)\,f(y_1,\ldots,y_n|\theta)$ given by Bayes' theorem. First draw a random sample $\theta_1,\ldots,\theta_N$ from the candidate density $g_0(\theta)$, which should have heavier tails than the unscaled posterior. (In sampling-importance-resampling the candidate distribution is sometimes called the starting distribution.) This can be determined by graphing the logarithms of the candidate and the unscaled posterior. Calculate the

Table 2.2 Table for performing sampling-importance-resampling

θ	Target	Candidate	Ratio	Weight
θ_1	$g(\theta_1)f(y_1,\ldots,y_n\|\theta_1)$	$g_0(\theta_1)$	r_1	$\frac{r_1}{\sum r_i}$
$\vdots$	$\vdots$	$\vdots$	$\vdots$	$\vdots$
θ_N	$g(\theta_N)f(y_1,\ldots,y_n\|\theta_N)$	$g_0(\theta_N)$	r_N	$\frac{r_N}{\sum r_i}$

unscaled posterior $g(\theta)\,f(y_1,\ldots,y_n|\theta)$ for each value in the sample and then form the importance weights

$$w_i = \frac{\frac{g(\theta_i)\,f(y_1\ldots,y_n|\theta_{(i)})}{g_0(\theta_i)}}{\left(\sum_{i=1}^{N}\frac{g(\theta_i)\,f(y_1\ldots,y_n|\theta_{(i)})}{g_0(\theta_i)}\right)}.$$

Now take a random draw from the values $\theta_1,\ldots,\theta_N$ where each value θ_i has the probability of being chosen given by its importance weight w_i. The steps are:

- Draw a random sample $\theta_1,\ldots,\theta_N$ from the starting density $g_0(\theta)$.
- Calculate the unscaled posterior $g(\theta)\times f(y_1\ldots,y_n|\theta)$ for each value of θ in the sample.
- Form the weights (probabilities)

$$w_i = \frac{g(\theta_i)\,f(y;\theta_{(i)})}{g_0(\theta_i)} \Big/ \sum_{i=1}^{N}\frac{g(\theta_i)\,f(y;\theta_{(i)})}{g_0(\theta_i)}.$$

- A random draw from the values $\theta_1,\ldots,\theta_N$ using the sampling probabilities $w_1,\ldots,w_N$ will be a random draw from the posterior distribution $g(\theta|y_1,\ldots,y_n)$.

Procedure for Drawing a Random Sample Using Sampling-Importance-Resampling

We can think of the first five steps as filling in the columns of the Table 2.2.

1. Draw a large sample $\theta_1,\ldots,\theta_N$ from the starting density $g_0(\theta)$. (First column)
2. Calculate the values of the unscaled posterior at each value. (Second column)
3. Calculate the value of the starting distribution at each value of θ in the sample. (Third column)
4. Calculate the ratio of the unscaled posterior to the starting distribution for each value of θ in the sample

$$r_i = \frac{g(\theta_i)f(y_1,\ldots,y_n|\theta_i)}{g_0(\theta_i)}.$$

(Fourth column)

5. Calculate the importance weights

$$w_i = \frac{r_i}{\sum r_i}\,.$$

(Fifth column)

The Resampling Step

6. Draw a random sample of size n (with replacement) from the first column using the weights in the fifth column as the sampling probabilities. This is the final sample. The final sample size n should be no more than 10% of N. Otherwise there will be too many repeated values in the final sample. The final sample will be approximately a random sample from the posterior.

Sampling-importance-resampling is sometimes called the *Bayesian bootstrap* since we are resampling. However, here we are resampling from the sample of parameters using the importance weights, not resampling from the data.

2.3 ADAPTIVE-REJECTION-SAMPLING FROM A LOG-CONCAVE DISTRIBUTION

Sometimes we do not have a candidate distribution that dominates the unscaled target density immediately available. Gilks and Wild (1992) have developed a method called adaptive-rejection-sampling (AdRS) that can be used for a univariate target which is log-concave. In other words, the log of the target is concave downward. For a log-concave target density, the log of the density can be bounded by tangent lines that never cross the log of the density. If point A and point B lie below the log of the density, all points on the line between the points also lie below the log of the density. Figure 2.6 shows the logs of two densities, one log-concave and one not log-concave. The second derivative of the log of a log-concave target density is always non-positive.

The form of adaptive rejection sampling we will use is known as the tangent method. First we construct an envelope function (upper bound) from piecewise exponential functions that dominate the log-concave unscaled target. The envelope function gives us the shape of the initial candidate density we use. The candidate is drawn from the candidate density and is either accepted or rejected by rejection sampling. If it is rejected, the envelope function is improved by adding another exponential piece that is tangent to the target at the rejected candidate value. Thus the improved candidate density is closer to the target density. This means the next candidate will have a better chance of being accepted. This process continues until a candidate is accepted. The envelope is updated at each sampled point, and hence the candidate density approaches the target. Thus the acceptance probability rapidly approaches 1.

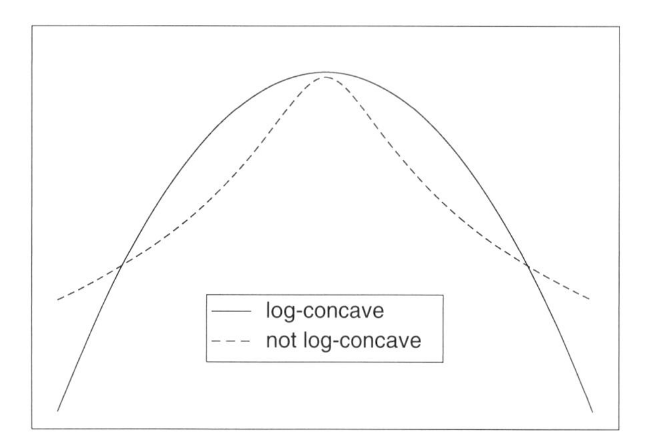

Figure 2.6 The logarithms of two densities.

Adaptive rejection sampling works for any log-concave posterior. Arbitrarily choose two points on the log posterior, one below the mode and the other above the mode. Because the posterior is log-concave, we can find the tangent lines to the log posterior through those points. Those tangents form the log of the first envelope function. The first envelope function will be found by exponentiating. Since the log of the first envelope function is the two tangent lines to the log of the posterior, the first envelope function will be made out of two exponential functions. This gives the shape of the first candidate density. Since the first candidate is piecewise exponential, the integrated first candidate density is easily obtained and will also be piecewise exponential. We can evaluate M, the limit of the integrated envelope as $\theta \to \infty$. Draw a *uniform*$(0, 1)$ and multiply it by M. Go up that amount, across to the integrated first candidate, and down to the θ axis. That will be the first candidate value θ. Then we decide whether to accept or reject the candidate. The acceptance weight, w, is the ratio of the unscaled target density to the envelope function at the candidate value. We draw u from *uniform*$(0, 1)$ distribution. If $u < w$ then we accept the candidate. It will be a random draw from the target.

However, if $u > w$ we reject the candidate. We calculate the tangent to the log of the unscaled target at the candidate value. The new log of the envelope function includes this tangent as well as the others. We exponentiate this to find the new envelope function, which now has one more exponential piece, and is closer to the target. We draw another candidate from the new envelope function and either reject it or accept it as before. The process is continued until we accept a candidate which will be a random draw from the target.

Example 3 *Suppose we wish to draw a random value from the target when all we know is that the unscaled target has the formula*

$$g(\theta) \quad = \quad \theta^4 \, e^{-\theta/2} \, .$$

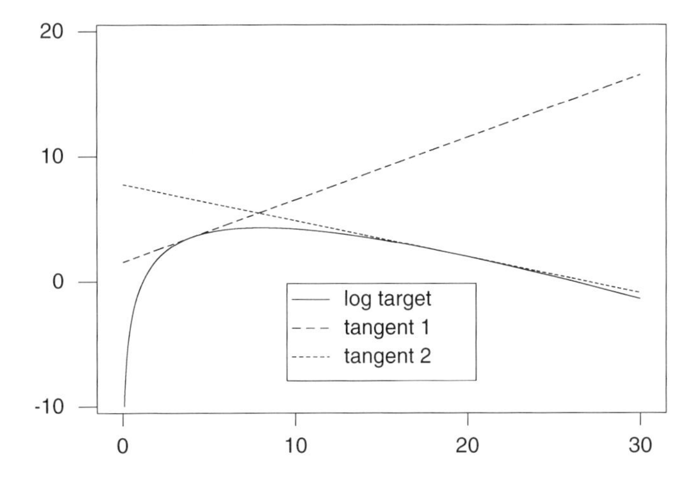

Figure 2.7 The logarithm of the target and two tangents.

We take the derivative of the logarithm of the unscaled target, set it equal to zero, and solve for the θ value that maximizes the log of the unscaled target. We find two tangents at arbitrarily chosen θ values, one below and one above the maximum. These are shown in Figure 2.7. We see that the two tangents lie entirely above the log of the target. This shows the target is log-concave. We can find an upper bound for the log of the target by the piecewise linear function made up of these two tangents. We know the point on the curve $(\theta, g(\theta))$ that each tangent goes through, and the slope $g'(\theta)$ of each tangent. We use these to determine the equation of the tangent. For our choice of tangents, the equation is

$$h(\theta) = \begin{cases} .500000\,\theta + 1.54518 & \text{for} \quad \theta < 7.89460 \\ -.289474\,\theta + 7.77776 & \text{for} \quad \theta > 7.89460 \end{cases}$$

and it is shown in Figure 2.8. We exponentiate to go back to the unscaled target and envelope density. The equation of the envelope function is

$$g_{env}(\theta) = \begin{cases} e^{+.500000\,\theta + 1.54518} & \text{for} \quad \theta < 7.89460 \\ e^{-.289474\,\theta + 7.77776} & \text{for} \quad \theta > 7.89460 \end{cases}.$$

We see that the envelope function is piecewise exponential, and dominates the unscaled target. It is shown in Figure 2.9. The integral of an exponential function is also exponential. So the integrated exponential function is given by

$$\int_{-\infty}^{\theta} g_{env}(\theta)\, d\theta = \begin{cases} \frac{1}{.50000} e^{+.500000\,\theta + 1.54518} & \text{for} \quad \theta < 7.89460 \\ \\ 1324.7 - \frac{1}{.289474} e^{-.289474\,\theta + 7.77776} & \text{for} \quad \theta > 7.89460 \end{cases}.$$

We draw u from the uniform$(0, 1)$ distribution, and scale it up by multiplying it by the maximum value of the integrated envelope. In this case, $u = .5232$ so we multiply

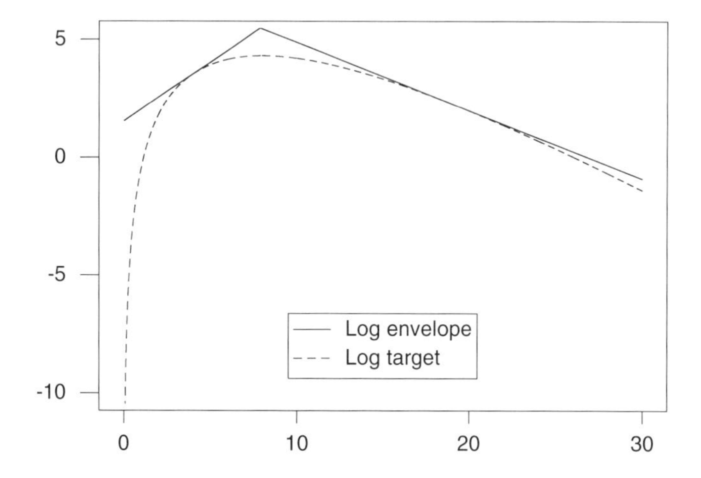

Figure 2.8 The logarithm of the target and the logarithm of the first envelope function.

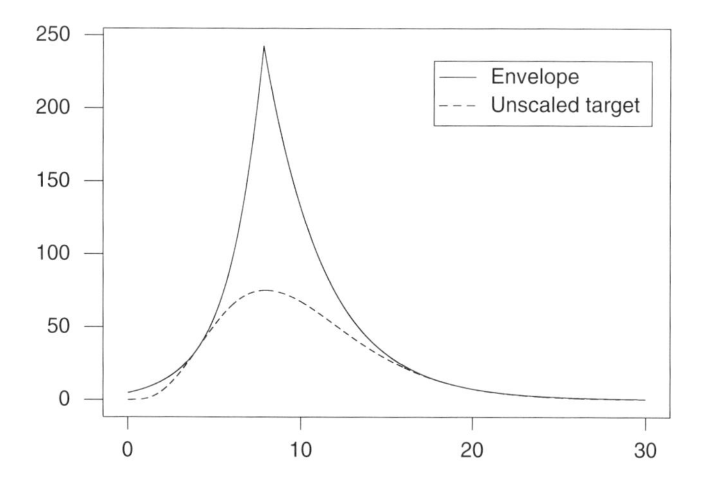

Figure 2.9 The target and the first envelope function.

it by 1324.7*, go up the vertical axis that amount, across to the integrated envelope, and down to horizontal axis to find the candidate value* $\theta = 8.875$*. This is shown in Figure 2.10. The unscaled target, first envelope, and the candidate are shown in Figure 2.11. The weight is the ratio of the height of the unscaled target to the first envelope at the candidate value. In this case the* $weight = .401229$*. We draw another* u *from the uniform*$(0, 1)$ *distribution. In this case we draw* $u = .787105$*, which is larger than the weight. Hence we do not accept this candidate. We now draw another tangent to the log of the unscaled target at the candidate value. This is shown in Figure 2.12. The log of the second envelope is made up of piecewise linear*

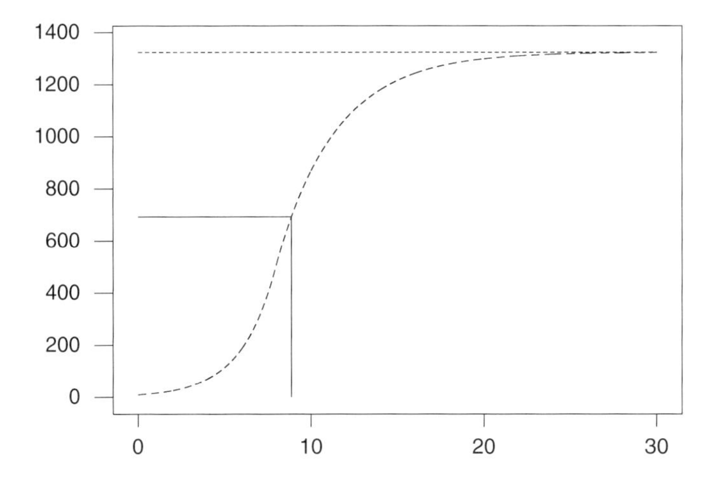

Figure 2.10 Finding the first candidate.

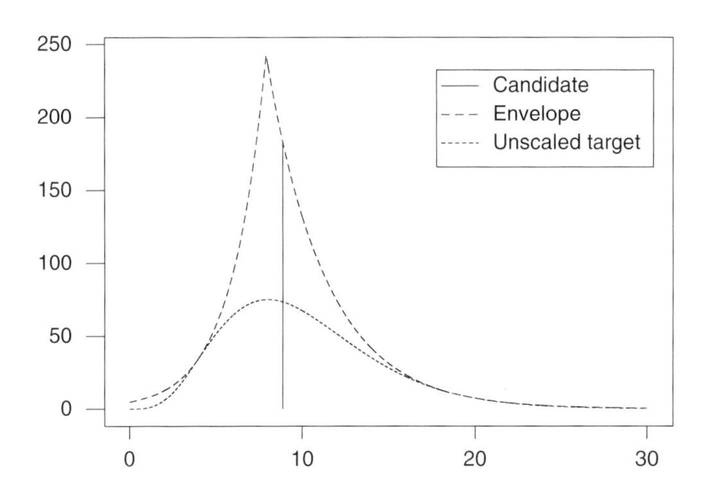

Figure 2.11 Accepting or rejecting the first candidate.

function based on all three tangents. Its equation will now be

$$h(\theta) = \begin{cases} .500000\,\theta + 1.54518 & \text{for} \quad \theta < 5.80339 \\ -.0492958\,\theta + 4.73295 & \text{for} \quad 5.80339 < \theta < 12.6773 \\ -.289474\,\theta + 7.77776 & \text{for} \quad \theta > 12.6773 \end{cases}.$$

It is shown in Figure 2.13. We exponentiate to get the second envelope function. Its equation is now

$$g_{env}(\theta) = \begin{cases} e^{+.500000\,\theta+1.54518} & \text{for} \quad \theta < 5.80339 \\ e^{-.0492958\,\theta+4.73295} & \text{for} \quad 5.80339 < \theta < 12.6773 \\ e^{-.289474\,\theta+7.77776} & \text{for} \quad 12.6773 < \theta \end{cases}.$$

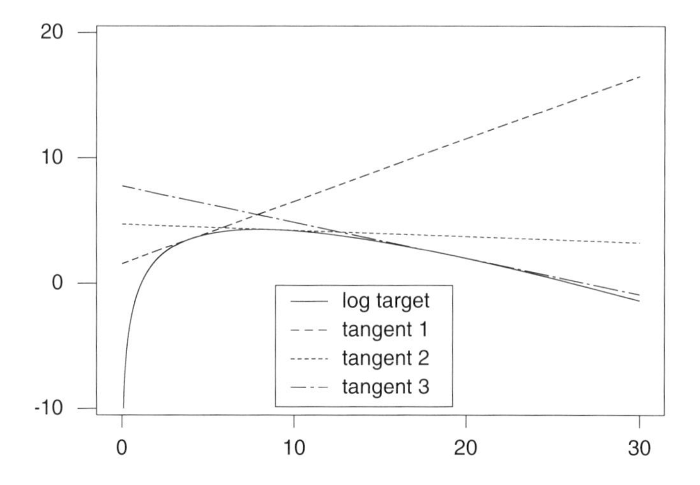

Figure 2.12 The logarithm of the target and three tangents.

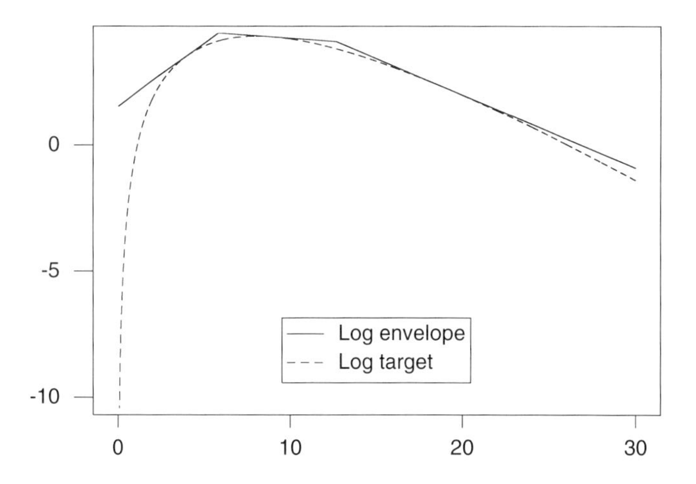

Figure 2.13 The logarithm of the target and the logarithm of the second envelope function.

The second envelope and the unscaled target are shown in Figure 2.14. We see that the second envelope function is closer to the unscaled target and still dominates it. We draw the next candidate from the second envelope by first integrating the second envelope function. That gives us the piecewise exponential with equation given by

$$\int_{-\infty}^{\theta} g_{env}(\theta)\,d\theta = \begin{cases} \frac{1}{.50000} e^{+.500000\,\theta+1.54518} & \text{for} \quad \theta < 5.80339 \\ 1731.57 - \frac{1}{.0492958} e^{-.0492958\,\theta+4.73295} & \text{for} \quad 5.80339 < \theta \\ & \quad < 12.6773 \\ 878.523 - \frac{1}{.289474} e^{-.289474*\,\theta+7.77776} & \text{for} \quad 12.6773 < \theta \end{cases} .$$

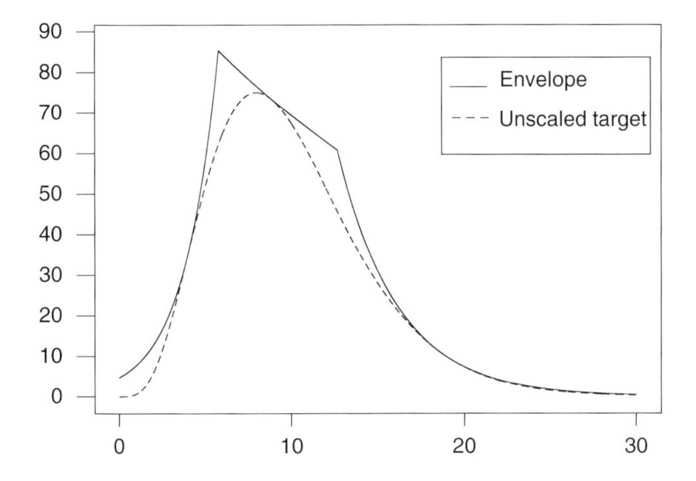

Figure 2.14 The unscaled target and the second envelope function.

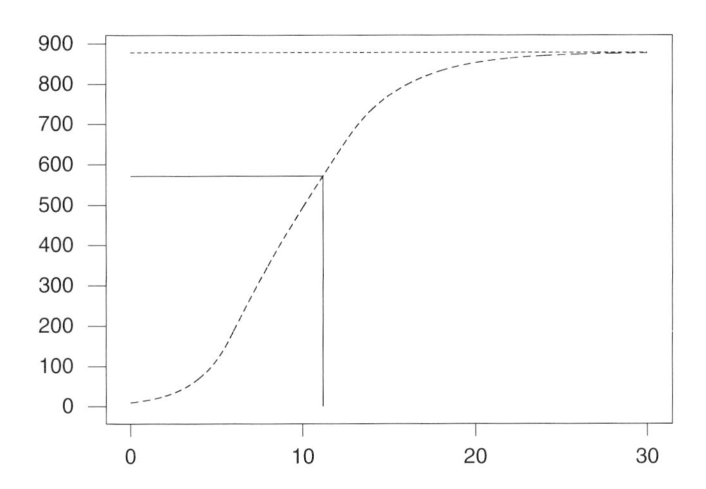

Figure 2.15 Finding the second candidate

We draw another u from the uniform$(0,1)$ distribution. In this case, we drew $u = .651319$. We multiply this by the maximum value of the integrated envelope which is 878.523 to get 572.199. We go up this far on the vertical axis, over to the integrated envelope function, and down to the horizontal axis to get the candidate value 11.1540. The second integrated envelope function with the second candidate value is shown in Figure 2.15. The unscaled target, second envelope, and the second candidate are shown in Figure 2.16. The weight is the ratio of the height of the unscaled target to the first envelope at the candidate value. In this case the $weight = .893226$. We draw another u from the uniform$(0,1)$ distribution. In this case we draw $u = .610158$,

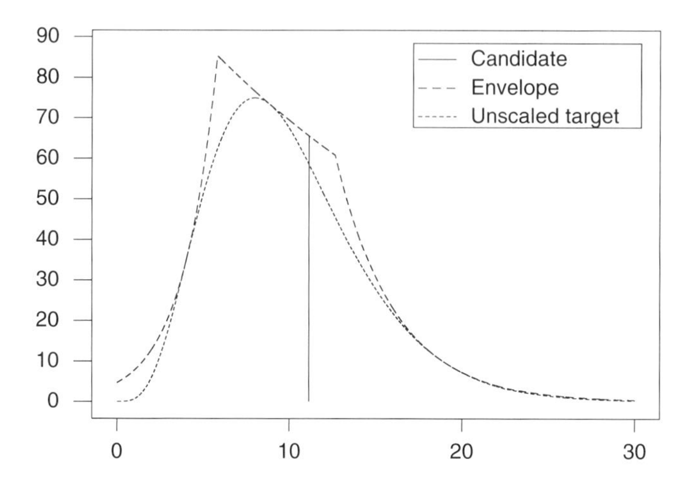

Figure 2.16 Accepting or rejecting the first candidate

which is less than the weight. Hence we accept this candidate. It is a random draw from the target distribution.

2.4 WHY DIRECT METHODS ARE INEFFICIENT FOR HIGH-DIMENSION PARAMETER SPACE

Both acceptance-rejection-sampling and sampling-importance-resampling methods can be used for multiple parameters. The candidate distribution must have heavier tails in all the parameter dimensions. This means that only a fraction of the candidates will be accepted into the final sample. As the number of parameters increases, the acceptance proportion goes down very quickly. Most of the candidate values will be drawn from a region of the parameter space that has very low posterior probability. So, the number of accepted values (in AR) goes down quite rapidly and (in SIR) the same few values get included over and over again. This makes these approaches very inefficient for high-dimensional parameter spaces.

Adaptive-rejection-sampling for multiple parameters would be based on the idea of bounding the logarithm of a log-concave target by tangent hyperplanes. While this would work in theory, the boundaries of the envelope function would become extremely complicated. In practice, adaptive-rejection-sampling only works for a single dimension parameter.

Thus, direct methods will not be used as stand-alone methods for drawing random samples from the posterior when there are multiple parameters. In later chapters we will develop Markov chain Monte Carlo methods such as Gibbs sampling and the Metropolis-Hastings algorithm for drawing samples from the posterior when there are multiple parameters. These methods set up a Markov chain that has the desired posterior as its long-run distribution. After the chain has been running for a while, a draw from the chain can be considered a draw from the desired posterior. We will sometimes use acceptance-rejection-sampling or adaptive-rejection-sampling for drawing a single parameter as part of a larger Gibbs sampling process.

Main points

- The unscaled posterior $g(\theta|y) \propto g(\theta)\, f(y|\theta)$ has all the shape information about the posterior.
- The exact posterior is found by dividing the unscaled posterior by its integral over the whole range of the parameter θ.
- A closed form for this integral only occurs for a few cases. In other cases the integral has to be found numerically. This may be hard to do.
- Acceptance-rejection-sampling is a method for drawing a random sample from the exact posterior (the target), even though all we know is the unscaled posterior.
- It draws a random sample of candidates from an easily sampled candidate distribution, $g_0(\theta)$, and reshapes this into a random sample from the exact posterior by only accepting some of the candidates into the final sample.
- The candidate density $g_0(\theta)$ must dominate the scaled up target $g(\theta)\, f(y|\theta)$. This means we can find a value M such that $M \times g_0(\theta) > g(\theta)\, f(y|\theta)$ for all values of θ. It is especially important that the candidate density have heavier tails than the unscaled target.
- Visual inspection of their graphs is not enough to make sure the candidate dominates the unscaled target. Information about this may be hidden by the axis as both are going to 0 as parameter goes to infinity.
- Take logarithms of the densities and graph them. If the log(candidate density) lies above the log(target density) in the tails, then the candidate dominates the target.
- We can check the relative tail weights exactly. Since we have formulas for the target and the starting densities, we can take the derivative of their ratio. Using the limit theorems of calculus and L'Hospital's rule we can find limits of this ratio for θ approaching $\pm\infty$. The candidate distribution has heavier tails if the limits of the ratio are greater than 1.
- Sampling-importance-resampling is a two-stage process that works by first drawing an initial sample of candidates from an easily sampled candidate distribution. Then we calculate the r ratio, which is the ratio of the unscaled posterior to the candidate density for each value in the initial sample. The sampling weight for each value is its r ratio divided by the sum of the r ratios, summed over the entire initial sample. The final sample is drawn from the sample of candidates using the calculated sampling weights.

- The second sample should be drawn with replacement. It should be no larger than 10% of the original sample size.

- Adaptive-rejection-sampling is a method for finding a random draw from a unscaled posterior (target) distribution that is log-concave. Log-concave means that the second derivative of its logarithm is always non-positive.

- For a log-concave target density, the logarithm of the unscaled target can be bounded by tangent lines. Transforming back to the unscaled target means that an envelope function can be constructed that is piecewise exponential. This envelope function is used as the candidate distribution.

- If a candidate is drawn, but not accepted, a new tangent is added to the log of the unscaled target at the unaccepted candidate value. This means the envelope function is revised by adding a new exponential part and is now closer to the target.

- Acception-rejection-sampling and sampling-importance-resampling will work for multiple parameters, but they become inefficient as the number of parameters increases.

- Adaptive-rejection-sampling would be extremely complicated for multiple parameters, so it is used only for single parameters.

- The main use of acceptance-rejection-sampling and adaptive-rejection-sampling will be to sample single parameters as part of a larger Markov-chain Monte Carlo process.

Exercises

2.1 Let the unscaled target density be given by

$$g(\theta|y) = e^{\frac{1}{2}(\theta-4)^2}\,.$$

(a) Show the *Laplace*$(3,1)$ candidate density given by

$$g_0(\theta) = \frac{1}{2}\,e^{-|\theta-3|}$$

dominates the target.

(b) Draw a sample of 10000 candidates from the candidate density, and reshape it into a sample from the target using acceptance-rejection-sampling.

(c) Calculate the mean and variance from the target sample.

2.2 Let the unscaled target density be given by

$$g(\theta|y) \propto .75\,e^{-\frac{1}{2}(\theta-3)^2} + .25\,e^{-\frac{1}{4}(\theta-6)^2}\,.$$

(a) Show the *normal*$(4, 3^2)$ candidate density dominates the target.

(b) Draw a sample of 10000 candidates from the candidate density, and reshape it into a sample from the target using acceptance-rejection-sampling.

(c) Calculate the mean and variance from the target sample.

2.3 Let the unscaled target density be given by

$$g(\theta|y) \propto e^{-|\theta|} \times e^{-\frac{1}{2}(\theta-.6)^2} .$$

(a) Show the *Laplace*$(0, 1^2)$ candidate density dominates the target.

(b) Draw a sample of 10000 candidates from the candidate density, and reshape it into a sample from the target using acceptance-rejection-sampling.

(c) Calculate the mean and variance from the target sample.

2.4 Let the unscaled target density be given by

$$g(\theta|y) \propto e^{-|\theta|} \times e^{-\frac{1}{2}(\theta-.6)^2} .$$

(a) Show the *Student's t* with 1 degree of freedom candidate density dominates the target.

(b) Draw a sample of 10000 candidates from the candidate density, and reshape it into a sample from the target using acceptance-rejection-sampling.

(c) Calculate the mean and variance from the target sample.

2.5 Let the unscaled target density be given by

$$g(\theta|y) \propto e^{-\frac{1}{2}(\theta-5)^2} .$$

(a) Show the *Laplace*$(4, 1)$ starting density candidate density given by

$$g_0(\theta) = \frac{1}{2} e^{-|\theta-4|}$$

dominates the target.

(b) Draw a sample of 100000 candidates from the starting density. Then use sampling-importance-resampling to obtain a sample of size 10000 from the target.

(c) Calculate the mean and variance from the target sample.

2.6 Let the unscaled target density be given by

$$g(\theta|y) \propto .8\, e^{-\frac{1}{2}(\theta-5)^2} + .2\, e^{-\frac{1}{2}(\theta-1)^2} .$$

(a) Show the *Normal*$(4, 3^2)$ starting density candidate density given by

$$g_0(\theta) = \frac{1}{2}e^{-|\theta-4|}$$

dominates the target.

(b) Draw a sample of 100000 candidates from the starting density. Then use sampling-importance-resampling to get a sample of size 10000 from the target.

(c) Calculate the mean and variance from the target sample.

2.7 Let the unscaled target density be given by

$$g(\theta|y) \propto e^{-|\theta-2|} \times e^{-\frac{1}{2}(\theta-1.8)^2}\,.$$

(a) Show the *Laplace*$(2, 1^2)$ starting density dominates the target.

(b) Draw a sample of 100000 candidates from the starting density. Then use sampling-importance-resampling to get a sample of size 10000 from the target.

(c) Calculate the mean and variance from the target sample.

2.8 Let the unscaled target density be given by

$$g(\theta|y) \propto e^{-|\theta|} \times e^{-\frac{1}{2}(\theta-.4)^2}\,.$$

(a) Show the *Student's t* with 1 degree of freedom starting density dominates the target.

(b) Draw a sample of 100000 candidates from the starting density. Then use sampling-importance-resampling to obtain a sample of size 10000 from the target.

(c) Calculate the mean and variance from the target sample.

3

Bayesian Inference

The posterior distribution summarizes all we believe about the parameter(s) after analyzing the data. It incorporates both our prior belief through the prior distribution and the information from the data through the likelihood function. In the Bayesian framework, the posterior is our entire inference about the parameter given the data. However, in the frequentist framework, there are several different types of inference: point estimation, interval estimation, and hypothesis testing. They are deeply embedded in statistical practice. Using the Bayesian approach, we can do each of these inferences based on the posterior distribution. In fact, under the Bayesian approach these inferences are more straightforward than under the frequentist approach. This is because, under the frequentist approach, the parameter is fixed but unknown and the only source of randomness is the distribution of the sample data given the unknown parameter which is the sampling distribution of the data. We are evaluating the parameter in its dimension based on a probability distribution in the data dimension. Under the Bayesian approach, we evaluate the parameter using a probability distribution in the parameter dimension, the posterior distribution. Bolstad (2007) gives a thorough discussion of the similarities and differences between Bayesian and frequentist inferences.

3.1 BAYESIAN INFERENCE FROM THE NUMERICAL POSTERIOR

In some cases we have a formula for the exact posterior. In other cases we only know the shape of the posterior using Bayes' theorem. The *posterior* is proportional to *prior* times *likelihood*. In those cases we can find the posterior density numerically by dividing through by the scale factor needed to make the integral of the posterior over its whole range of values equal to one. This scale factor is found by integrating the prior times likelihood over the whole range of parameter values. Thus the posterior is given by

$$g(\theta|y_1,\ldots,y_n) = \frac{g(\theta)\, f(y_1,\ldots,y_n|\theta)}{\int g(\theta)\, f(y_1,\ldots,y_n|\theta)\, d\theta} \tag{3.1}$$

where we have performed numerical integration in the denominator. The posterior density summarizes everything we know about the parameter given the data. The inferences are summaries of the posterior taken to answer particular questions.

Bayesian Point Estimation

The first type of inference is where a single statistic is calculated from the sample data and is used to estimate the unknown parameter. From the Bayesian perspective, point estimation is choosing a value to summarize the posterior distribution. The most important summary number of a distribution is its location. The posterior mean and the posterior median are good measures of location and hence would be good Bayesian estimators of the parameter. Generally we will use the posterior mean as our Bayesian estimator since it minimizes the posterior mean squared error

$$PMS(\hat{\theta}) = \int (\theta - \hat{\theta})^2\, g(\theta|y_1,\ldots,y_n) d\theta\,. \tag{3.2}$$

The posterior mean is the first moment (or balance point) of the posterior distribution. We find it by

$$\hat{\theta} = \int_{-\infty}^{\infty} \theta g(\theta|y_1,\ldots,y_n)\, d\theta. \tag{3.3}$$

The posterior median could also be used as a Bayesian estimator since it minimizes the posterior mean absolute deviation

$$PMAD(\hat{\theta}) = \int |\theta - \hat{\theta}|\, g(\theta|y_1,\ldots,y_n) d\theta\,. \tag{3.4}$$

We find the posterior median by numerically integrating the posterior. The posterior median $\tilde{\theta}$ is found by solving

$$.5 \quad = \quad \int_{-\infty}^{\tilde{\theta}} g(\theta|y_1,\ldots,y_n)\, d\theta\,. \tag{3.5}$$

The posterior median has half the posterior probability below it, and half the posterior probability above it. The posterior median can also be found from the numerical cumulative distribution function (CDF) of the posterior $G(\theta|y_1,\ldots,y_n) = \int_{-\infty}^{\theta} g(\theta|y_1,\ldots,y_n)\, d\theta$. The posterior median is the value $\tilde{\theta}$ that gives the numerical CDF value equal to .5.

Example 4 *Suppose the unscaled posterior density has shape given by*

$$g(\theta|y_1,\ldots,y_n) = .8 \times e^{-\frac{1}{2}(\theta)^2} + .2 \times \frac{1}{2}\, e^{-\frac{1}{2\times 2^2}(\theta-3)^2}\,.$$

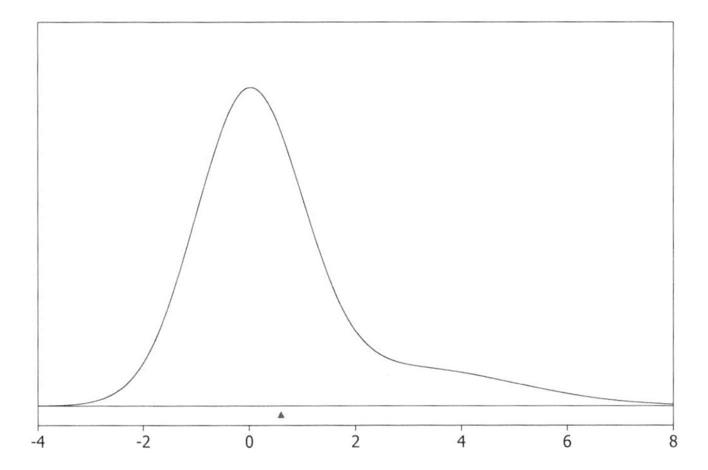

Figure 3.1 The posterior mean is the first moment or balance point of the posterior distribution.

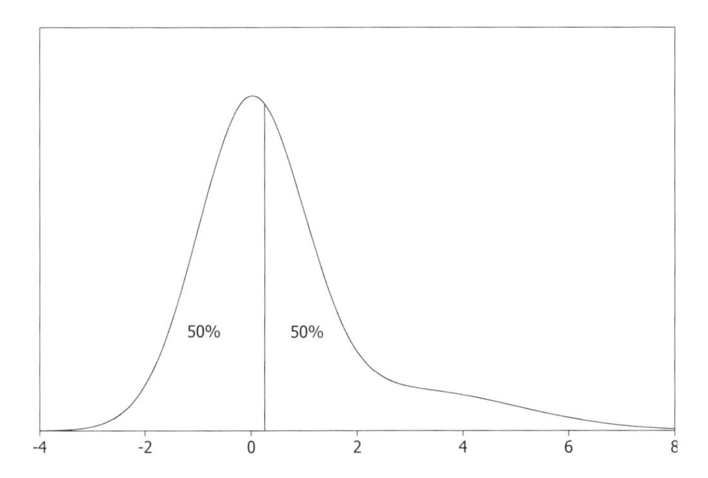

Figure 3.2 The posterior median is the point of the posterior distribution which has equal area above and below.

This is a mixture of a normal$(0, 1^2)$ *and a normal*$(3, 2^2)$. *We find the integral of the unscaled target by integrating it over the whole range*

$$\int_{-\infty}^{\infty} .8 \times e^{-\frac{1}{2}(\theta)^2} + .2 \times \frac{1}{2} e^{-\frac{1}{2\times 2^2}(\theta-3)^2} \, d\theta = 2.50633 \, .$$

Thus the numerical density is given by

$$g(\theta|y_1, \ldots, y_n) = .39899 \times \left(.8 \times e^{-\frac{1}{2}(\theta)^2} + .2 \times \frac{1}{2} e^{-\frac{1}{2\times 2^2}(\theta-3)^2} \right) .$$

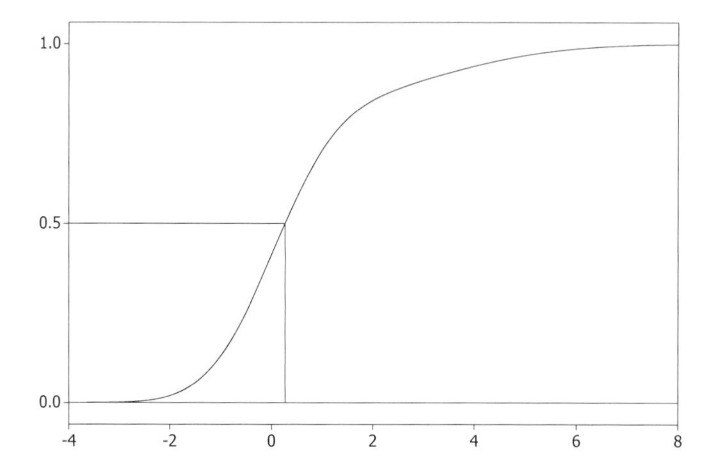

Figure 3.3 The posterior median is the θ value that has numerical CDF value equal to .5.

We find the posterior mean , $\hat{\theta}$, by integrating θ times the numerical posterior density.

$$
\begin{aligned}
\hat{\theta} &= E(\theta|y_1, \ldots, y_n) \\
&= \int_{-\infty}^{\infty} \theta \times .39899 \times \left(.8 \times e^{-\frac{1}{2}(\theta)^2} + .2 \times \frac{1}{2} e^{-\frac{1}{2\times 2^2}(\theta-3)^2} \right) d\theta \\
&= .5999
\end{aligned}
$$

The target density with the posterior mean as its balance point is shown in Figure 3.1. We find the posterior median $\tilde{\theta}$ from the numerical posterior density. It is the solution of

$$
\int_{-\infty}^{\tilde{\theta}} .39899 \times \left(.8 \times e^{-\frac{1}{2}(\theta)^2} + .2 \times \frac{1}{2} e^{-\frac{1}{2\times 2^2}(\theta-3)^2} \right) d\theta = .5000
$$

The posterior median is found to be $\tilde{\theta} = .2627$. The target density with the posterior median as the point that has equal area above and below is shown in Figure 3.2. The posterior median is shown on the graph of the CDF in Figure 3.3.

The loss-function is the cost for estimating with estimator $\hat{\theta}$ when the true parameter value is θ. The posterior mean is the Bayesian estimator that minimizes the squared-error loss-function, while the posterior median is the Bayesian estimator that minimizes the absolute value loss-function. One of the strengths of Bayesian statistics, is that we could decide on any particular loss function, and find the estimator that minimizes it. This is covered in the field of statistical decision theory. We will not pursue this topic further in this book. Readers are referred to Berger (1980) and DeGroot (1970).

Bayesian Interval Estimation

The second type of inference is where we find an interval of possible values that has a specific probability of containing the true parameter value. In the Bayesian approach, we have the posterior distribution of the parameter given the data. Hence we can calculate an interval that has the specified posterior probability of containing the random parameter θ. These are called *credible intervals.*

When we want to find a $(1-\alpha) \times 100\%$ credible interval for θ from the posterior we are looking for an interval (θ_l, θ_u) such that the posterior probability

$$\begin{aligned}(1-\alpha) &= P(\theta_l < \theta < \theta_u) \\ &= \int_{\theta_l}^{u} g(\theta|y_1, \ldots, y_n)\, d\theta\,.\end{aligned}$$

There are many possible intervals that have the required coverage probability. The shortest interval (θ_l, θ_u) with the required probability will have equal density values. That is $g(l|y_1, \ldots, y_n) = g(u|y_1, \ldots, y_n)$. However, often it is easier to find the interval (θ_l, θ_u) that has equal tail areas. This will be the interval (θ_l, θ_u) where we find θ_l and θ_u by

$$\int_{-\infty}^{\theta_l} g(\theta|y_1, \ldots, y_n)\, d\theta = \frac{\alpha}{2} \quad \text{and} \quad \int_{\theta_u}^{\infty} g(\theta|y_1, \ldots, y_n)\, d\theta = \frac{\alpha}{2}$$

respectively.

Example 4 (continued) *We find the equal tail area 95% credible interval using the numerical posterior density. The lower limit is the solution of*

$$\int_{-\infty}^{\theta_l} .39899 \times \left(.8 \times e^{-\frac{1}{2}(\theta)^2} + .2 \times \frac{1}{2}\, e^{-\frac{1}{2\times 2^2}(\theta-3)^2} \right) d\theta = .025$$

and the upper limit is the solution of

$$\int_{\theta_u}^{\infty} .39899 \times \left(.8 \times e^{-\frac{1}{2}(\theta)^2} + .2 \times \frac{1}{2}\, e^{-\frac{1}{2\times 2^2}(\theta-3)^2} \right) d\theta = .025\,.$$

We find the 95% credible interval is $(-1.888,\ 5.299)$. *The density with the credible interval is shown in Figure 3.4.*

Bayesian Testing of a One-Sided Hypothesis

The third type of inference that we want to do is to decide whether the parameter value for the new treatment is greater than the historical value of the parameter for the standard treatment. When we want to decide whether or not the treatment effect makes the parameter greater than the historical value it had for the standard treatment, we set this up as a one-sided hypothesis test. The historical value of the parameter for

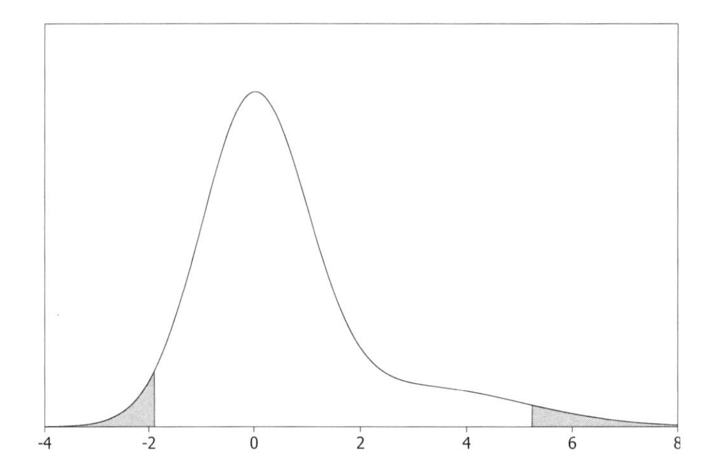

Figure 3.4 The 95% credible interval for the mean with equal tail areas.

the standard treatment is called the null value. For instance, if we want to determine if θ is greater than the null value θ_0 we decide to test the hypothesis

$$H_0 : \theta \leq \theta_0 \quad \text{versus} \quad H_1 : \theta > \theta_0 \,.$$

We note that the alternative hypothesis shows the effect we want to find out about. The null hypothesis is that there is no effect or possibly a negative effect. We only want to reject the null hypothesis when the evidence shows that it is unlikely. We decide on the significance level α. In Bayesian statistics, we test the one-sided hypothesis by calculating the posterior probability of the null hypothesis

$$\int_{-\infty}^{\theta_0} g(\theta|y_1, \ldots, y_n \, d\theta$$

using the posterior. If this probability is less than the level of significance α, we can reject the null hypothesis, and conclude the alternative hypothesis is correct.

Example 4 (continued) *Suppose we wish to determine if the parameter $\theta > 0$ at the 5% level of significance. We calculate the posterior probability of the null hypothesis using the numerical posterior density.*

$$\int_{-\infty}^{0} .39899 \times \left(.8 \times e^{-\frac{1}{2}(\theta)^2} + .2 \times \frac{1}{2} e^{-\frac{1}{2\times 2^2}(\theta-3)^2} \right) d\theta = .4134$$

This is greater than the level of significance, so we cannot reject the null hypothesis. The posterior probability of the null hypothesis is shown in Figure 3.5.

The Bayesian one-sided hypothesis test calculates the posterior probability of the null hypothesis. That is, its probability given the observed data. The null hypothesis is rejected when this calculated probability is less than the chosen level of significance

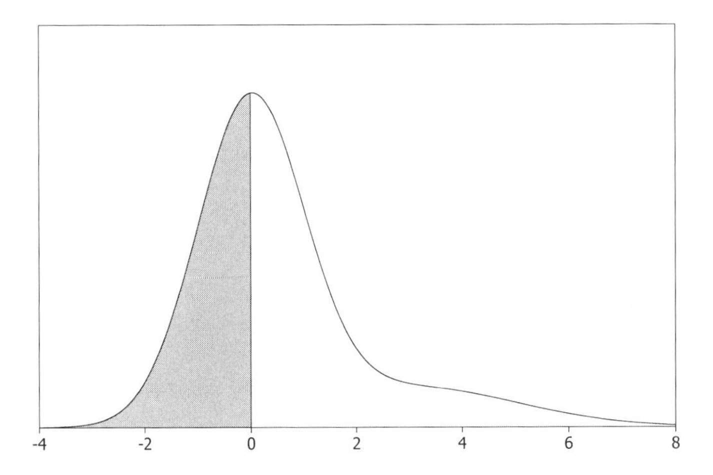

Figure 3.5 The posterior probability of null hypothesis $H_0 : \theta \leq 0$.

α. It is settling a question about where we are in the parameter dimension given a probability calculated in the parameter dimension. This contrasts with the frequentist *p-value*, which calculates the tail probability of the data, or something more extreme, from the sampling distribution of the data given the null hypothesis is true. The frequentist approach is settling the question about where we are in the parameter dimension with a probability calculated in the observation dimension. Clearly the Bayesian approach is more straightforward.

Bayesian Testing of a Two-Sided Hypothesis

On the other hand, we may want to test

$$H_0 : \theta = \theta_0 \quad \text{vs} \quad H_1 : \theta \neq \theta_0 \,.$$

The null hypothesis is true only at the single point $\theta = \theta_0$ so it is called a point null hypothesis. Its negation, the alternative hypothesis is two-sided. This would be called a two-sided hypothesis test. Under the Bayesian approach with a continuous prior, the posterior will also be continuous, so the posterior probability of the point null hypothesis will always be 0. So we can't test a two-sided hypothesis by calculating the posterior probability of the null hypothesis. Clearly we can't test the "truth" of the null hypothesis. Our choice of a continuous prior means we do not believe the null hypothesis can be literally true. Instead we calculate a $(1 - \alpha) \times 100\%$ credible interval for the parameter θ. If the null value θ_0 lies in the credible interval, then we cannot reject the null hypothesis and θ_0 remains a credible value. Note: we are not testing the "truth" of the null hypothesis, but rather its "credibility."

The Predictive Distribution of a New Observation

The predictive distribution of a new observation is the conditional distribution of the new observation y_{n+1} given the previous observations $y_1, \ldots, y_n$. It can be found by integrating the parameter out of the joint posterior of the parameter and new observations

$$\begin{aligned} f(y_{n+1}|y_1,\ldots,y_n) &\propto \int g(\theta) \times f(y_{n+1}|\theta) \times \ldots \times f(y_1|\theta)\, d\theta \\ &\propto \int f(y_{n+1}|\theta) g(\theta) \times f(y_{n+1}|\theta) \times \ldots \times f(y_1|\theta)\, d\theta \\ &\propto \int f(y_{n+1}|\theta) g(\theta|y_1,\ldots,y_n)\, d\theta\,. \end{aligned} \quad (3.6)$$

We are treating the parameter as a nuisance parameter, and marginalizing it out of the joint posterior. One of the advantages of the Bayesian approach is that we have a clear-cut way to find the predictive distribution that works in all circumstances.

3.2 BAYESIAN INFERENCE FROM POSTERIOR RANDOM SAMPLE

When we find a random sample from the posterior using one of the methods we developed in Chapter 2, we can use this sample as the basis for our inferences. We find the sample equivalent to what we calculated from the numerical posterior.

Bayesian Point Estimation

If we want to use the posterior mean as our point estimator, we calculate the mean of the posterior sample instead of calculating the mean from the numerical posterior density. Similarly, if we want to use the posterior median as our point estimator, we find the median of the posterior sample instead of calculating the median from the numerical CDF. We find the value that has 50% of the sample above and 50% below it.

Example 4 (continued) *Suppose the unscaled posterior has the shape given by*

$$g(\theta|y_1,\ldots,y_n) = .8 \times e^{-\frac{1}{2}(\theta)^2} + .2 \times \frac{1}{2} e^{-\frac{1}{2\times 2^2}(\theta-3)^2}\,.$$

This is the same shape as before. We generate a random sample from this density by acceptance-rejection-sampling. The candidate density we will use is normal$(2, 3^2)$. *Out of 100000 candidates, 31502 are accepted into the final sample. The mean for this sample equals .6039. The histogram of the random sample from the posterior with the sample mean as its balance point is shown in Figure 3.6. We see this is the sample analog of Figure 3.1. The median for this sample equals .2671. The histogram of the random sample from the posterior with the sample median showing 50% of the area of the histogram above and 50% below is given in Figure 3.7. This is the sample analog of Figure 3.2.*

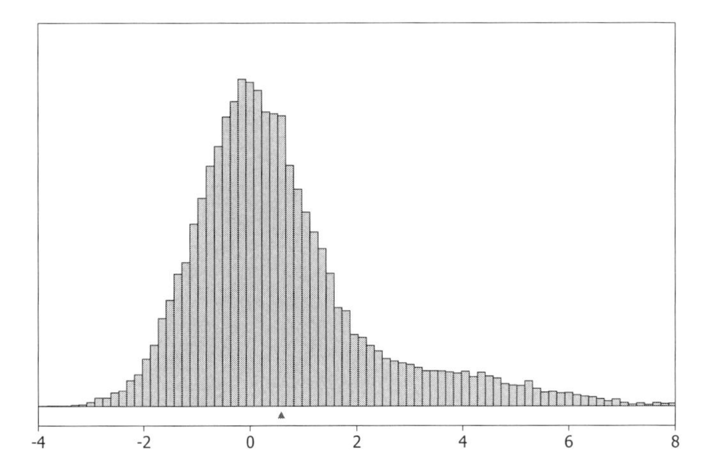

Figure 3.6 The sample mean is the first moment or balance point of the histogram of the sample from the posterior distribution.

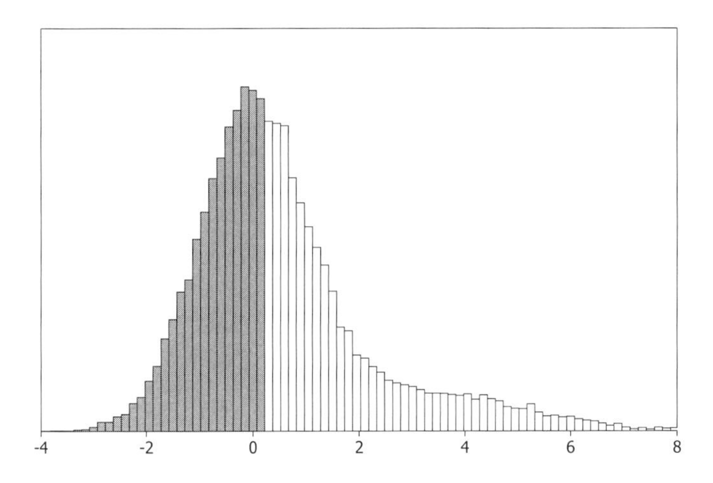

Figure 3.7 The histogram of the sample from the posterior distribution with the sample median.

Bayesian Interval Estimation

The random sample from the posterior can be used to calculate an equal-tail credible interval. If we had the exact posterior, we would find the value θ_l and θ_u such that $P(\theta \leq \theta_l) = \frac{\alpha}{2}$ and $P(\theta \geq \theta_u) = \frac{\alpha}{2}$, respectively. Since we are using the random sample from the posterior instead of the exact posterior, we find the θ_l and θ_u so the proportion of the sample values less than θ_l is $\frac{\alpha}{2}$ and the proportion of sample values greater than θ_u is also $\frac{\alpha}{2}$.

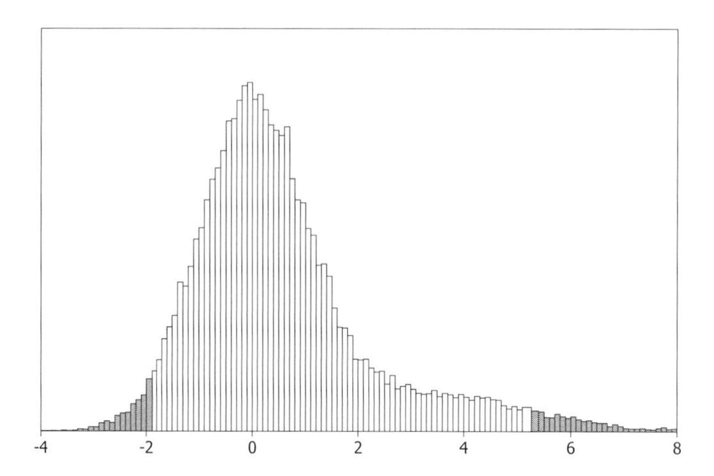

Figure 3.8 The histogram of the sample from the posterior distribution showing the 95% credible interval.

Example 4 (continued) *We calculate the equal tail 95% credible interval using the random sample from the unscaled posterior having shape given by*

$$g(\theta|y_1,\ldots,y_n) = .8 \times e^{-\frac{1}{2}(\theta)^2} + .2 \times \frac{1}{2} e^{-\frac{1}{2\times 2^2}(\theta-3)^2}\,.$$

It is $(-1.88407,\ 5.26849)$. *The histogram of the random sample with the 95% credible interval are given in Figure 3.8. This is the sample analog of Figure 3.3.*

Testing a One-Sided Hypothesis

When we have the exact posterior, $g(\theta|y_1,\ldots,y_n)$, we test a one-sided hypothesis

$$H_0 : \theta \leq \theta_0 \quad \text{versus} \quad H_1 : \theta > \theta_0$$

by calculating the posterior probability of the null hypothesis and comparing it to the chosen level of significance α. When we have a random sample from the posterior, we calculate the proportion of the sample values for which the null hypothesis is true, and compare it to α. This is the area for which the null hypothesis is true under the density histogram of the sample.

Example 4 (continued) *We wish to test the hypothesis* $H_0 : \mu \leq 0$ *versus the alternative* $H_1 : \mu > 0$ *at the 5% level of significance. We calculate the proportion of values from the random sample from the posterior that are less than or equal 0. In this case, the proportion of values less than or equal to the null value equals .418449. This is greater than the level of significance* $\alpha = .05$ *so we cannot reject the null hypothesis. The histogram with the posterior proportion is shown in Figure 3.9. This is the sample analog of Figure 3.4.*

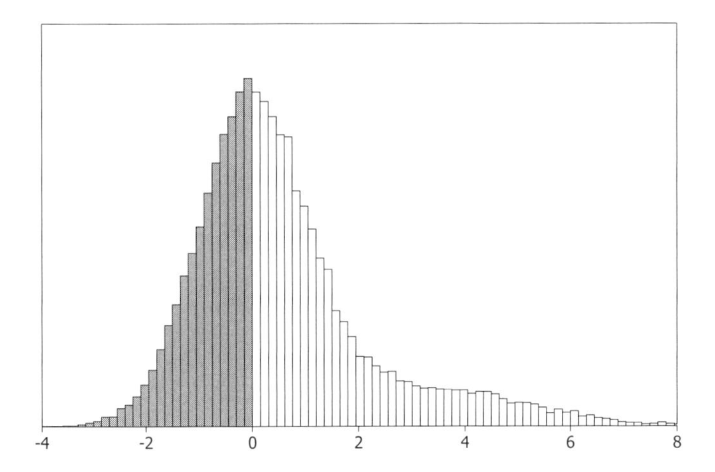

Figure 3.9 The histogram of the sample from the posterior distribution showing the proportion of sample cases that satisfy the null hypothesis $H_0 : \mu \leq 0$.

The Predictive Distribution

A random sample can be drawn from the predictive distribution by drawing a random value from $f(y_{n+1}|\theta_j)$ for each value θ_j in the posterior sample.

Summary

Smith and Roberts (1992) point out we can explore the posterior using exploratory data analysis techniques on the random sample from the posterior. This is statistics at the most basic level. However, the data analysis techniques are used on the sample from the posterior, not on the observed data. Inferences, including point estimates, credible intervals, and hypothesis tests, can be made from a random sample from the posterior. They are the sample analogs to the procedures we would use to make that inference from the exact numerical posterior. These inferences are approximations, since they come from a random sample the posterior. These approximations can be made as accurate as we need by taking a large enough sample size.

Main Points

- In the Bayesian framework the posterior distribution summarizes what we believe about the parameter given the observed data.

- Statistical inferences such as point estimation, confidence intervals, and hypothesis testing developed under the frequentist framework use the sampling distribution of the statistic given the unknown parameter. They answer questions about where we are in the parameter dimension using a probability distribution in the observation dimension.

- These inferences can be performed in a Bayesian manner using probabilities calculated from the posterior distribution of the parameter given the observed data. They answer questions about where we are in the parameter dimension using a probability distribution in the parameter dimension.
  - Posterior mean. $\hat{\beta} = \int_{-\infty}^{\infty} \theta g(\theta|y_1, \ldots, y_n)\, d\theta$.
  - Posterior median. Solution of $\int_{-\infty}^{\tilde{\theta}} g(\theta|y_1, \ldots, y_n)\, d\theta = .5$
  - Equal tail $(1 - \alpha) \times 100\%$ credible interval is an interval (θ_l, θ_u) such that

    $$\int_{-\infty}^{\theta_l} g(\theta|y_1, \ldots, y_n)\, d\theta = \frac{\alpha}{2} \quad \text{and} \quad \int_{\theta_u}^{\infty} g(\theta|y_1, \ldots, y_n)\, d\theta = \frac{\alpha}{2}\,.$$

    There is $(1 - \alpha)$ posterior probability that θ lies in the interval (θ_l, θ_u). We can test the two sided hypothesis $H_0 : \theta = \theta_0$ vs $H_1 : \theta \neq \theta_0$ at the α level of significance by observing whether or not the null value θ_0 lies in the $(1 - \alpha) \times 100\%$ credible interval for θ. If it does we reject the null hypothesis at the level α and conclude θ_0 remains a credible value. If it does not, we cannot reject the null hypothesis.
  - One-sided hypothesis test. To test $H_0 : \theta \leq \theta_0$ vs. $H_1 : \theta > \theta_0$ we calculate the posterior probability of the null hypothesis

    $$P(\theta \leq \theta_0) = \int_{-\infty}^{\theta_0} g(\theta|y_1, \ldots, y_n)\, d\theta\,.$$

    If this is less than α, then we can reject the null hypothesis at the α level of significance.

- When we have a random sample from the posterior instead of the exact numerical posterior, we do the Bayesian inferences using the analogous procedure on the posterior sample.
  - Posterior mean is the first moment or the balance point of the posterior sample. $\hat{\theta} = \frac{1}{N} \sum_{j=1}^{N} \theta_j$.
  - Posterior median is the median of the posterior sample.
  - Equal tail $(1 - \alpha) \times 100\%$ credible interval for θ is the interval (θ_l, θ_u) where the proportion of the posterior sample less than θ_l is $\frac{\alpha}{2}$ and the proportion of the posterior sample is greater than θ_u is $\frac{\alpha}{2}$. We can test a two-sided hypothesis at the α level of significance by seeing whether the null value θ_0 lies inside or outside the $(1 - \alpha) \times 100\%$ credible interval for θ.
  - One-sided hypothesis test. To test $H_0 : \theta \leq \theta_0$ vs. $H_1 : \theta > \theta_0$ we calculate the proportion of the random sample from the posterior that satisfies the null hypothesis. If this proportion is less than α, we reject the null hypothesis at that level.

Exercises

3.1 Let the unscaled posterior density be given by the formula

$$g(\theta|y_1,\ldots,y_n) \propto e^{-|\theta|} \times e^{-\frac{1}{2}(\theta-.9)^2}\,.$$

(a) Set θ for $-5 \leq \theta \leq 5$ in steps of .0001 into column 1. Calculate the values of the unscaled posterior into column 2.

(b) Find the exact numerical posterior density using the Minitab macro *tintegral.mac* or the R-function `sintegral`.

(c) Find the posterior mean and standard deviation using the Minitab macro *tintegral.mac* or the R-function `sintegral`.

(d) Find the CDF of the posterior using the Minitab macro *tintegral.mac* or the R-function `sintegral`.

(e) Find an equal tail area 95% credible interval for θ using the Minitab macro *CredIntNum.mac* or the R-function `credint`.

(f) Test the hypothesis

$$H_0 : \theta \leq 0 \quad \text{vs.} \quad H_0 : \theta > 0$$

at the 5% level of significance using the Minitab macro *PNullHypNum.mac* or the R-function `pnull`.

3.2 Let the unscaled posterior density be given by the formula

$$g(\theta|y_1,\ldots,y_n) = .3\,e^{-\frac{1}{2}((\theta+1.1)^2)} + 7 \times \frac{1}{1.5}\,e^{-\frac{1}{2\times 1.5^2}(\theta-2)^2}\,.$$

(a) Set θ for $-6 \leq \theta \leq 8$ in steps of .0001 into column 1. Calculate the values of the unscaled posterior into column 2.

(b) Find the exact numerical posterior density using the Minitab macro *tintegral.mac* or the R-function `sintegral`.

(c) Find the posterior mean and standard deviation using the Minitab macro *tintegral.mac* or the R-function `sintegral`.

(d) Find the CDF of the posterior using the Minitab macro *tintegral.mac* or the R-function `sintegral`.

(e) Find an equal tail area 95% credible interval for θ using the Minitab macro *CredIntNum.mac* or the R-function `credint`.

(f) Test the hypothesis

$$H_0 : \theta \geq 3 \quad \text{vs.} \quad H_1 : \theta < 3$$

at the 5% level of significance using the Minitab macro *PNullHypNum.mac* or the R-function `pnull`.

3.3 We want to generate a random sample from the posterior given in Exercise 3.1. First we draw a random sample of size 100000 from a *Laplace*$(0, 1)$ candidate density. Use acceptance-rejection-sampling to reshape it to be a random sample from the posterior.

(a) Calculate the mean and standard deviation of the posterior sample. How do they compare to the exact numerical solution found in Exercise 3.1?

(b) Calculate a 95% equal tail area credible interval for θ from the posterior sample using the Minitab macro *CredIntSamp.mac* or the R-function `credint`. How does it compare to the exact numerical credible interval found in Exercise 3.1?

(c) Test the hypothesis

$$H_0 : \theta \leq 0 \quad \text{vs.} \quad H_0 : \theta > 0$$

at the 5% level of significance using the Minitab macro *PNullHypSamp.mac* or the R-function `pnull`. How do the results compare with the results of the exact numerical hypothesis test in Exercise 3.1?

3.4 We want to generate a random sample from the posterior given in Exercise 3.2. First we draw a random sample of size 100000 from a *Normal*$(0, 3)$ candidate density. Use acceptance-rejection-sampling to reshape it to be a random sample from the posterior.

(a) Calculate the mean and standard deviation of the posterior sample. How do they compare to the exact numerical solution found in Exercise 3.2?

(b) Calculate a 95% equal tail area credible interval for θ from the posterior sample using the Minitab macro *CredIntSamp.mac* or the R-function `credint`. How does it compare to the exact numerical credible interval found in Exercise 3.2?

(c) Test the hypothesis

$$H_0 : \theta \leq 0 \quad \text{vs.} \quad H_0 : \theta > 0$$

at the 5% level of significance using the Minitab macro *PNullHypSamp.mac* or the R-function `pnull`. How do the results compare with the results of the exact numerical hypothesis test in Exercise 3.2?

4

Bayesian Statistics Using Conjugate Priors

In this chapter we go over the cases where the posterior distribution can be found easily, without having to do any numerical integration. In these cases, the observations come from a distribution that is a member of the exponential family of distributions, and a conjugate prior is used. The methods developed in this chapter will be used in later chapters as steps to help us draw samples from the posterior for more complicated models having many parameters.

4.1 ONE-DIMENSIONAL EXPONENTIAL FAMILY OF DENSITIES

When we have a random sample of observations $y_1, \ldots, y_n$ from a density $f(y|\theta)$ and we have a prior density $g(\theta)$, the posterior distribution is given by Bayes' theorem

$$g(\theta|y_1, \ldots, y_n) = \frac{g(\theta) \times f(y_1, \ldots, y_n|\theta)}{k}$$

where $k = \int g(\theta) \times f(y_1, \ldots, y_n|\theta)d\theta$ is the scale factor needed to make this a density function.

For most observation densities, and most prior densities, the denominator integral can't be found analytically, and instead has to be evaluated numerically. However, there is one important class of problems where the integral can be found analytically. For problems in this class, the density the random samples come from is a member of the *one-dimensional exponential family* of distributions, and the prior density comes from the *conjugate family* for that sampling density. The posterior distribution is easily found analytically for this class of problem using only simple updating formulas. There is no need for any numerical integration.

Understanding Computational Bayesian Statistics. By William M. Bolstad

The density $f(y|\theta)$ is a member of the one-dimensional exponential family of densities if and only if the observation density function can be written

$$f(y|\theta) = A(\theta)B(y)e^{C(\theta)\times T(y)} \tag{4.1}$$

for some functions $A(\theta)$, $B(y)$, $C(\theta)$, and $T(y)$. In other words, we see that when the density factors into a part depending only on the parameter times a part only dependent on the statistic times e raised to the product of a part depending on the parameter times a part depending on the statistic, then the density is a member of the one-dimensional exponential family. We will see that many of the common densities are members of this family, although their densities may not usually be written in this form. We will show how finding the posterior distribution using Bayes theorem is very simple in this case.

The conjugate family. The likelihood function has the same formula as the observation density, only with the observation held fixed and the parameter varying over all possible values. For a one-dimensional exponential family likelihood, we can absorb the factor $B(y)$ into the constant of proportionality since it is only a scale factor and does not affect the shape. The conjugate family of priors for a member of the one-dimensional exponential family of densities has the same form as the likelihood. It is given by

$$g(\theta) \propto A(\theta)^k e^{C(\theta)\times l} \tag{4.2}$$

where k and l are the constants that determine its shape. The posterior will be given by

$$\begin{aligned} g(\theta|y) &\propto g(\theta) \times f(y|\theta) && (4.3) \\ &\propto A(\theta)^k e^{C(\theta)\times l} \times A(\theta)e^{C(\theta)\times T(y)} && (4.4) \\ &\propto A(\theta)^{k+1} e^{C(\theta)\times(l+T(y))} \,. && (4.5) \end{aligned}$$

We recognize this is the member of the conjugate family with constants k' and l' where the new values of the constants are given by

$$k' = k + 1 \quad \text{and} \quad l' = l + T(y)\,.$$

Thus when the observations are from a one-dimensional exponential family, and the prior is from the conjugate family, the posterior is easily found without any need for integration. All that is needed is the simple updating formulas for the constants. We will now look at some common distributions that are members of the one-dimensional exponential family of densities.

4.2 DISTRIBUTIONS FOR COUNT DATA

In this section we will look at two common distributions for count data. The observations are numbers of occurrences, so the only possible values are non-negative integers.

Binomial Distribution

The binomial distribution arises when we are counting the number of "successes" that occur in n independent trials, where each trial results in either a "success" or a "failure" and the probability of "success" remains constant over all n trials. Let π be the probability of "success" on an individual trial, and Y be the total number of successes observed over the n trials. We know n, so it is a known constant, not a parameter. Then Y has the *binomial*(n, π) distribution with probability function given by

$$f(y|\pi) = \binom{n}{y} \pi^y (1-\pi)^{n-y}$$

for $y = 0, \ldots, n$ and $0 \leq \pi \leq 1$. We see that we can rearrange this formula to give

$$f(y|\pi) = (1-\pi)^n \binom{n}{y} \left(\frac{\pi}{1-\pi}\right)^y$$

so this distribution is a member of the one-dimensional exponential family where

$$A(\pi) = (1-\pi)^n, \quad B(y) = \binom{n}{y}, \quad C(\pi) = \log\left(\frac{\pi}{1-\pi}\right), \quad \text{and} \quad T(y) = y\,.$$

Mean and variance of binomial. Let W_i be the number of successes on trial i, for $i = 1, \ldots, n$. W_i has the binomial$(1, \pi)$ distribution and has possible values 0 and 1. Its mean equals

$$\begin{aligned} E(W_i|\pi) &= \sum_{w=0}^{1} w \times f(w|\pi) \\ &= 0 \times \pi^0(1-\pi)^1 + 1 \times \pi^1(1-\pi)^0 \\ &= \pi\,. \end{aligned}$$

Its variance equals

$$\begin{aligned} Var(W_i|\pi) &= \sum_{w=0}^{1} (w-\pi)^2 \times f(w|\pi) \\ &= (0-\pi)^2 \times \pi^0(1-\pi)^1 + (1-\pi)^2 \times \pi^1(1-\pi)^0 \\ &= \pi^2(1-\pi) + \pi(1-\pi)^2 \\ &= \pi(1-\pi)\,. \end{aligned}$$

The total number of successes $Y = \sum_{i=1}^{n} W_i$, and the W_i are all independent of each other. Thus the mean and variance of Y are given by

$$E(Y|\pi) = n\pi \quad \text{and} \quad Var(Y|\pi) = n\pi(1-\pi)\,.$$

Conjugate prior for binomial(n, π) distribution is the beta(a, b) distribution. The conjugate prior will have the form

$$\begin{aligned} g(\pi) &\propto (1-\pi)^k e^{\log\left(\frac{\pi}{1-\pi}\right)\times l} \\ &\propto \pi^l (1-\pi)^{k-l} \\ &\propto \pi^{a-1}(1-\pi)^{b-1} \end{aligned}$$

where $a = l + 1$ and $b = k - l + 1$. The density having this shape is known as the *beta*(a, b) distribution and has the exact density function given by

$$\frac{\Gamma(a+b)}{\Gamma(a)\Gamma(b)}\pi^{a-1}(1-\pi)^{b-1}$$

for $0 \le \pi \le 1$, where the $\Gamma(c)$ function is a generalization of the factorial function.[1] The Gamma function always satisfies the equation $\Gamma(c) = (c-1) \times \Gamma(c-1)$. When we use a *beta*$(a, b)$ prior for π the shape of the posterior distribution of $\pi|y$ will be

$$\begin{aligned} g(\pi|y) &\propto g(\pi)f(y|\pi) \\ &\propto \pi^{a-1}(1-\pi)^{b-1} \times \pi^y(1-\pi)^{n-y} \\ &\propto \pi^{a-1+y}(1-\pi)^{b-1+n-y} \end{aligned}$$

which we recognize as the *beta*(a', b') where the constants are updated by the simple formulas $a' = a + y$ and $b' = b + n - y$. We add the number of "successes" to a, and add the number of "failures" to b. The mean of the *beta*(a', b') distribution is found by

$$\begin{aligned} E(\pi) &= \int_0^1 \pi \times \frac{\Gamma(a'+b')}{\Gamma(a')\Gamma(b')}\pi^{a'-1}(1-\pi)^{b'-1}d\pi \\ &= \frac{\Gamma(a'+b')}{\Gamma(a')\Gamma(b')} \times \frac{\Gamma(a'+1)\Gamma(b')}{\Gamma(a'+b'+1)} \times \int_0^1 \frac{\Gamma(a'+b'+1)}{\Gamma(a'+1)\Gamma(b')}\pi^{a'}(1-\pi)^{b'-1}d\pi \\ &= \frac{a'}{a'+b'}. \end{aligned}$$

Similarly,

$$\begin{aligned} E(\pi^2) &= \int_0^1 \pi^2 \times \frac{\Gamma(a'+b')}{\Gamma(a')\Gamma(b')}\pi^{a'-1}(1-\pi)^{b'-1}d\pi \\ &= \frac{\Gamma(a'+b')}{\Gamma(a')\Gamma(b')} \times \frac{\Gamma(a'+2)\Gamma(b')}{\Gamma(a'+b'+2)} \times \int_0^1 \frac{\Gamma(a'+b'+2)}{\Gamma(a'+2)\Gamma(b')}\pi^{a'+1}(1-\pi)^{b'-1}d\pi \\ &= \frac{(a'+1)\times a'}{(a'+b'+1)\times(a'+b)}, \end{aligned}$$

[1] When c is a non-negative integer, $\Gamma(c) = (c-1)!$

so the variance of the *beta*(a', b') distribution is given by

$$\begin{aligned} Var(\pi) &= E(\pi^2) - [E(\pi)]^2 \\ &= \frac{(a'+1) \times a'}{(a'+b'+1) \times (a'+b')} - \left(\frac{a'}{a'+b'}\right)^2 \\ &= \frac{a' \times b'}{(a'+b')^2 \times (a'+b'+1)} . \end{aligned}$$

Bernoulli Process

Observations come from a *Bernoulli process*[2] when they are from a sequence of independent *Bernoulli* trials. For Bernoulli trials, each trial has two possible outcomes which we label "success" and "failure." The probability of success, π, remains constant over all the trials. The *binomial*(n, π) distribution arises when Y is the number of "successes" in a sequence of n Bernoulli trials with success probability π.

Poisson Distribution

The Poisson distribution is another distribution for counts.[3] Specifically the Poisson is a distribution which counts the number of occurrences of rare events. Unlike the binomial which counts the number of events (successes) in a known number of independent trials, the number of trials in the Poisson is so large that it isn't known. However looking at the binomial gives us way to start our investigation of the Poisson. Let Y be a binomial random variable where n is very large, and π is very small. The binomial probability function is

$$\begin{aligned} P(Y = y|\pi) &= \binom{n}{y} (\pi)^y (1-\pi)^{n-y} \\ &= \frac{n!}{(n-y)!y!} (\pi)^y (1-\pi)^{n-y} \end{aligned}$$

for $y = 0, \ldots, n$. Since π is small, the only terms that have appreciable probability are those where y is much smaller than n. We will look at the probabilities for those small values of y. Let $\mu = n\pi$. The probability function is

$$P(Y = y|\mu) = \frac{n!}{(n-y)!y!} \left(\frac{\mu}{n}\right)^y \left(1 - \frac{\mu}{n}\right)^{n-y} .$$

Rearranging the terms we get

$$P(Y = y|\mu) = \frac{n}{n} \times \frac{n-1}{n} \times \ldots \times \frac{n-y+1}{n} \times \frac{\mu^y}{y!} \left(1 - \frac{\mu}{n}\right)^n \left(1 - \frac{\mu}{n}\right)^{-y} .$$

[2]First studied by Jacob Bernoulli (1654–1705).

[3]First studied by Simeon Poisson (1781–1840).

But all the values $\frac{n}{n}, \frac{n-1}{n}, \ldots, \frac{n-y+1}{n}$ are approximately equal to 1 since y is much smaller than n. We let n approach infinity, and π approach 0 in such a way that $\mu = n\pi$ approaches a constant. We know that

$$\lim_{n\to\infty}\left(1-\frac{\mu}{n}\right)^n = e^{-\mu} \quad \text{and} \quad \lim_{n\to\infty}\left(1-\frac{\mu}{n}\right)^{-y} = 1$$

so the probability function is given by

$$P(Y = y|\mu) = \frac{\mu^y e^{-\mu}}{y!} \tag{4.6}$$

for $y = 0, 1, \ldots$. Thus, the *Poisson*(μ) distribution can be used to approximate a *binomial*(n, π) when n is large, π is very small, and $\mu = n\pi$. Clearly the *Poisson*(μ) distribution is a member of the one-dimensional exponential family with

$$A(\mu) = e^{-\mu}, \quad B(y) = \frac{1}{y!}, \quad C(\mu) = \log\mu, \quad \text{and} \quad T(y) = y\,.$$

Mean and variance of Poisson. The mean of the *Poisson*(μ) can be found by

$$\begin{aligned} E(Y|\mu) &= \sum_{y=0}^{\infty} y\frac{\mu^y e^{-\mu}}{y!} \\ &= \sum_{y=1}^{\infty} \frac{\mu^y e^{-\mu}}{(y-1)!}\,. \end{aligned}$$

We let $y' = y - 1$, and factor out μ

$$E(Y|\mu) = \mu\sum_{y'=0}^{\infty}\frac{\mu^{y'} e^{-\mu}}{y'!}\,.$$

The sum equals one since it is the the sum over all possible values of a Poisson distribution, so the mean of the *Poisson*(μ) is

$$E(Y|\mu) = \mu\,.$$

Similarly we can evaluate

$$\begin{aligned} E(Y \times (Y-1)|\mu) &= \sum_{y=0}^{\infty} y \times (y-1) \times \frac{\mu^y e^{-\mu}}{y!} \\ &= \sum_{y=2}^{\infty} \frac{\mu^y e^{-\mu}}{(y-2)!}\,. \end{aligned}$$

We let $y' = y - 2$, and factor out μ^2

$$E(Y \times (Y-1)|\mu) = \mu^2 \sum_{y'=0}^{\infty} \frac{\mu^{y'} e^{-\mu}}{y'!} .$$

The sum equals one since it is the the sum is over all possible values of a Poisson distribution, so $E(y \times (y-1))$ for a *Poisson*(μ) is given by

$$E(Y \times (Y-1)|\mu) = \mu^2 .$$

The variance is given by

$$\begin{aligned} var(Y|\mu) &= E(Y^2) - [E(Y)]^2 \\ &= E(Y \times (Y-1)) + E(Y) - [E(Y)]^2 \\ &= \mu^2 + \mu - \mu^2 \\ &= \mu . \end{aligned}$$

Thus, we see the mean and variance of a *Poisson*(μ) are both equal to μ. Note: the mean and variance of *binomial*(n, π) when n is large and π is small are $n\pi$ and $n\pi(1-\pi)$ which is approximately $n\pi$.

Conjugate prior for Poisson(μ) is the gamma(r, v) distribution. The conjugate prior for the *Poisson*(μ) will have the same form as the likelihood. Hence it has shape given by

$$\begin{aligned} g(\mu) &\propto e^{-k\mu} e^{\log \mu \times l} \\ &\propto \mu^l e^{-k\mu} . \end{aligned}$$

The distribution having this shape is known as the *gamma*(r, v) distribution and has density given by

$$g(\mu; r, v) = \frac{v^r \mu^{r-1} e^{-v\mu}}{\Gamma(r)}$$

where $r - 1 = l$ and $v = k$. When we have a *Poisson*(μ) observation, and use *gamma*(r, v) prior for μ, the shape of the posterior is given by

$$\begin{aligned} g(\mu|y) &\propto g(\mu) \times f(y|\mu) \\ &\propto \frac{v^r \mu^{r-1} e^{-v\mu}}{\Gamma(r)} \times \frac{\mu^y e^{-\mu}}{y!} \\ &\propto \mu^{r-1+y} e^{-(v+1)\mu} . \end{aligned}$$

We recognize this to be a $gamma(r', v')$ density where the constants are updated by the simple formulas $r' = r + y$ and $v' = v + 1$. We add y to r, and we add 1 to v. The mean of the $gamma(r', v')$ distribution is found by

$$\begin{aligned} E(\mu) &= \int_0^\infty \mu \times \frac{(v')^{r'} \mu^{r'-1} e^{-v'\mu}}{\Gamma(r')} d\mu \\ &= \int_0^\infty \frac{(v')^{r'} \mu^{r'} e^{-v'\mu}}{\Gamma(r')} d\mu \\ &= \frac{r'}{v'} \int_0^\infty \frac{(v')^{r'+1} \mu^{r'} e^{-v'\mu}}{\Gamma(r'+1)} d\mu \\ &= \frac{r'}{v'} . \end{aligned}$$

Similarly

$$\begin{aligned} E(\mu^2) &= \int_0^\infty \mu^2 \times \frac{(v')^{r'} \mu^{r'-1} e^{-v'\mu}}{\Gamma(r')} d\mu \\ &= \int_0^\infty \frac{(v')^{r'} \mu^{r'+1} e^{-v'\mu}}{\Gamma(r')} d\mu \\ &= \frac{r'(r'+1)}{(v')^2} \int_0^\infty \frac{(v')^{r'+2} \mu^{r'+1} e^{-v'\mu}}{\Gamma(r'+2)} d\mu \\ &= \frac{r'(r'+1)}{(v')^2} . \end{aligned}$$

Thus the variance of the $gamma(r', v')$ distribution is

$$\begin{aligned} Var(\mu) &= E(\mu^2) - [E(\mu)]^2 \\ &= \frac{r'(r'+1)}{(v')^2} - \left[\frac{r'}{v'}\right]^2 \\ &= \frac{r'}{(v')^2} . \end{aligned}$$

Poisson Process

The *Poisson process* counts the number of occurrences of events that are occurring randomly through time and have *independent increments*. This means that what happens (the occurrence or non-occurrence of events) in time intervals that do not overlap are independent of each other. While the events themselves are occurring at random times, the intensity rate at which they are occurring is constant. This means the process is *time-homogeneous*.[4] The events are occurring one at a time. The *Poisson process* would be a suitable model for traffic accidents on a particular highway, but not for deaths from traffic accidents on the highway, since some accidents cause multiple deaths.

Under the following assumptions, Y_t number of occurrences in the time interval $(0, t)$ has the *Poisson* distribution with mean $\mu = \lambda t$.

1. Occurrences of an event happen randomly through time at a constant rate λ.
2. Y_t counts the number of occurrences of the event in the time interval. $(0, t)$.
3. What happens in one time interval is independent of what happens in another non-overlapping time interval. (*independent increments*)
4. The probability of exactly one event occurring in a very short time interval $(t, t + h)$ is approximately proportional to the length of the interval h.

$$P(\text{exactly one event occurs in}(t, t + h)) = \lambda h + o(h)$$

where $o(h)$ is a function that approaches 0 faster than h, i.e.,

$$\lim_{h \to 0} \left(\frac{o(h)}{h} \right) = 0.$$

5. The probability that more than one event occurs in a very short time interval $(t, t + h)$ approaches 0 faster than h. i.e.,

$$\begin{aligned} \lim_{h \to 0} P(\text{more than one event occurs in}(t, t + h)) &= \lim_{h \to 0} \left(\frac{o(h)}{h} \right) \\ &= 0\,. \end{aligned}$$

Then Y_t number of occurrences in the time interval $(0, t)$ has the Poisson distribution with mean $\mu = \lambda t$. When this is true, we say the process follows the Poisson process with intensity rate λ. The details of the proof are given in the Appendix at the end of this chapter for readers that want the mathematical details. Some readers may prefer to skip over the proof and just accept the conclusion.

4.3 DISTRIBUTIONS FOR WAITING TIMES

Sometimes the random variable we are interested in is the time we have to wait until some event occurs. In this section we look at four distributions for waiting time, two discrete distributions associated with Bernoulli processes, and two continuous distributions associated with the Poisson process. In each situation the first distribution is a special case of the second.

[4] Non-time-homogeneous Poisson processes with non-constant intensity rates are possible, but beyond the scope of this text.

The Geometric (π) Distribution

A sequence of independent trials where each trial has two possible outcomes which we call "success" or "failure", and the probability of success remains constant for each trial are called *Bernoulli* trials. The *binomial*(n, π) distribution gives the probability that we have Y successes in a sequence of n *Bernoulli* trials.

Suppose we look at the number of "failures" that occur before the first success. So we will let Y be the number of "failures" that occur before the first success. The possible values will be the non-negative integers $y = 0, 1, 2, \ldots$ as it is not possible to set an upper limit for the number of failures before the first success is certain. The event $Y = y$ corresponds to the sequence $F\,F\,\ldots\,F\,S$ where there are y consecutive failures followed by a success. Since we have independent Bernoulli trials which all have the same probability of success π,

$$\begin{aligned} P(Y = y) &= (1-\pi) \times \ldots \times (1-\pi) \times \pi \\ &= (1-\pi)^y \times \pi \end{aligned}$$

for $y = 0, 1, 2, \ldots$. The mean and variance of the *geometric*(π) are easily found using the relationships

$$\sum_1^\infty kab^k = \frac{ab}{(1-b)^2} \quad \text{and} \quad \sum_1^\infty k^2 ab^k = \frac{ab(1+b)}{(1-b)^3}$$

where $0 < b < 1$. The mean and variance of the *geometric*(π) are

$$E(Y|\pi) = \frac{1-\pi}{\pi} \quad \text{and} \quad Var(Y|\pi) = \frac{1-\pi}{\pi^2}$$

respectively.

Conjugate prior for the geometric (π) is the beta(a, b) distribution. The *geometric*(π) probability function can be written as

$$\begin{aligned} f(y) &= \pi\,(1-\pi)^y \\ &= \pi e^{\log(1-\pi) \times y} \end{aligned}$$

so clearly it is a member of the one-dimensional exponential family with $a(\pi) = \pi$, $c(\pi) = \log(1-\pi)$, and $T(y) = y$. Its conjugate family will have the same form as the likelihood. Its shape is given by

$$\begin{aligned} g(\pi) &= \pi^k e^{\log(1-\pi) \times l} \\ &= \pi^k\,(1-\pi)^l \\ &\propto \pi^{a-1}(1-\pi)^{b-1} \end{aligned}$$

where $a = k + 1$ and $b = l + l + 1$. The density having this shape is known as the *beta*(a, b) distribution and has the exact density function given by

$$\frac{\Gamma(a+b)}{\Gamma(a)\Gamma(b)}\pi^{a-1}(1-\pi)^{b-1}$$

for $0 \leq \pi \leq 1$. When we have a *geometric*(π) observation y, and use a *beta*(a, b) prior for π, the posterior is given by

$$\begin{aligned} g(\pi|y) &\propto g(\pi) \times f(y|\pi) \\ &\propto \pi^{a-1}(1-\pi)^{b-1} \times \pi(1-\pi)y \\ &\propto \pi^{a}(1-\pi)^{b+y-1} \end{aligned}$$

which we recognize as a *beta*(a', b') where $a' = a + 1$ and $b' = b + y$. The simple updating rules are add number of "successes" (which equals 1) to a, and add number of "failures" to b. In words, this is the same simple updating rule as for the binomial observation, but for geometric, the observed y is number of failures, not number of successes as it is in the binomial.

Negative Binomial (r, π) Distribution

Again, suppose we have a sequence of independent *Bernoulli* trials. Each trial has two possible outcomes, "success" and "failure" and the probability of success remains constant for each trial. The *binomial*(n, π) distribution gives the probability that we have y successes in a sequence of n *Bernoulli* trials. Instead of counting the number of successes in a fixed number of trials, y is the number of failures that occur before the rth success. So r is a positive integer.[5] Clearly, the *geometric*(π) distribution is a special case of the *negative binomial*(r, π) distribution where $r = 1$. For the negative binomial random variable Y to have the value y, there must be $y - 1$ failures in the first $r + y - 1$ trials, followed by the rth success. The trials are independent so the probability $P(Y = y)$ is the product of these two probabilities. The first is the binomial probability of $y - 1$ successes in $r + y - 1$ trials, and the second is the binomial probability of a success in a single *Bernoulli* trial. Thus the probability equals

$$\begin{aligned} P(Y = y) &= \binom{y+r-1}{r-1}\pi^{r-1}(1-\pi)^{y} \times \pi \\ &= \binom{y+r-1}{r-1}\pi^{r}(1-\pi)^{y} \end{aligned}$$

for $y = 0, 1, 2, \ldots$.

[5]The *negative binomial* can be defined for values of r that are not integers, but that is beyond the scope of this book.

Mean and variance of the negative binomial. The number of failures until the rth success is the sum of r geometric random variables, the number of failures before the first success and the number of failures between each success and the next one. The sequence of *Bernoulli* trials are all independent. Hence, the mean and variance of the *negative binomial*(r, π) are given by

$$E(Y|\pi) = \frac{r(1-\pi)}{\pi} \quad \text{and} \quad Var(Y|\pi) = \frac{r(1-\pi)}{\pi^2}. \tag{4.7}$$

Conjugate prior for negative binomial(r, π) is the beta(a, b). Since the likelihood of *negative binomial*(r, π) has the same form as the likelihood of the *binomial*(n, π), it will also have the same conjugate prior. When we have a *negative binomial*(r, π) observation y, and are using a *beta*(a, b) prior for π, the shape of the posterior is given by

$$\begin{aligned} g(\pi|y) &\propto g(\pi) \times f(y|\pi) \\ &\propto \pi^{a-1}(1-\pi)^{b-1} \times \pi^{r}(1-\pi)^{y} \\ &\propto \pi^{a-1+r}(1-\pi)^{b-1+y} \end{aligned}$$

which we recognize as a *beta*(a', b') where the constants are updated by the equations $a' = a + r$ and $b' = b + y$. In words, this is the same simple updating rules as for the binomial. Add the number of "successes" to a, and add the number of "failures" to b. But here, y is the number of "failures" not the number of "successes" as it is in the binomial case.

Exponential(λ) Waiting Time Distribution

Suppose events are occurring according to a *Poisson process*. This means they occur singly (one-at-a-time) and randomly in time at a constant rate λ. We saw previously that Y_t, the number of events that occur in the time interval $(0, t)$ has the *Poisson* distribution with parameter λt. Let T be the time until the first arrival. The two events "the waiting time for the first arrival is greater than t" and "the number of arrivals in the interval $(0, t)$ equals 0" are equivalent. Hence

$$\begin{aligned} P(T > t) &= P(Y_t = 0) \\ &= \frac{(\lambda t)^0 \, e^{\lambda t}}{0\,!} \end{aligned}$$

since Y_t has the *Poisson* distribution with parameter λt. The density function $f(t)$ is found by taking derivative of $P(T \leq t)$ with respect to t. Clearly

$$\begin{aligned} f(t|\lambda) &= \frac{d}{dt}(1 - e^{-\lambda t}) \\ &= \lambda e^{-\lambda t}\,. \end{aligned}$$

This is the *exponential*(λ) density. So the waiting time for the first occurrence of a *Poisson process* with intensity rate λ has the *exponential*(λ) distribution. Let $T_1, T_2, \ldots$ be the times of the first, second, ... occurrence of the *Poisson process* respectively. The interarrival times $T_2 - T_1, T_3 - T_2, \ldots$ between occurrences of the *Poisson process* with intensity rate λ will all have the *exponential*(λ) distribution. The "arrivals occurring according to a *Poisson process* with rate λ" is equivalent to "the interarrival times have the *exponential*(λ) distribution."

Exponential distribution is "memoryless." Suppose we look at the probability that $T > t + s$ given we know $T > t$.

$$\begin{aligned} P(T > t + s | T > t) &= \frac{P(T > t + s \cap T > t)}{P(T > t)} \\ &= \frac{P(T > t + s)}{P(T > t)} \\ &= \frac{e^{-\lambda(t+s)}}{e^{-\lambda t}} \\ &= e^{-\lambda s}. \end{aligned}$$

The distribution of the waiting time until first occurrence, given that we have already waited until time t is the same as the distribution of the waiting time starting from 0. The exponential distribution is the only continuous distribution that has this memoryless property.

Conjugate prior for exponential(λ) is gamma(r, v). The conjugate prior for λ will have the same form as the likelihood. Thus its shape is given by

$$g(\lambda) \propto \lambda^k e^{-\lambda l}.$$

We recognize this to be the gamma(r, v) distribution which has exact density given by

$$g(\lambda) = \frac{\lambda^{r-1} v^r}{\Gamma(r)} e^{-v\lambda}$$

where $r - 1 = k$ and $v = l$. If t is the waiting time from the *exponential*(λ) distribution and we use the gamma(r, v) prior distribution for λ the shape of the posterior is given by

$$\begin{aligned} g(\lambda | t) &\propto \frac{\lambda^{r-1} v^r}{\Gamma(r)} e^{-v\lambda} \times \lambda e^{-\lambda t} \\ &\propto \lambda^{r-1+1} e^{-(v+t)\lambda} \\ &\propto \lambda^{r'-1} e^{v'\lambda} \end{aligned}$$

where the constants are updated by $r' = r + 1$ and $v' = v + t$. We recognize this to be the *gamma*(r', v') distribution.

Gamma(n, λ) Waiting Time Distribution

Suppose arrivals are occurring according to a *Poisson process* with intensity rate λ and we want to find the waiting time distribution until the nth arrival. Let W_n be the waiting time until the nth arrival.

$$\begin{aligned} P(W_n > t) &= P(Y_t < n) \\ &= \sum_{k=0}^{n-1} P(Y_t = k)\,. \end{aligned}$$

The density function of W_n is found by taking the derivative of $P(W_n < t)$ with respect to t.

$$\begin{aligned} f(t|\lambda) &= \frac{d}{dt}(P(W_n < t)) \\ &= \frac{d}{dt}\left[\sum_{k=n}^{\infty} \frac{(\lambda t)^k e^{-\lambda t}}{k\,!}\right] \\ &= \sum_{k=n}^{\infty}\left(\frac{\lambda(\lambda t)^{k-1} e^{-\lambda t}}{k-1\,!} - \frac{\lambda(\lambda t)^k e^{-\lambda t}}{k\,!}\right). \end{aligned}$$

Since the second term for each k cancels out with the first term for the next k, and the terms are going to zero as k increases without bound all we will be left with is the first term for $k = n$. This gives

$$\begin{aligned} f(t|\lambda) &= \frac{\lambda(\lambda t)^{n-1} e^{-\lambda t}}{n-1\,!} \\ &= \frac{\lambda^n t^{n-1} e^{-\lambda t}}{n-1\,!}\,. \end{aligned}$$

This is known as the *gamma*(n, λ) density. We see clearly that the *exponential*(λ) density is the *gamma*$(1, \lambda)$ density.

Conjugate prior for gamma(n, λ) distribution is the gamma(r, v) distribution. The conjugate prior for λ will have the same form as the likelihood. Thus its shape is given by

$$g(\lambda) \propto \lambda^k e^{-\lambda l}\,.$$

We recognize this to be the gamma(r, v) distribution, which has exact density given by

$$g(\lambda) = \frac{\lambda^{r-1} v^r}{\Gamma(r)} e^{-v\lambda}$$

where $r-1=k$ and $v=l$. If t is the waiting time from the *gamma*(n, λ) distribution and we use the gamma(r, v) prior distribution for λ the shape of the posterior is given by

$$\begin{aligned} g(\lambda|t) &\propto \frac{\lambda^{r-1}v^r}{\Gamma(r)}e^{-v\lambda} \times \lambda^n t^{n-1}e^{-\lambda t} \\ &\propto \lambda^{r-1+n}e^{-(v+t)\lambda} \\ &\propto \lambda^{r'-1}e^{v'\lambda} \end{aligned}$$

where the constants are updated by $r' = r + n$ and $v' = v + t$. We recognize this to be the *gamma*(r', v') distribution.

4.4 NORMALLY DISTRIBUTED OBSERVATIONS WITH KNOWN VARIANCE

Let us look at where we have a random sample of normally distributed observations. We will look at the two parameters, the mean μ and the variance σ^2, separately. In this section we will find the posterior distribution of the mean μ given the variance σ^2 is a known constant. This way the observations come from a one-dimensional exponential family making the analysis relatively simple.

Single Draw from a Normal Distribution with Known Variance

Suppose y is a random draw from a *normal*(μ, σ^2) distribution where σ^2 is a known constant. The likelihood is a function of the single parameter μ and is given by

$$\begin{aligned} f(y|\mu) &= \frac{1}{\sqrt{2\pi}\sigma}e^{-\frac{1}{2\sigma^2}(y-\mu)2} \\ &= e^{-\frac{\mu^2}{2\sigma^2}} \times \frac{1}{\sqrt{2\pi}\sigma} \times e^{-\frac{y2}{2\sigma^2}} \times e^{\frac{\mu}{\sigma^2}\times y} \\ &= A(\mu) \times B(y) \times e^{C(\mu)\times T(y)} . \end{aligned}$$

Thus it is a one-dimensional exponential family. The conjugate family of priors will have the same form as the likelihood.

$$\begin{aligned} g(\mu; k, l) &\propto [A(\mu)]^k e^{C(\mu)\times l} \\ &\propto e^{-\frac{1}{2\sigma^2}(k\mu^2 - 2l\mu)} . \end{aligned}$$

Factoring out k, completing the square (and absorbing the extra bit into the constant) gives

$$\begin{aligned} g(\mu;k,l) &\propto e^{-\frac{1}{2\sigma^2/k}\left[\mu^2-2\frac{l\mu}{k}+\left(\frac{l}{k}\right)^2-\left(\frac{l}{k}\right)^2\right]} \\ &\propto e^{-\frac{1}{2\sigma^2/k}\left[\mu-\frac{l}{k}\right]^2} \\ &\propto e^{-\frac{1}{2s^2}\left[\mu-m\right]^2} \end{aligned}$$

where $s^2 = \sigma^2/k$ and $m = \frac{l}{k}$. We recognize this has the *normal* shape with mean m and variance s^2. The posterior will be proportional to the prior times likelihood. This will be

$$\begin{aligned} g(\mu|y) &\propto g(\mu) \times f((y|\mu) \\ &\propto [A(\mu)]^{k'} e^{C(\mu)\times l'} . \end{aligned}$$

This is the conjugate posterior with $k' = k+1$ and $l' = l+y$. When we put this back in the usual form, this is normal with mean $m' = \frac{l+y}{k+1}$ and variance $(s')^2 = \frac{\sigma^2}{k+1}$.

Simple updating formulas for single observation. When we put this in terms of the precisions, which are the reciprocals of the variance,

$$\frac{1}{(s')^2} = \frac{1}{s^2} + \frac{1}{\sigma^2} \quad \text{and} \quad \frac{\frac{1}{s^2}}{\frac{1}{(s')^2}} \times m + \frac{\frac{1}{\sigma^2}}{\frac{1}{(s')^2}} \times y .$$

We can summarize the first updating formula by "the posterior precision equals the sum of the prior precision plus the observation precision," and the second updating formula by "the posterior mean is the weighted average of the prior mean and the observation, where the weights are the proportions of their precisions to the posterior precision."

Random sample from a normal distribution with known variance. Suppose $y_1, \ldots, y_n$ are a random sample from a *normal*(μ, σ^2) distribution where the variance σ^2 is a known constant. For a random sample, the joint likelihood is the product of the individual likelihoods, so it is given by

$$\begin{aligned} f(y_1,\ldots,y_n|\mu) &= f(y_1|\mu) \times \ldots \times f(y_n|\mu) \\ &\propto \prod_{i=1}^{n} e^{-\frac{1}{2\sigma^2}(y_i-\mu)^2} \\ &\propto e^{-\frac{1}{2\sigma^2}\sum(y_i-\mu)^2} . \end{aligned}$$

Multiply out the terms in the exponent and collect like terms

$$\begin{aligned}\sum_{i=1}^{n}(y_i-\mu)^2 &= \sum_{i=1}^{n} y_i^2 - \sum_{i=1}^{n} 2\mu y_i + \sum_{i=1}^{n}\mu^2 \\ &= n\mu^2 - 2\mu n\bar{y} + \sum_{i=1}^{n} y_i^2 .\end{aligned}$$

Factor out n from the first terms and complete the square

$$\begin{aligned}\sum_{i=1}^{n}(y_i-\mu)^2 &= n(\mu^2 - 2\mu\bar{y}) + \sum_{i=1}^{n} y_i^2 \\ &= n(\mu-\bar{y})^2 - n(\bar{y})^2 + \sum_{i=1}^{n} y_i^2 .\end{aligned}$$

Put this back in the likelihood, and absorb the part that does not affect the shape into the constant

$$\begin{aligned}f(y_1,\ldots,y_n|\mu) &\propto e^{-\frac{1}{2\sigma^2/n}(\mu-\bar{y})^2} \times e^{-\frac{1}{2\sigma^2}[n(\bar{y}^2)+\sum_{i=1}^{n} y_i^2]} \\ &\propto e^{-\frac{1}{2\sigma^2/n}(\mu-\bar{y})^2}\end{aligned}$$

We recognize this is the likelihood of $\bar{y}$. It is a normal distribution with mean μ and variance $\sigma_{\bar{y}}2 = \frac{\sigma^2}{n}$. Thus the likelihood of the whole random sample is proportional to the likelihood of $\bar{y}$, a single draw from a *normal*$(\mu, \sigma_{\bar{y}}^2)$, where $\sigma_{\bar{y}}^2 = \frac{\sigma^2}{n}$.

Simple updating formulas for random sample. The random sample can be summarized by $\bar{y}$, a single draw from a *normal*$(\mu, \sigma^2/n)$ distribution.[6] Hence

$$\frac{1}{(s')^2} = \frac{1}{s^2} + \frac{n}{\sigma^2} \quad \text{and} \quad \frac{\frac{1}{s^2}}{\frac{1}{(s')^2}} \times m + \frac{\frac{n}{\sigma^2}}{\frac{1}{(s')^2}} \times \bar{y} . \tag{4.8}$$

We can summarize the first updating formula by "the posterior precision of the sample mean equals the sum of the prior precision plus the precision," and the second updating formula by "the posterior mean is the weighted average of the prior mean and the sample mean, where the weights are the proportions of their precisions to the posterior precision."

More realistically, both parameters are unknown, and then the observations would be from a two-dimensional exponential family. Usually we want to do our inference on the mean μ and regard the variance σ^2 as a nuisance parameter. Using a joint prior, we could find the joint posterior. The marginal distribution of μ would be found by marginalizing the nuisance parameter σ^2 out of the joint posterior. This will be done in Section 4.6.

[6]We could also justify these updating formula by updating the posterior one observation at a time, and using the posterior after one observation as the prior for the next observation. After analyzing all the observations we would end up with the same posterior.

4.5 NORMALLY DISTRIBUTED OBSERVATIONS WITH KNOWN MEAN

In this section we will consider that we have normal observations where the mean μ is a known constant, and the variance σ^2 is the unknown parameter. Again, the observations come from a one-dimensional exponential family so the analysis is simple.

Single Draw from a Normal Distribution with Known Mean

Suppose y is a random draw from a *normal*(μ, σ^2) distribution where the mean μ is assumed to be a known constant. This is a draw from a one-dimensional exponential family with single parameter σ^2, since we can write the likelihood function as

$$\begin{aligned} f(y|\sigma^2) &= \frac{1}{\sqrt{2\pi}\sigma} e^{-\frac{1}{2\sigma^2}(y-\mu)^2} \\ &= \frac{1}{\sqrt{\sigma^2}} \times \frac{1}{\sqrt{2\pi}} \times e^{-\frac{1}{2\sigma^2}\times(y-\mu)^2}. \end{aligned}$$

We see that this is a member of the one-dimensional exponential family with

$$A(\sigma^2) = \frac{1}{\sqrt{\sigma^2}} \quad \text{and} \quad B(y) = \frac{1}{\sqrt{2\pi}}$$

$$C(\sigma^2) = -\frac{1}{2\sigma^2} \quad \text{and} \quad T(y) = (y-\mu)^2.$$

The conjugate family will have the same form as the likelihood. It is given by

$$\begin{aligned} g(\sigma^2; k, l) &\propto [A(\sigma^2)]^k e^{C(\sigma^2)\times l} \\ &\propto (\sigma^2)^{-\frac{k}{2}} e^{-\frac{1}{2\sigma^2} l}. \end{aligned}$$

Its shape is given by

$$g(\sigma^2) \propto (\sigma^2)^{-\frac{\kappa}{2}-1} e^{-\frac{S}{2\sigma^2}}$$

where $S = l$ and $\kappa = k - 2$. This is S times an *inverse chi-squared* distribution with κ degrees of freedom. The exact prior density is

$$g(\sigma^2) = \frac{S^{\frac{\kappa}{2}}}{2^{\frac{\kappa}{2}}\Gamma(\frac{\kappa}{2})} \times (\sigma^2)^{-\frac{\kappa}{2}-1} e^{-\frac{S}{2\sigma^2}}$$

where $\frac{S^{\frac{\kappa}{2}}}{2^{\frac{\kappa}{2}}\Gamma(\frac{\kappa}{2})}$ is the constant scale factor needed to make it a density. The density is sometimes written as

$$g(\sigma^2) = \frac{S^{\frac{\kappa}{2}}}{2^{\frac{\kappa}{2}}\Gamma(\frac{\kappa}{2})} \times \frac{1}{(\sigma^2)^{\frac{\kappa}{2}+1}} e^{-\frac{S}{2\sigma^2}}.$$

The posterior will be proportional to the prior times likelihood. This will be

$$g(\sigma^2|y) \quad \propto \quad [A(\sigma^2)]^{k'} \times e^{C(\sigma^2)\times l'}$$

where $k' = k + 1$ and $l' = l + (y - \mu)^2$. This will be S' times an *inverse chi-squared* distribution with κ' where $S' = S + (y - \mu)^2$ and $\kappa' = \kappa + 1$.

Simple updating rules for a single observation. The first updating rule is "add $(y - \mu)^2$ to the constant S" and the second updating rule is "add 1 to the degrees of freedom."

Random Sample from a Normal Distribution with Known Mean

Suppose $y_1, \ldots, y_n$ is a random sample from a *normal*(μ, σ^2) distribution where the mean μ is assumed to be a known constant. The likelihood of the random sample equals the product of the individual likelihoods

$$\begin{aligned} f(y_1, \ldots, y_n|\sigma^2) &= \prod_{i=1}^{n} f(y_i|\sigma^2) \\ &\propto \prod_{i=1}^{n} \left((\sigma^2)^{-\frac{1}{2}}\, e^{-\frac{(y_i-\mu)^2}{2\sigma^2}} \right) \\ &\propto (\sigma^2)^{-\frac{n}{2}}\, e^{-\frac{1}{2\sigma^2}\sum(y_i-\mu)^2} \\ &\propto \frac{1}{(\sigma^2)^{\frac{n}{2}}}\, e^{-\frac{SS_t}{2\sigma^2}} \end{aligned}$$

where $SS_t = \sum_{i=1}^{n}(y_i - \mu)^2$ is the total sum of squares around the known mean μ. The posterior is proportional to prior times likelihood, so the shape of the posterior will be given by

$$\begin{aligned} g(\sigma^2|y) &\propto g(\sigma^2) \times f(y_1, \ldots, y_n|\sigma^2) \\ &\propto (\sigma^2)^{-\frac{\kappa}{2}-1}\, e^{-\frac{S}{2\sigma^2}} \times (\sigma^2)^{-\frac{n}{2}}\, e^{-\frac{SS_t}{2\sigma^2}} \\ &\propto (\sigma^2)^{-\frac{\kappa'}{2}-1} e^{-\frac{S'}{2\sigma^2}} \end{aligned}$$

where $\kappa' = \kappa + n$ and $S' = S + SS_t$. We recognize this to be S' times an *inverse chi-squared* distribution with κ' degrees of freedom.

Simple updating rules for a random sample. The first updating rule is "add $SS_t = \sum(y - \mu)^2$ to the constant S" and the second updating rule is "add the sample size n to the degrees of freedom."

Change of variable to standard deviation σ. The variance σ^2 is in squared units. It is more difficult to understand our prior belief about the variance. Bolstad (2007) shows how to chose an *inverse chi-squared* prior for the variance σ^2 that matches our prior belief about the median of the standard deviation σ. Also, the variance cannot be compared directly to the mean because of the different units. Often, we reparameterize to σ, the standard deviation which can be directly compared to the mean. The corresponding (prior or posterior) density for σ is found from the corresponding (prior or posterior) density of σ^2 using the change of variable formula

$$g_\sigma(\sigma) = g_{\sigma^2}(\sigma^2) \times 2\sigma\,.$$

Generally, we do the calculations to find the posterior for the variance σ^2, then graph the corresponding posterior for the standard deviation σ since it is more easily understood.

We have now looked at each of the parameters of a *normal*(μ, σ^2) separately, assuming the other parameter is a known constant. In fact, this is all we need to do computational Bayesian statistics using the Gibbs sampler since knowing the conditional distribution of each parameter given the other parameters are known is all that is necessary to implement it. This will be shown in Chapter 10. However, we will look at the case where we have *normal*(μ, σ^2) observations where both parameters are unknown in the next section.

4.6 NORMALLY DISTRIBUTED OBSERVATIONS WITH UNKNOWN MEAN AND VARIANCE

A more realistic model when we have a random sample of observations from a *normal*(μ, σ^2) distribution is that both parameters are unknown. The parameter μ is usually the only parameter of interest, and σ^2 is a nuisance parameter. We want to do inference on μ while taking into account the additional uncertainty caused by the unknown value of σ^2. The Bayesian approach allows us to do this by marginalizing out the nuisance parameter from the joint posterior of the parameters given the data. The joint posterior will be proportional to the joint prior times the joint likelihood.

The Joint Likelihood Function

The likelihood function of the random sample is the product of the individual observation likelihoods. It is given by

$$\begin{aligned} f(y_1,\ldots,y_n|\mu,\sigma^2) &\propto \prod_{i=1}^{n} f(y_i|\mu,\sigma^2) \\ &\propto \frac{1}{(\sigma^2)^{\frac{n}{2}}} e^{-\frac{1}{2\sigma^2}\sum(y_i-\mu)^2} \\ &\propto \frac{1}{(\sigma^2)^{\frac{n}{2}}} e^{-\frac{1}{2\sigma^2}\sum(y_i-\bar{y}+\bar{y}-\mu)^2} \\ &\propto \frac{1}{(\sigma^2)^{\frac{n}{2}}} e^{-\frac{1}{2\sigma^2}[n(\bar{y}-\mu)^2+\sum(y_i-\bar{y})^2]} \\ &\propto \frac{1}{(\sigma^2)^{\frac{n}{2}}} e^{-\frac{1}{2\sigma^2}[n(\bar{y}-\mu)^2+SS_y]} \end{aligned}$$

where $SS_y = \sum(y_i - \bar{y})$ is the sum of square away from the sample mean.

Finding the posterior using independent Jeffrey's priors. Jeffrey's prior for μ is the improper prior

$$g_\mu(\mu) = 1 \quad \text{for} \quad \infty < \mu < \infty$$

and Jeffrey's prior for σ^2 is the improper prior

$$g_{\sigma^2}(\sigma^2) = \frac{1}{\sigma^2} \quad \text{for} \quad 0 < \sigma^2 < \infty\,.$$

The joint prior will be their product of the independent Jeffrey's priors

$$g_{\mu,\sigma^2}(\mu,\sigma^2) = \frac{1}{\sigma^2} \quad \text{for} \quad \left\{ \begin{array}{c} 0 < \sigma^2 < \infty \\ -\infty < \mu < \infty \end{array} \right. .$$

The joint posterior density will be proportional to the joint prior times the likelihood and is given by

$$\begin{aligned} g_{\mu,\sigma^2}(\mu,\sigma^2|y_1,\ldots,y_n) &\propto g_{\mu,\sigma^2}(\mu,\sigma^2) \times f(y_1,\ldots,y_n|\mu,\sigma^2) \\ &\propto \frac{1}{\sigma^2} \times \frac{1}{(\sigma^2)^{\frac{n}{2}}} e^{-\frac{1}{2\sigma^2}[n(\bar{y}-\mu)^2+SS_y]} \\ &\propto \frac{1}{(\sigma^2)^{\frac{n}{2}+1}} e^{-\frac{1}{2\sigma^2}[n(\bar{y}-\mu)^2+SS_y]}\,. \end{aligned}$$

We find the marginal posterior density for μ by marginalizing the nuisance parameter σ^2 out of the joint posterior.

$$\begin{aligned} g_\mu(\mu|y_1,\ldots,y_n) &\propto \int_0^\infty g_{\mu,\sigma^2}(\mu,\sigma^2|y_1,\ldots,y_n)\,d\sigma^2 \\ &\propto \int_0^\infty \frac{1}{(\sigma^2)^{\frac{n}{2}+1}} e^{-\frac{1}{2\sigma^2}[n(\bar{y}-\mu)^2+SS_y]}\,d\sigma^2 . \end{aligned}$$

We change the variable by letting

$$u = \frac{n(\mu-\bar{y})^2 + SS_y}{\sigma^2}$$

and when we do the integration we find that

$$\begin{aligned} g_\mu(\mu|y_1,\ldots,y_n) &\propto [n(\mu-\bar{y})^2 + SS_y]^{-\frac{n}{2}} \int_\infty^0 -u^{\frac{n}{2}-1} e^{-\frac{u}{2}}\,du \\ &\propto [n(\mu-\bar{y})^2 + SS_y]^{-\frac{n}{2}} . \end{aligned}$$

Suppose we change the variables to

$$t = \frac{\mu-\bar{y}}{\sqrt{\frac{SS_y}{n(n-1)}}}$$

and apply the change of variable formula. We find the posterior density of t to be

$$g(t|y_1,\ldots,y_n) \propto \left(\frac{t^2}{n-1}+1\right)^{-\frac{n}{2}} .$$

This is the density of the *Student's t* distribution with $n-1$ degrees of freedom. Note that the maximum likelihood estimate of the variance is

$$\hat{\sigma}^2 = \frac{SS_y}{n-1} .$$

This means that the posterior density of μ is $\bar{y}$ plus $\frac{\hat{\sigma}}{\sqrt{n}}$ times a *Student's t* with $n-1$ degrees of freedom. This means we can do the inferences for μ treating the unknown variance σ^2 as if it had the value $\hat{\sigma}^2$ but using the *Student's t* table instead of the standard normal table.

The Joint Conjugate Prior

We might think that the product of independent conjugate priors for each of the two parameters μ and σ^2 would be jointly conjugate for the two parameters (μ,σ^2) considered together. Surprisingly, this is not the case. The product of an *normal*(m,s^2) conjugate prior for μ and S times an *inverse chi-squared* conjugate prior for σ^2 will not be the joint conjugate prior when both parameters are unknown. See DeGroot (2004) and O'Hagan and Forster (1994). The observations come from a two-dimensional exponential family and the conjugate prior will have the same form as the likelihood. The likelihood can be written

$$\begin{aligned} f(y_1, \ldots, y_n|\mu, \sigma^2) &\propto f_{\bar{y}}(\bar{y}|\mu, \sigma^2) \times f_{SS_y}(SS_y|\sigma^2) \\ &\propto \frac{1}{(\sigma^2)^{\frac{1}{2}}} e^{-\frac{1}{2\sigma^2}[n(\bar{y}-\mu)^2]} \times \frac{1}{(\sigma^2)^{\frac{n-1}{2}}} e^{-\frac{1}{2\sigma^2}SS_y} . \end{aligned} \tag{4.9}$$

We recognize the first is a *normal*$(\mu, \frac{\sigma^2}{n})$, which depends on both parameters, and the second is SS_y times an *inverse chi-squared* with $n-1$ degrees of freedom. We will let $g_{\sigma^2}(\sigma^2)$, the prior for σ^2, be S times an *inverse chi-squared* with κ degrees of freedom, and let $g_\mu(\mu|\sigma^2)$, the prior for μ conditional on σ^2 be a *normal*$(m, \frac{\sigma^2}{n_0})$. The parameter n_0 is the equivalent sample size of the prior. It says that our prior knowledge about μ is equivalent to the knowledge we would obtain from n_0 observations. The joint conjugate prior will be their product

$$\begin{aligned} g_{\mu,\sigma^2}(\mu, \sigma^2) &= g_\mu(\mu|\sigma^2) \times g_{\sigma^2}(\sigma^2) \\ &\propto \frac{1}{(\sigma^2)^{\frac{1}{2}}} e^{-\frac{n_0}{2\sigma^2}(\mu-m)^2} \times \frac{1}{(\sigma^2)^{\frac{\kappa}{2}+1}} e^{-\frac{S}{2\sigma^2}} \\ &\propto \frac{1}{(\sigma^2)^{\frac{\kappa+1}{2}+1}} e^{-\frac{1}{2\sigma^2}[n_0(\mu-m)^2+S]} . \end{aligned}$$

Finding the posterior using joint conjugate prior. The joint posterior distribution of the two parameters will be proportional to the joint prior times likelihood. It will be

$$\begin{aligned} g_{\mu,\sigma^2}(\mu, \sigma^2|y_1, \ldots, y_n) &\propto g_{\mu,\sigma^2}(\mu, \sigma^2) \times f(y_1, \ldots, y_n|\mu, \sigma^2) \\ &\propto \frac{1}{(\sigma^2)^{\frac{\kappa+1}{2}+1}} e^{-\frac{1}{2\sigma^2}[n_0(\mu-m)^2+S]} \\ &\quad \times \frac{1}{(\sigma^2)^{\frac{n}{2}}} e^{-\frac{1}{2\sigma^2}[n(\bar{y}-\mu)^2+SS_y]} \\ &\propto \frac{1}{(\sigma^2)^{\frac{\kappa+n+1}{2}+1}} e^{-\frac{1}{2\sigma^2}[n_0(\mu-m)^2+S+n(\bar{y}-\mu)^2+SS_y]} \\ &\propto \frac{1}{(\sigma^2)^{\frac{\kappa'+1}{2}+1}} e^{-\frac{1}{2\sigma^2}[n'(\mu-m')^2+S'+\left(\frac{n_0 n}{n_0+n}\right)(\bar{y}-m)^2]} \end{aligned}$$

where $S' = S + SS_y$ and $\kappa' = \kappa + n$ and $m' = \frac{n\bar{y}+n_0 m}{n+n_0}$ and $n' = n_0 + n$. We find the marginal posterior of μ by marginalizing σ^2 out of the joint posterior.

$$
\begin{aligned}
g_\mu(\mu|y_1,\ldots,y_n) &\propto \int_0^\infty g_{\mu,\sigma^2}(\mu,\sigma^2|y_1,\ldots,y_n)\,d\sigma^2 \\
&\propto \int_0^\infty \frac{1}{(\sigma^2)^{\frac{\kappa'+1}{2}+1}} e^{-\frac{1}{2\sigma^2}[n'(\mu-m')^2+S'+\left(\frac{n_0 n}{n_0+n}\right)(\bar{y}-m)^2]}\,d\sigma^2 .
\end{aligned}
$$

When we do the integration we find that

$$
g_\mu(\mu|y_1,\ldots,y_n) \propto [(n')(\mu-m')^2 + S' + \left(\frac{n_0 n}{n_0+n}\right)(\bar{y}-m)^2]^{-\frac{\kappa'+1}{2}} .
$$

Suppose we change the variables to

$$
t = \frac{\mu - m'}{\sqrt{\frac{S'+\left(\frac{n_0 n}{n_0+n}\right)(\bar{y}-m)^2}{n'\kappa'}}}
$$

and apply the change of variable formula. We find the posterior density of t to be

$$
g(t|y_1,\ldots,y_n) \propto \left(\frac{t^2}{\kappa'}+1\right)^{-\frac{\kappa'+1}{2}} .
$$

This is the density of the *Student's t* distribution with κ' degrees of freedom. Note that

$$
\begin{aligned}
\frac{S' + \left(\frac{n_0 n}{n_0+n}\right)(\bar{y}-m)^2}{n'} &= \frac{S + SS_y + \left(\frac{n_0 n}{n_0+n}\right)(\bar{y}-m)^2}{n_0+n} \\
&= \left(\frac{n_0}{n_0+n}\right)\left(\frac{S}{n_0} + \frac{n(\bar{y}-m)^2}{n_0+n}\right) + \left(\frac{n}{n_0+n}\right)\frac{SS_y}{n} \\
&= \hat{\sigma}_B^2
\end{aligned}
$$

which is the weighted average of two estimates of the variance. The first incorporates the prior and the distance $\bar{y}$ is from its prior mean m, and the second is the maximum likelihood estimator of the variance. This means the posterior density of μ is m' plus $\frac{\hat{\sigma}_B}{\sqrt{\kappa'}}$ times a *Student's t* with κ' degrees of freedom. Again, we can do the inference treating the unknown variance σ^2 as if it had the value $\hat{\sigma}_B^2$ but using the *Student's t* table to find the critical values instead of the standard normal table.

4.7 MULTIVARIATE NORMAL OBSERVATIONS WITH KNOWN COVARIANCE MATRIX

In this section we look at the case where we have a random sample of observations that come from a *multivariate normal* distribution. Let the ith observation be

$$\mathbf{y_i} = \begin{pmatrix} y_{1\,i} \\ \vdots \\ y_{k\,i} \end{pmatrix}$$

for $i = 1, \ldots, n$. Each observation is a random vector that comes from a *multivariate normal* distribution with mean vector and covariance matrix

$$\boldsymbol{\mu} = \begin{pmatrix} \mu_1 \\ \vdots \\ \mu_k \end{pmatrix} \quad \text{and} \quad \boldsymbol{\Sigma} = \begin{bmatrix} \sigma_{1\,1} & \ldots & \sigma_{1\,k} \\ \vdots & \vdots\vdots\vdots & \vdots \\ \sigma_{k\,1} & \ldots & \sigma_{k\,k} \end{bmatrix}$$

where the covariance matrix $\boldsymbol{\Sigma}$ is known.

Single Multivariate Normal Observation

The likelihood of the single observation $\mathbf{y}$ is given by

$$f(\mathbf{y}|\boldsymbol{\mu}) \quad \propto \quad e^{-\frac{1}{2}(\mathbf{y}-\boldsymbol{\mu})'\boldsymbol{\Sigma}^{-1}(\mathbf{y}-\boldsymbol{\mu})} .$$

Suppose we use a *multivariate normal*$(\mathbf{m_0}, \mathbf{V_0})$ prior for $\boldsymbol{\mu}$. The prior density is given by

$$g(\boldsymbol{\mu}) \quad \propto \quad e^{-\frac{1}{2}(\boldsymbol{\mu}-\mathbf{m_0})'\mathbf{V_0}^{-1}(\boldsymbol{\mu}-\mathbf{m_0})} .$$

Using Bayes' theorem, the posterior will be

$$g(\boldsymbol{\mu}|\mathbf{y}) \quad \propto \quad e^{-\frac{1}{2}(\boldsymbol{\mu}-\mathbf{m_0})'\mathbf{V_0}^{-1}(\boldsymbol{\mu}-\mathbf{m_0})} \times e^{-\frac{1}{2}(\boldsymbol{\mu}-\mathbf{y})'\boldsymbol{\Sigma}^{-1}(\boldsymbol{\mu}-\mathbf{y})}$$

$$\propto \quad e^{-\frac{1}{2}[(\boldsymbol{\mu}-\mathbf{m_0})'\mathbf{V_0}^{-1}(\boldsymbol{\mu}-\mathbf{m_0})+(\boldsymbol{\mu}-\mathbf{y})'\boldsymbol{\Sigma}^{-1}(\boldsymbol{\mu}-\mathbf{y})]} .$$

We expand both terms, combine like terms, and absorb the part not containing the parameter $\boldsymbol{\mu}$ into the constant. This yields

$$g(\boldsymbol{\mu}|\mathbf{y}) \quad \propto \quad e^{-\frac{1}{2}[\boldsymbol{\mu}'(\mathbf{V_1^{-1}})\boldsymbol{\mu}-\boldsymbol{\mu}'(\boldsymbol{\Sigma}^{-1}\mathbf{y}+\mathbf{V_0^{-1}}\mathbf{m_0})-(\mathbf{y}'\boldsymbol{\Sigma}^{-1}+\mathbf{m_0}'\mathbf{V_0^{-1}})\boldsymbol{\mu}]}$$

where $\mathbf{V_1^{-1}} = \mathbf{V_0^{-1}} + \boldsymbol{\Sigma}^{-1}$. The posterior precision matrix equals the prior precision matrix plus the precision matrix of the observation. This is similar to the rule for an ordinary univariate normal observation but adapted to the multivariate normal situation. Let $\mathbf{U}'\mathbf{U} = \mathbf{V_1^{-1}}$ where $\mathbf{U}$ is an orthogonal matrix. We are assuming $\mathbf{V_1^{-1}}$

is of full rank so that $\mathbf{U}$ and $\mathbf{U}'$ are of full rank so their inverses exist. Simplifying the posterior we get

$$g(\boldsymbol{\mu}|\mathbf{y}) \propto e^{-\frac{1}{2}[\boldsymbol{\mu}'(\mathbf{U}'\mathbf{U})\boldsymbol{\mu}-\boldsymbol{\mu}'(\mathbf{U}'(\mathbf{U}')^{-1})(\boldsymbol{\Sigma}^{-1}\mathbf{y}+\mathbf{V}_0^{-1}\mathbf{m}_0)-(\mathbf{y}'\boldsymbol{\Sigma}^{-1}+\mathbf{m}_0{}'\mathbf{V}_0^{-1})(\mathbf{U}^{-1}\mathbf{U})\boldsymbol{\mu}]} .$$

Completing the square and absorbing the part that does not contain the parameter into the constant the posterior simplifies to

$$g(\boldsymbol{\mu}|\mathbf{y}) \propto e^{-\frac{1}{2}[\boldsymbol{\mu}'\mathbf{U}'-(\mathbf{y}'\boldsymbol{\Sigma}^{-1}+\mathbf{m}_0{}'\mathbf{V}_0{}^{-1})\mathbf{U}^{-1}][\mathbf{U}\boldsymbol{\mu}-(\mathbf{U}'^{-1})(\boldsymbol{\Sigma}^{-1}\mathbf{y}+\mathbf{V}_0^{-1}\mathbf{m}_0)]} .$$

Since both $\mathbf{U}$ and $\mathbf{U}'$ are of full rank, $(\mathbf{U}')^{-1}\mathbf{U}'$ and $\mathbf{U}\mathbf{U}^{-1}$ both equal the k dimensional identity matrix. Hence

$$g(\boldsymbol{\mu}|\mathbf{y}) \propto e^{-\frac{1}{2}[\boldsymbol{\mu}'-(\mathbf{y}'\boldsymbol{\Sigma}^{-1}+\mathbf{m}_0{}'\mathbf{V}_0{}^{-1})\mathbf{V}_1](\mathbf{V}_1)^{-1}[\boldsymbol{\mu}-\mathbf{V}_1'(\boldsymbol{\Sigma}^{-1}\mathbf{y}+\mathbf{V}_0^{-1}\mathbf{m}_0)]}$$

$$\propto e^{-\frac{1}{2}[\boldsymbol{\mu}'-\mathbf{m}_1'](\mathbf{V}_1)^{-1}[\boldsymbol{\mu}-\mathbf{m}_1]}$$

where $\mathbf{m}_1 = \mathbf{V}_1[\boldsymbol{\Sigma}^{-1}]\mathbf{y} + \mathbf{V}_1[\mathbf{V}_0^{-1}]\mathbf{m}_0$ is the posterior mean. It is given by the rule "posterior mean vector equals the inverse of posterior precision matrix times prior precision matrix times prior mean vector plus inverse of posterior precision matrix times precision matrix of observation vector times the observation vector." This is similar to the rule for single normal observation, but adapted to vector observations. We recognize that the posterior distribution of $\boldsymbol{\mu}|\mathbf{y}$ is *multivariate normal*$(\mathbf{m}_1, \mathbf{V}_1)$.

Random Sample of Multivariate Normal Observations

Let $\mathbf{y}_j$ for $j = 1, \ldots, J$ be a random sample from a *multivariate normal* distribution having mean vector $\boldsymbol{\mu}$ and known covariance matrix $\boldsymbol{\Sigma}$. The likelihood of the random sample will be equal to the product of the individual likelihoods. It is given by

$$f(\mathbf{y}_1, \ldots, \mathbf{y}_n|\boldsymbol{\mu}) \propto \prod_{j=1}^{n} e^{-\frac{1}{2}(\mathbf{y}_j-\boldsymbol{\mu})'\boldsymbol{\Sigma}^{-1}(\mathbf{y}_j-\boldsymbol{\mu})}$$

$$\propto e^{-\frac{1}{2}\sum(\mathbf{y}_j-\boldsymbol{\mu})'\boldsymbol{\Sigma}^{-1}(\mathbf{y}_j-\boldsymbol{\mu})}$$

$$\propto e^{-\frac{n}{2}(\bar{\mathbf{y}}-\boldsymbol{\mu})'\boldsymbol{\Sigma}^{-1}(\bar{\mathbf{y}}-\boldsymbol{\mu})} .$$

Thus the likelihood of the random sample from the *multivariate normal* distribution is proportional to the likelihood of the mean vector, which is like a single observation from a *multivariate normal* distribution with mean vector $\boldsymbol{\mu}$ and covariance matrix $\boldsymbol{\Sigma}/n$.

Simple updating formulas. We condense the random sample of *multivariate normal*$(\mu, \boldsymbol{\Sigma})$ down to a single observation of the sample mean vector, $\bar{\mathbf{y}}$, which comes from a *multivariate normal*$(\mu, \Sigma/n)$ distribution. Hence the posterior precision matrix is the sum of prior precision matrix plus the precision matrix of the sample mean vector $\bar{\mathbf{y}}$. It given by

$$\mathbf{V_1^{-1}} = \mathbf{V_0^{-1}} + n\mathbf{\Sigma^{-1}}. \tag{4.10}$$

The posterior mean vector is the weighted average of the prior mean vector and the sample mean vector, where the weights are the proportions of their precision matrices to the posterior precision matrix. It is given by

$$\mathbf{m_1} = \mathbf{V_1}[\mathbf{V_0^{-1}}]\mathbf{m_0} + \mathbf{V_1}[n\mathbf{\Sigma^{-1}}]\bar{\mathbf{y}}. \tag{4.11}$$

The posterior distribution of $\boldsymbol{\mu}|\mathbf{y}$ is *multivariate normal*$(\mathbf{m_1}, \mathbf{V_1})$.

4.8 OBSERVATIONS FROM NORMAL LINEAR REGRESSION MODEL

In the normal linear regression model, we have n independent observations $y_1, \ldots, y_n$ where each observation y_i has its own mean μ_i, and all observations have the same variance σ^2. The means are unknown linear functions of the p predictor variables $x_1, \ldots, x_p$. The values of the predictor variables are known for each observation. Hence we can write the mean as

$$\mu_i = \sum_{j=0}^{p} X_{ij}\beta_j\,,$$

where the unknown parameters $\beta_0, \ldots, \beta_p$ are the intercept and the slope coefficients of the predictor variables. We write the observations as the vector

$$\mathbf{y} = \begin{pmatrix} y_1 \\ \vdots \\ y_n \end{pmatrix},$$

the parameters as the vector

$$\boldsymbol{\beta} = \begin{pmatrix} \beta_0 \\ \vdots \\ \beta_p \end{pmatrix},$$

and the values of the predictor variables as a matrix

$$\mathbf{X} = \begin{bmatrix} 1 & X_{11} & \ldots & X_{1p} \\ \vdots & \vdots & \vdots & \vdots \\ 1 & X_{n1} & \ldots & X_{np} \end{bmatrix}.$$

Suppose we use a *multivariate normal* prior for $\boldsymbol{\beta}$ having mean vector $\mathbf{b_0}$ and covariance matrix $\mathbf{V_0}$. Let $\mathbf{b_{LS}} = (\mathbf{X}'\mathbf{X})^{-1}\mathbf{X}'\mathbf{y}$ and $\mathbf{V_{LS}} = \sigma^2(\mathbf{X}'\mathbf{X})^{-1}$ be the vector of least squares coefficients and the covariance matrix of the least squares coefficients, respectively. The posterior will be

$$\begin{aligned}
g(\boldsymbol{\beta}|\mathbf{y}) &\propto g(\boldsymbol{\beta}) \times f(\mathbf{y}|\boldsymbol{\beta}) \\
&\propto e^{-\frac{1}{2}(\boldsymbol{\beta}-\mathbf{b_0})'\mathbf{V_0}^{-1}(\boldsymbol{\beta}-\mathbf{b_0})} \times e^{-\frac{1}{2}(\boldsymbol{\beta}-\mathbf{b_{LS}})'\mathbf{V_{LS}^{-1}}(\boldsymbol{\beta}-\mathbf{b_{LS}})} \\
&\propto e^{-\frac{1}{2}[(\boldsymbol{\beta}-\mathbf{b_0})'\mathbf{V_0}^{-1}(\boldsymbol{\beta}-\mathbf{b_0})+(\boldsymbol{\beta}-\mathbf{b_{LS}})'\mathbf{V_{LS}}^{-1}(\boldsymbol{\beta}-\mathbf{b_{LS}})]} \\
&\propto e^{-\frac{1}{2}\left[\boldsymbol{\beta}'(\mathbf{V_0^{-1}}+\mathbf{V_{LS}^{-1}})\boldsymbol{\beta}-\boldsymbol{\beta}'\left(\mathbf{V_{LS}^{-1}b_{LS}}+\mathbf{V_0^{-1}b_0}\right)-\left(\mathbf{b_{LS}'V_{LS}^{-1}}+\mathbf{b_0'V_0^{-1}}\right)\boldsymbol{\beta}\right.} \\
&\qquad {}^{\left.+\left(\mathbf{b_{LS}'V_{LS}^{-1}}+\mathbf{b_0'V_0^{-1}}\right)\left(\mathbf{V_{LS}^{-1}b_{LS}}+\mathbf{V_0^{-1}b_0}\right)\right]}.
\end{aligned}$$

The last term does not contain $\boldsymbol{\beta}$ so it does not enter the likelihood and can be absorbed into the proportionality constant. We let $\mathbf{V_1^{-1}} = \mathbf{V_0^{-1}}+\mathbf{V_{LS}^{-1}}$. The posterior becomes

$$\propto e^{-\frac{1}{2}\left[\boldsymbol{\beta}'\mathbf{V_1^{-1}}\boldsymbol{\beta}-\boldsymbol{\beta}'\left(\mathbf{V_{LS}^{-1}b_{LS}}+\mathbf{V_0^{-1}b_0}\right)-\left(\hat{\boldsymbol{\beta}}_{LS}'\mathbf{V_{LS}^{-1}}+\mathbf{b_0'V_0^{-1}}\right)\boldsymbol{\beta}\right]}.$$

Let $\mathbf{U'U} = \mathbf{V_1^{-1}}$ where $\mathbf{U}$ an orthogonal matrix. We are assuming $\mathbf{V_1^{-1}}$ is of full rank so both $\mathbf{U}$ and $\mathbf{U'}$ are also full rank, and their inverses exist. We complete the square by adding $(\mathbf{b_{LS}'V_{LS}^{-1}} + \mathbf{b_0'V_0^{-1}})\mathbf{U}(\mathbf{U'})^{-1}(\mathbf{V_{LS}^{-1}b_{LS}} + \mathbf{V_0^{-1}b_0})$. We subtract it as well, but since that does not contain the parameter $\boldsymbol{\beta}$, that part gets absorbed into the constant. The posterior becomes

$$\begin{aligned}
&\propto e^{-\frac{1}{2}\left[\boldsymbol{\beta}'\mathbf{U'U}\boldsymbol{\beta}-\boldsymbol{\beta}'\mathbf{U'}(\mathbf{U'})^{-1}\left(\mathbf{V_{LS}^{-1}b_{LS}}+\mathbf{V_0^{-1}b_0}\right)-\left(\mathbf{b_{LS}'V_{LS}^{-1}}+\mathbf{b_0'V_0^{-1}}\right)\mathbf{U}^{-1}\mathbf{U}\boldsymbol{\beta}+\right.} \\
&\qquad {}^{\left.(\mathbf{b_0'V_0^{-1}}+\mathbf{b_{LS}V_{LS}^{-1}})\mathbf{U}^{-1}(\mathbf{U'})^{-1}(\mathbf{V_0^{-1}b_0}+\mathbf{V_{LS}^{-1}b_{LS}})\right]}.
\end{aligned}$$

We factor the exponent. The posterior becomes

$$\propto e^{-\frac{1}{2}\left(\boldsymbol{\beta}'\mathbf{U'}-(\mathbf{V_{LS}^{-1}b_{LS}}+\mathbf{V_0^{-1}b_0})\mathbf{U}^{-1}\right)\left(\mathbf{U}\boldsymbol{\beta}-(\mathbf{U'})^{-1}(\mathbf{V_0^{-1}b_0}+\mathbf{V_{LS}^{-1}b_{LS}})\right)}.$$

When we factor $\mathbf{U'}$ out of the first factor and $\mathbf{U}$ out of the second factor in the product we get

$$= e^{-\frac{1}{2}\left[\boldsymbol{\beta}-(\mathbf{b_0'V_0^{-1}}+\mathbf{b_{LS}'V_{LS}^{-1}})\mathbf{U}^{-1}(\mathbf{U'})^{-1}\right]'(\mathbf{U'U})\left[\boldsymbol{\beta}-\mathbf{U}^{-1}(\mathbf{U'})^{-1}(\mathbf{V_0^{-1}b_0}+\mathbf{V_{LS}^{-1}b_{LS}})\right]}.$$

Since $\mathbf{U'U} = \mathbf{V}^{-1}$ and they are all of full rank we have $(\mathbf{U'})^{-1}\mathbf{U}^{-1} = \mathbf{V_1}$. When we substitute back into the posterior we get

$$g(\boldsymbol{\beta}|\mathbf{y}) \propto e^{-\frac{1}{2}(\boldsymbol{\beta}-\mathbf{b_1})'\mathbf{V_1^{-1}}(\boldsymbol{\beta}-\mathbf{b_1})}$$

where $\mathbf{b_1} = \mathbf{V_1}[\mathbf{V_0^{-1}}]\mathbf{b_0} + \mathbf{V_1}[\mathbf{V_{LS}^{-1}}]\mathbf{b_{LS}}$. The posterior distribution of $\boldsymbol{\beta}|\mathbf{y}$ will be *multivariate normal*$(\mathbf{b_1}, \mathbf{V_1})$.

The updating formulas. Under the normal linear regression assumptions, the least squares estimates maximize the likelihood function. This makes them the maximum likelihood estimates and their covariance matrix the covariance matrix of the maximum likelihood estimates. Thus the posterior has the *multivariate normal* where the constants are found by the updating formulas "the posterior precision matrix equals the sum of the prior precision matrix plus the precision matrix of the "maximum likelihood estimates"

$$\begin{aligned} \mathbf{V}_1^{-1} &= \mathbf{V}_0^{-1} + \mathbf{V}_{\mathbf{LS}}^{-1} \\ &= \mathbf{V}_0^{-1} + \mathbf{V}_{\mathbf{ML}}^{-1} \end{aligned} \tag{4.12}$$

and "the posterior mean vector is the weighted average of the prior mean vector and the maximum likelihood estimate vector where their weights are the inverse of the posterior precision matrix (which is the posterior covariance matrix) multiplied by the inverse of their respective precision matrices"

$$\begin{aligned} \mathbf{b}_1 &= \mathbf{V}_1[\mathbf{V}_0^{-1}]\mathbf{b}_0 + \mathbf{V}_1[\mathbf{V}_{\mathbf{LS}}^{-1}]\mathbf{b}_{\mathbf{LS}} \\ &= \mathbf{V}_1[\mathbf{V}_0^{-1}]\mathbf{b}_0 + \mathbf{V}_1[\mathbf{V}_{\mathbf{ML}}^{-1}]\hat{\beta}_{ML}\,. \end{aligned} \tag{4.13}$$

Main Points

- The posterior can be found easily when the observations come from a one-dimensional exponential family. The observation density has the form

  $$f(y|\theta) = A(\theta)B(y)e^{C(\theta)\times T(y)}$$

  for some functions $A(\theta)$, $B(y)$, $C(\theta)$, and $T(y)$.

- The conjugate family of prior densities has the same form as the likelihood.

- The posterior will be another member of the same family. Simple rules are all that is necessary to determine the constants.

- Many commonly used distributions are members of the one-dimensional exponential family. These include the *binomial*(n, π) and *Poisson*(μ) distributions for count data, the *geometric*(π) and *negative binomial*(r, π) distributions for waiting time in a *Bernoulli process*, the *exponential*(λ) and *gamma*(n, λ) distributions for waiting times in a *Poisson process*, and the *normal*(μ, σ^2) where σ^2 is a known constant and the *normal*(μ, σ^2) where μ is a known constant.

- When the observation comes from the *binomial*(n, π) distribution, the conjugate prior distribution for π is *beta*(a, b). The posterior is *beta*(a', b') where the constants are found by

  $$a' = a + y \quad \text{and} \quad b' = b + n - y\,.$$

The simple updating rules are "add number of successes to a" and "add number of failures to b."

- When the observation comes from the *Poisson*(μ) distribution, the conjugate prior distribution for μ is *gamma*(r, v). The posterior is *gamma*(r', v') where the constants are found by

$$r' = r + y \quad \text{and} \quad v' = v + 1\,.$$

The simple updating rules are "add the observation to r" and "add 1 to v."

- When the observation comes from the *negative binomial*(r, π) distribution, the conjugate prior distribution for π is *beta*(a, b). The posterior is *beta*(a', b') where

$$a' = a + r \quad \text{and} \quad b' = b + y\,.$$

The simple updating rules are "add number of successes to a" and "add number of failures to b."

- When the observation comes from the *gamma*(n, λ) distribution, the conjugate prior distribution for λ is *gamma*(r, v). The posterior is *gamma*(r', v') where

$$r' = r + n \quad \text{and} \quad v' = v + t\,.$$

The simple updating rules are "add n to r" and "add observation to v."

- When the observations come from a *normal*(μ, σ^2) distribution where the variance σ^2 is known, the conjugate prior for $\boldsymbol{\mu}$ is *normal*(m, s^2). The posterior is *normal*$(m', (s')^2)$ where

$$\frac{1}{(s')^2} = \frac{1}{s^2} + \frac{n}{\sigma^2} \quad \text{and} \quad m' = \frac{\frac{1}{s^2}}{\frac{1}{(s')^2}} \times m + \frac{\frac{n}{\sigma^2}}{\frac{1}{(s')^2}} \times \bar{y}$$

The simple updating rules are "the posterior precision is the sum of the prior precision plus the precision of the sample mean" and "the posterior mean is the weighted average of the prior mean and the sample mean where the weights are the proportions of their respective precisions relative to the posterior precision."

- When the observations come from a *normal*(μ, σ^2) distribution where the mean μ is known, the conjugate prior for σ^2 is S times an *inverse chi-squared* distribution with κ degrees of freedom. The posterior is S' times an *inverse chi-squared* distribution with κ' degrees of freedom where

$$S' = S + SS_t \quad \text{and} \quad \kappa' = \kappa + n\,.$$

where $SS_t = \sum(y_i - \mu)^2$. The simple updating rules are "add sum of squares away from the mean μ to S" and "add sample size to the degrees of freedom κ."

- The *normal*(μ, σ^2) where both parameters are unknown is a member of a two-dimensional exponential family. Surprisingly, the product of independent conjugate priors for μ and σ^2 is not the conjugate prior for the case where μ and σ^2 are both unknown.

- When the observations come from a *normal*(μ, σ^2) distribution where both parameters are unknown, the joint conjugate prior is the product of S times an *inverse chi-squared* distribution with κ degrees of freedom for σ^2 times a *normal*$(m, \frac{\sigma^2}{n_0})$ prior for μ given σ^2. Note n_0 is the equivalent sample size of the prior for μ.

$$S' = S + SS_y \quad \text{and} \quad \kappa' = \kappa + n\,.$$

where $SS_y = \sum(y_i - \bar{y})^2$ and

$$n' = n + n_0 \quad \text{and} \quad m' = \frac{n\bar{y} + n_0 m}{n + n_0}\,.$$

The simple updating rules are "add sum of squares away from sample mean $\bar{y}$ to S" and "add sample size to κ." followed by "the posterior precision is the sum of the prior precision plus the precision of the sample mean" and "the posterior mean is the weighted average of the prior mean and the sample mean where the weights are the proportions of their respective precisions relative to the posterior precision."

- When we treat the variance σ^2 as a nuisance parameter, we find the marginal posterior density of μ is m' plus $\frac{S'}{\kappa'}$ time a *Student's t* with κ' degrees of freedom.

- When we have a random sample of *multivariate normal*$(\boldsymbol{\mu})$ observations with known covariance matrix $\boldsymbol{\Sigma}$, the conjugate prior for $\boldsymbol{\mu}$ is *multivariate normal*$(\mathbf{m_0}, \mathbf{V_0})$. The posterior is *multivariate normal*$(\mathbf{m_1}, \mathbf{V_1})$ where

$$\mathbf{V_1}^{-1} = \mathbf{V_0}^{-1} + n\,\boldsymbol{\Sigma}^{-1} \quad \text{and} \quad \mathbf{m_1} = \mathbf{V_1}\mathbf{V_0}^{-1}\mathbf{m_0} + \mathbf{V_1}[n\,\boldsymbol{\Sigma}^{-1}]\bar{\mathbf{y}}$$

The simple updating rules are "posterior precision matrix is the sum of the prior precision matrix plus the precision matrix of the sample mean vector" and "the posterior mean vector is the weighted sum of the prior mean vector and the sample mean vector where the weights are the proportions of their precision matrices to the posterior precision matrix."

- When we have n independent observations from the normal linear regression model where the observations all have the same known variance, the conjugate prior distribution for the regression coefficient vector $\boldsymbol{\beta}$ is *multivariate normal*$(\mathbf{b_0}, \mathbf{V_0})$. The posterior distribution of $\boldsymbol{\beta}$ will be *multivariate normal*$(\mathbf{b_1}, \mathbf{V_1})$, where

$$\mathbf{V_1}^{-1} = \mathbf{V_0}^{-1} + \mathbf{V_{LS}}^{-1} \quad \text{and} \quad \mathbf{b_1} = \mathbf{V_1}\mathbf{V_0}^{-1}\mathbf{b_0} + \mathbf{V_1}\mathbf{V_{LS}}^{-1}\mathbf{b_{LS}}$$

where $\mathbf{b}_{\mathbf{LS}} = (\mathbf{X}'\mathbf{X})^{-1}\mathbf{X}'\mathbf{y}$ is the vector of least squares estimates and $\mathbf{V}_{\mathbf{LS}} = \sigma^2(\mathbf{X}'\mathbf{X})^{-1}$ is its covariance matrix. The simple updating rules are "posterior precision matrix is the sum of the prior precision matrix plus the precision matrix of the least squares estimates" and "the posterior mean vector is the weighted average of the prior mean vector plus the least squares vector where the weights are the relative proportions of their precision matrices to the posterior precision matrix."

Exercises

4.1 Suppose we observe $y = 15$ successes out of $n = 22$ independent *Bernoulli* trials, each having success probability π.

(a) Find the posterior distribution when a *beta*$(2, 2)$ prior is used for π.

(b) Find the posterior mean and standard deviation of π.

4.2 Suppose we observe $y = 3$ successes out of $n = 15$ independent *Bernoulli* trials, each having success probability π.

(a) Find the posterior distribution when a *beta*$(2, 1)$ prior is used for π.

(b) Find the posterior mean and standard deviation. of π.

4.3 Suppose that y has the *Poisson*(μ) distribution. We observe $y = 12$.

(a) Find the posterior distribution for μ when we use a *gamma*$(10, 1)$ prior.

(b) Find the posterior mean and standard deviation of μ.

4.4 Suppose that we observe a random sample of eight *Poisson*(μ) random variables. The random sample is given by

5	5	2	3	3	4	1	4

(a) Find the posterior distribution for μ when we use a *gamma*$(12, 2)$ prior.

(b) Find the posterior mean and standard deviation of μ.

4.5 Suppose we observe $y = 8$, the number of failures until the first success in a sequence of indendent *Bernoulli* trials where π is the probability of a success on a single trial.

(a) Find the posterior distribution for π when we use a *beta*$(1, 1)$ prior.

(b) Find the posterior mean and standard deviation of π.

4.6 Suppose we observe $y = 25$, the number of failures until the sixth success in a sequence of indendent *Bernoulli* trials where π is the probability of a success on a single trial.

(a) Find the posterior distribution for π when we use a *beta*$(3, 1)$ prior.

(b) Find the posterior mean and standard deviation of π.

4.7 Suppose we observe $t = 5.3$, the waiting time until the first occurrence of a *Poisson process* with intensity rate λ.

(a) Find the posterior distribution for λ given we use a $gamma(10, 1)$ prior.

(b) Find the posterior mean and standard deviation of λ.

4.8 Suppose we observe $t = 21.6$, the waiting time until the fourth occurrence of a *Poisson process* with intensity rate λ.

(a) Find the posterior distribution for λ given we use a $gamma(10, 1)$ prior.

(b) Find the posterior mean and standard deviation of λ.

4.9 We observe a random sample of ten $normal(\mu, \sigma^2)$ observations where $\sigma = .4$ is known. The random sample is

2.98	2.19	3.25	3.23	3.25
3.13	2.68	2.50	2.64	2.43

(a) Find the posterior distribution for μ when we use a $normal(5, 3)$ prior.

(b) Find the posterior mean and standard deviation of μ.

4.10 We observe a random sample of twelve $normal(\mu, \sigma^2)$ observations where $\sigma = 10$ is known. The random sample is

76.0	61.5	91.7	79.0	90.6	75.8
60.5	91.5	108.1	107.0	73.8	89.8

(a) Find the posterior distribution for μ when we use a $normal(5, 3)$ prior.

(b) Find the posterior mean and standard deviation of μ.

4.11 We observe a random sample of ten $normal(\mu, \sigma^2)$ observations where $\mu = 305$ is known. The random sample is

336	302	246	307	286	289	303	283

(a) Find the posterior distribution for σ^2 when we use a 410 times an *inverse chi-squared* prior with 1 degree of freedom.

(b) Find the posterior mean and standard deviation of σ.

4.12 We observe a random sample of twelve $normal(\mu, \sigma^2)$ observations where $\mu = .1000$ is known. The random sample is

0.105	0.095	0.102	0.099	0.099	0.098
0.095	0.099	0.100	0.100	0.097	0.095

(a) Find the posterior distribution for μ when we use a .000007278 times an *inverse chi-squared* prior with 1 degree of freedom.

(b) Find the posterior mean and standard deviation of σ.

4.13 We observe a random sample of size 10 from a *normal*(μ, σ^2) distribution where both the mean μ and the variance σ^2 are unknown. We consider the variance σ^2 to be a nuisance parameter. The random sample is

422.3	425.6	446.9	439.7	414.4
496.4	389.5	454.1	445.3	502.3

(a) Suppose we use a flat prior for μ and the Jeffrey's prior for σ^2. Find the equation for the joint posterior.

(b) Find the marginal posterior distribution of μ.

4.14 We observe a random sample of size 10 from a *normal*(μ, σ^2) distribution where both the mean μ and the variance σ^2 are unknown. We consider the variance σ^2 to be a nuisance parameter. The random sample is

296.3	285.7	302.5	281.4	278.2
295.8	282.5	279.2	324.6	310.3

(a) Suppose we use a flat prior for μ and the Jeffrey's prior for σ^2. Find the equation for the joint posterior.

(b) Find the marginal posterior distribution of μ.

4.15 We observe a random sample of size 15 from a *normal*(μ, σ^2) distribution where both the mean μ and the variance σ^2 are unknown. We consider the variance σ^2 to be a nuisance parameter. The random sample is

778.1	748.8	732.4	743.4	733.7
779.1	758.0	755.7	776.1	765.5
774.5	774.0	797.5	761.4	766.8

(a) Suppose we want to use the joint conjugate prior for μ and σ^2. It is the product of a *normal*$(m, \sigma^2/n_0)$ distribution for μ given σ^2 where n_0 is the equivalent sample size, and an S_0 time an *inverse chi-squared* distribution with κ_0 degrees of freedom for σ^2. Let the equivalent sample size $n_0 = 1$, $S_0 = 400$, and $\kappa_0 = 1$. Find the equation for the joint posterior.

(b) Find the marginal posterior distribution of μ.

4.16 We observe a random sample of size 20 from a *normal*(μ, σ^2) distribution where both the mean μ and the variance σ^2 are unknown. We consider the variance σ^2 to be a nuisance parameter. The random sample is

1.193	1.200	1.201	1.207	1.202
1.204	1.200	1.201	1.203	1.202
1.204	1.200	1.199	1.197	1.202
1.201	1.193	1.193	1.197	1.197

(a) Suppose we want to use the joint conjugate prior for μ and σ^2. It is the product of a *normal*$(m, \sigma^2/n_0)$ distribution for μ given σ^2 where n_0 is the equivalent sample size, and an S_0 time an *inverse chi-squared* distribution with κ_0 degrees of freedom for σ^2. Let the equivalent sample size $n_0 = 1$, $S_0 = 1$, and $\kappa_0 = 3$. Find the equation for the joint posterior.

(b) Find the marginal posterior distribution of μ.

4.17 Suppose

$$\begin{aligned} \mathbf{y}' &= (y_1, y_2, y_3) \\ &= (10.1, 6.6, 8.8) \end{aligned}$$

is a random draw from a multivariate normal with mean vector $\boldsymbol{\mu}$ and known covariance matrix

$$\mathbf{V_y} = \begin{bmatrix} 4 & 1 & 1 \\ 1 & 3 & 2 \\ 1 & 2 & 4 \end{bmatrix}.$$

(a) Let the prior distribution for $\boldsymbol{\mu}$ be *multivariate normal* with mean vector and covariance matrix equal to

$$\mathbf{m_0} = \begin{pmatrix} 8 \\ 5 \\ 6.8 \end{pmatrix} \quad \text{and} \quad \mathbf{V_0} = \begin{bmatrix} 6 & 1 & 2 \\ 1 & 5 & 2 \\ 2 & 2 & 6 \end{bmatrix}$$

respectively. Find the posterior distribution of $\boldsymbol{\mu}|\mathbf{y}$.

4.18 Suppose we have the ten observations from the multivariate normal with mean vector $\boldsymbol{\mu}$ and known covariance matrix

$$\boldsymbol{\Sigma} = \begin{bmatrix} 10000 & 2000 & 5000 \\ 2000 & 2500 & 1600 \\ 5000 & 1600 & 6400 \end{bmatrix}$$

given below.

y_1	y_2	y_3	y_1	y_2	y_3
383.6	188.7	339.8	488.5	221.0	503.8
520.3	189.2	442.3	509.6	137.5	376.5
502.6	236.4	461.3	703.3	221.2	457.2
426.6	202.1	286.9	601.5	215.9	506.4
495.1	131.1	424.4	400.1	214.7	422.3

(a) Let the prior distribution for $\boldsymbol{\mu}$ be *multivariate normal* with mean vector and covariance matrix equal to

$$\mathbf{m_0} = \begin{pmatrix} 500 \\ 200 \\ 400 \end{pmatrix} \quad \text{and} \quad \mathbf{V_0} = \begin{bmatrix} 20000 & 6000 & 8500 \\ 6000 & 5000 & 4000 \\ 8500 & 4000 & 5000 \end{bmatrix}$$

respectively. Find the posterior distribution of $\mu|\mathbf{y}$.

4.19 Suppose we have the following nine observations from the simple linear regression model where we know the standard deviation $\sigma = 1$.

x	21	22	23	24	25	26	27	28	29
y	10.18	8.82	8.48	9.38	10.39	10.71	10.93	11.09	11.45

(a) Recode the x variable so it is centered around mean zero. Find the the vector of maximum likelihood estimates $\hat{\beta}_{ML}$ and its covariance matrix $\mathbf{V_{ML}}$ by least squares.

(b) Suppose we use a *multivariate normal*$(\mathbf{b_0}, \mathbf{V_0})$ prior where

$$\mathbf{b_0} = \begin{pmatrix} 10 \\ 0 \end{pmatrix} \quad \text{and} \quad \mathbf{V_0} = \begin{bmatrix} 10 & 0 \\ 0 & 4 \end{bmatrix}.$$

i. Find the posterior covariance matrix $\mathbf{V_1}$.

ii. Find the posterior mean $\mathbf{b_1}$.

4.20 Suppose we have the following sixteen observations of the response variable y together with two predictor variables x_1 and x_2. We want to do a multiple regression of y onto the two predictor variables, where we know the standard deviation $\sigma = 2$.

y	x_1	x_2	y	x_1	x_2
35.8	20	0	39.1	40	0
36.1	20	0	37.9	40	0
32.5	20	0	39.4	40	0
33.4	20	0	37.6	40	0
32.7	30	1	36.2	50	1
30.3	30	1	34.1	50	1
31.1	30	1	41.1	50	1
32.0	30	1	34.2	50	1

(a) Recode the predictor variables so each is centered around mean zero. Find the the vector of maximum likelihood estimates $\hat{\beta}_{ML}$ and its covariance matrix $\mathbf{V_{ML}}$ by least squares.

(b) Suppose we use a *multivariate normal*$(\mathbf{b_0}, \mathbf{V_0})$ prior where

$$\mathbf{b_0} = \begin{pmatrix} 30 \\ 0 \\ 0 \end{pmatrix} \quad \text{and} \quad \mathbf{V_0} = \begin{bmatrix} 25 & 0 & 0 \\ 0 & 1 & 0 \\ 0 & 0 & 16 \end{bmatrix}.$$

i. Find the posterior covariance matrix $\mathbf{V_1}$.

ii. Find the posterior mean $\mathbf{b_1}$.

Appendix: Proof of Poisson Process Theorem

Theorem 1 *Under the following assumptions, Y_t number of occurrences in the time interval $(0, t)$ has the* Poisson *distribution with mean $\mu = \lambda t$.*

Assumptions of Poisson Process

1. *Y_t counts the number of occurrences of the event in the time interval $(0, t)$.*

2. *Occurrences of an event happen randomly through time at a constant rate λ.*

3. *What happens in one time interval is independent of what happens in another non-overlapping time interval. (independent increments)*

4. *The probability of exactly one event occurring in a very short time interval $(t, t+h)$ is approximately proportional to the length of the interval h.*

$$P(\text{exactly one event occurs in } (t, t+h) = \lambda h + o(h)$$

where $o(h)$ is a function that approaches 0 faster than h, i.e.,

$$\lim_{h \to 0} \left(\frac{o(h)}{h} \right) = 0.$$

5. *The probability that more than one event occurs in a very short time interval $(t, t+h)$ approaches 0 faster than h, i.e.,*

$$\begin{aligned} \lim_{h \to 0} P(\text{more than one event occurs in} (t, t+h) &= \lim_{h \to 0} \left(\frac{o(h)}{h} \right) \\ &= 0\,. \end{aligned}$$

Then Y_t number of occurrences in the time interval $(0, t)$ has the Poisson distribution with mean $\mu = \lambda t$. When this is true, we say the process follows the Poisson process with intensity rate λ.

Proof: *Let $p_{i\,j}(h) = P(Y_{t+h} = j \mid Y_t = i)$ be the probability of a transition from i to j in a short time interval h. Note: this does not depend on t the starting time since the events occur at constant rate λ.*

Zero occurrences. *First we look at the case for zero occurrences of the event in the interval. Since there are independent increments*

$$p_{0\,0}(t+h) \quad = \quad p_{0\,0}(t) \times p_{0\,0}(h).$$

Subtracting $p_{0\,0}(t)$ from both sides and dividing both sides by h

$$\begin{aligned}\frac{p_{0\,0}(t+h)-p_{0\,0}(t)}{h} &= -p_{0\,0}(t)\times\frac{1-p_{0\,0}(h)}{h}\\ &= -p_{0\,0}(t)\times\frac{1-(1-\lambda h-o(h))}{h}\\ &= = -p_{0\,0}(t)\times\frac{\lambda h+o(h)}{h}\,.\end{aligned}$$

When we take the limit of both sides as $h\to 0$, the left-hand side is the derivative of the probability function. This gives the differential equation that the probability function must satisfy.

$$p'_{0\,0}(t) = -p_{0\,0}(t)\times\lambda\,.$$

We let $u = log(p_{0\,0}(t))$. Then $du = \frac{1}{p_{0\,0}(t)}\,p'_{0\,0}(t)$. The equation becomes $u' = -\lambda$ so the solution is $u = -\lambda t$. Transform back to get

$$p_{0\,0}(t) = e^{-\lambda t}\,.$$

This is the Poisson probability of zero occurrences with parameter λt.

More than one occurrence. *Now we look at the case where there is more than one occurrence of the event in the interval. Suppose there are j occurrences of the event*

$$\begin{aligned}p_{0\,j}(t+h) &= \sum_{k<j} p_{0k}(t)\times p_{k\,j}(h)+p_{0\,j}(t)\times p_{j\,j}(h))\\ &= \sum_{k<j-1} p_{0k}(t)\times p_{k\,j}(h)+p_{0\,j-1}(t)\times p_{j-1\,j}(h)\\ &\quad +p_{0\,j}(t)\times p_{j\,j}(h))\,.\end{aligned}$$

Subtract $p_{0\,j}(t)$ from both sides and divide both sides by h

$$\begin{aligned}\frac{p_{0\,j}(t+h)-p_{0\,j}(t)}{h} &= \sum_{k<j-1} p_{0k}(t)\times\frac{p_{k\,j}(h)}{h}+p_{0\,j-1}(t)\times\frac{p_{j-1\,j}(h)}{h}\\ &\quad +p_{0\,j}(t)\times\frac{p_{j\,j}(h)}{h}\\ &= \sum_{k<j-1} p_{0k}(t)\times\frac{o(h)}{h}+p_{0\,j-1}(t)\times\frac{\lambda h+o(h)}{h}\\ &\quad +p_{0\,j}(t)\times\frac{1-(1-\lambda h+o(h))}{h}\,.\end{aligned}$$

We take limits of both sides of the equation as $h\to 0$. The left-hand side is the derivative of $p_{0\,j}(t)$. We note that $p_{j\,j}(t) = p_{0\,0}(t) = e^{-\lambda t}$ as all cases have 0 *occurrences in the interval. This simplifies to the differential equation*

$$p'_{0\,j}(t) \quad = \quad \lambda p_{0\,j-1}(t) - \lambda p_{0\,j}(t)$$

that $p_{0\,j}(t)$ must satisfy.

***Solve for** $j = 1$. We can write the equation for $j = 1$ as*

$$p'_{0\,1}(t) = \lambda p_{0\,0}(t) - \lambda p_{0\,1}(t)\,.$$

We let D be the differential operator $D(f) = f'$. The equation becomes

$$(D + \lambda)P_{0\,1}(t) \quad = \quad \lambda e^{-\lambda t}\,.$$

Let $P_{0\,1}(t) = V(t)e^{-\lambda t}$. Using the rule for derivative of a product the equation becomes

$$\begin{aligned} -\lambda V(t)e^{-\lambda t} + V'(t)e^{-\lambda t} + \lambda V(t)e^{-\lambda t} &= \lambda e^{-\lambda t} \\ V'(t)e^{-\lambda t} &= \lambda e^{-\lambda t} \end{aligned}$$

so $V'(t) = \lambda$ and hence $V(t) = \lambda t$. Hence the solution for $j = 1$ is

$$\begin{aligned} P_{0\,1}(t) &= V(t)e^{-\lambda t} \\ &= \lambda t e^{-\lambda t}\,. \end{aligned}$$

This is the Poisson probability of one occurrence with parameter λt.

***By mathematical induction.** We have seen that for $j = 0$ and $j = 1$ the probability of j occurrences is given by*

$$P_{0\,j}(t) = \frac{(\lambda t)^j \; e^{-\lambda t}}{j\,!}\,.$$

We assume that this holds true for all integers $j \leq k$. The differential equation that $p_{0\,k+1}(t)$ must satisfy is given by

$$p'_{0\,k+1}(t) \quad = \quad \lambda p_{0\,k}(t) - \lambda p_{0\,k+1}(t)\,.$$

Using the differential operator $D(f) = f'$ the equation becomes

$$(D + \lambda)P_{0\,k+1}(t) \quad = \quad \lambda p_{0\,k}(t)\,.$$

Let $P_{0\,k+1}(t) = V(t)\,e^{-\lambda t}$. Using the rule for derivative of a product the equation becomes

$$V'(t)e^{-\lambda t} - \lambda\,V(t)\,e^{-\lambda t} + \lambda\,V(t)\,e^{-\lambda t} \quad = \quad \frac{\lambda\,(\lambda t)^k e^{-\lambda t}}{k\,!}\,.$$

This simplifies to

$$V'(t)e^{-\lambda t} = \frac{\lambda\,(\lambda t)^k e^{-\lambda t}}{k\,!}.$$

Thus

$$V'(t) = \frac{\lambda\,(\lambda t)^k}{k\,!}$$

and hence

$$V(t) = \frac{(\lambda t)^{k+1}}{k+1!}.$$

Then

$$P_{0\,k+1}(t) = \frac{(\lambda t)^{k+1} e^{-\lambda t}}{k+1!}.$$

This is the Poisson probability of $k+1$ *occurrences with Poisson parameter* λt. *This holds true for all integers* n *by the principal of mathematical induction.*

So when we have a *Poisson process*, the random variable Y_t, the number of occurrences of the event in the time interval $(0, t)$ has the *Poisson* distribution with parameter $\mu = \lambda t$.

5

Markov Chains

In this chapter we introduce Markov chains. These are a special type of *stochastic process*, which are processes that move around a set of possible values where the future values can't be predicted with certainty. There is some chance element in the evolution of the process through time. The set of possible values is called the *state space* of the process. Markov chains have the "memoryless" property that, given the past and present states, the future state only depends on the present state. This chapter will give us the necessary background knowledge about Markov chains that we will need to understand Markov chain Monte Carlo sampling.

In Section 5.1 we introduce the stochastic processes. In Section 5.2 we will introduce Markov chains and define some terms associated with them. In Section 5.3 we find the n-step transition probability matrix in terms of one-step transition probability matrix for time invariant Markov chains with a finite state space. Then we investigate when a Markov chain has a long-run distribution and discover the relationship between the long-run distribution of the Markov chain and the steady state equation. In Section 5.4 we classify the states of a Markov chain with a discrete state space, and find that all states in an irreducible Markov chain are of the same type. In Section 5.5 we investigate sampling from a Markov chain. In Section 5.6 we look at time-reversible Markov chains and discover the detailed balance conditions, which are needed to find a Markov chain with a given steady state distribution. In Section 5.7 we look at Markov chains with a continuous state space to determine the features analogous to those for discrete space Markov chains.

In Chapter 6, we will let the *parameter space* be the state space for the Markov chains we will use. We will learn ways to find a Markov chain that has the posterior distribution as its long-run distribution. Then when we take a sample value from the

Markov chain after it has been running a long time it can be considered a random draw from the posterior. This method for drawing a sample from the posterior is known as Markov chain Monte Carlo sampling.

5.1 STOCHASTIC PROCESSES

A stochastic process is a process that evolves through time according to some probabilistic law. There is an element of random chance in the evolution of the process. If we repeated the process starting from the identical conditions, we would not get the same pattern of moves through the states as we observed the first time.

The first characteristic of a stochastic process is the time points when the process is allowed to change values. If the process can only change values at discrete values such as $t = 1, 2, \ldots$, it is called a *discrete time* stochastic process. If the process can change values at any time point $t \geq 0$ it is called a *continuous time* stochastic process. We only consider discrete time stochastic processes, where changes are only allowed at times $n = 1, 2, \ldots$.

The value that the process has at time t is called its *state* at time t. The set of possible values that the stochastic process has at any particular time is called the *state space*, and can be either discrete or continuous. The second thing that characterizes a stochastic process is the size and type of state space allowed. The state space may be one or more dimensions. In each dimension there may be a finite number of discrete possible values, an infinite number of discrete possible values, or all values in a continuous interval. If the possible values in each dimension are discrete, the process is equivalent to a process on a single dimensional discrete space, so the types of state space we need to consider include:

- Finite discrete number of states. $S = \{1, 2, \ldots, m\}$
- Infinite number of discrete states. $S = \{\ldots, -1, 0, 1, \ldots, \}$
- Continuous state space of one dimension. $S = \{x : (\infty \ < x < \infty)\}$
- Continuous state space of p dimensions. $S = \{(x_1, \ldots, x_p) : (\infty \ < x_i < \infty) \ \text{ for } \ i = 1, \ldots, p\}$

Initially we will consider processes having a finite discrete number of states. The properties of processes having more complicated state spaces will follow by analogy.

Mathematical Model for a Stochastic Process

The mathematical model for a discrete stochastic process is a sequence of random variables $X^{(n)}$ for $n = 0, 1, 2, \ldots$, where $X^{(n)}$ is the state at time n. Here the superscript (n) indicates time point n. All knowledge about a single random variable is in its probability distribution. Similarly, all knowledge about the probabilistic law for a stochastic process is contained in the joint probability distribution of every sequence of the random variables. The joint probabilities are given by

$$
\begin{gathered}
P(X^{(0)} = x_0) \\
P(X^{(1)} = x_1, X^{(0)} = x_0) \\
P(X^{(2)} = x_2, X^{(1)} = x_1, X^{(0)} = x_0) \\
etc.
\end{gathered}
$$

for all possible values x_n for $n = 0, 1, \ldots$ in the state space. The joint probabilities of the process at times $0, 1, \ldots, n$ can be built up sequentially from the previous joint probabilities and the conditional probabilities (transition probabilities) of the next observation in the process given the previous values of the process the by using the multiplication rule.

$$
\begin{aligned}
P(X^{(1)} = x_1, X^{(0)} = x_0) &= P(X^{(1)} = x_1 | X^{(0)} = x_0) \\
&\quad \times P(X^{(0)} = x_0) \\
P(X^{(2)} = x_12, X^{(1)} = x_1, X^{(0)} = x_0) &= P(X^{(2)} = x_2 | X^{(1)} = x_1, X^{(0)} = x_0) \\
&\quad \times P(X^{(1)} = x_1, X^{(0)} = x_0) \\
&\quad etc.
\end{aligned}
$$

for all possible values of x_n for $n = 0, 1, \ldots$. Note that the transition probabilities of a stochastic process generally depend on the entire past history of the process.

Occupation Probability Distribution

The occupation probabilities at each time can be found by summing the joint probabilities. The occupation probabilities are given by

$$
\begin{gathered}
P(X^{(1)} = x_1) = \textstyle\sum_{x_0} P(X^{(1)} = x_1, X^{(0)} = x_0) \\
P(X^{(2)} = x_2) = \textstyle\sum_{x_1} \sum_{x_0} P(X^{(2)} = x_2, X^{(1)} = x_1, X^{(0)} = x_0) \\
etc.
\end{gathered}
$$

for all possible values of x_n for $n = 0, 1, \ldots$.

5.2 MARKOV CHAINS

A Markov chain[1] is a special kind of stochastic process. The conditional probabilities of the process at time n given the states at all the previous times $n-1, \ldots, 0$ only depends on the single previous state at time $n-1$. Hence

$$
\begin{aligned}
P(X^{(2)} = x_2 | X^{(1)} = x_1, X^{(0)} = x_0) &= P(X^{(2)} = x_2 | X^{(1)} = x_1) \\
P(X^{(3)} = x_3 | X^{(2)} = x_2, X^{(1)} = x_1, X^{(0)} = x_0) &= P(X^{(3)} = x_3 | X^{(2)} = x_2) \\
&\quad etc.
\end{aligned}
$$

[1]These were first studied by the Russian mathematician A. A. Markov (1856–1922).

The future evolution of the Markov chain only depends on the present state. It does not depend on the path taken to get to the present state. This is known as the *Markov property*. The joint probability distributions of all the states up to time n can be built up sequentially by

$$\begin{aligned}
P(X^{(1)} = x_1, X^{(0)} = x_0) &= P(X^{(1)} = x_1 | X^{(0)} = x_0) \\
&\quad \times P(X^{(0)} = x_0) \\
P(X^{(2)} = x_2, X^{(1)} = x_1, X^{(0)} = x_0) &= P(X^{(2)} = x_2 | X^{(1)} = x_1) \\
&\quad \times P(X^{(1)} = x_1 | X^{(0)} = x_0) \\
&\quad \times P(X^{(0)} = x_0) \\
&\quad etc.
\end{aligned}$$

If we are only interested in the process at time n then we would sum both sides over all possible values of $x_{n-1}, x_{n-2}, \ldots x_0$. This gives the probability distribution over the possible states at time n

$$P(X^{(n)} = x_n) = \sum P(X^{(n)} = x_n | X^{(n-1)} = x_{n-1}) \times P(X^{(n-1)} = x_{n-1})$$

where the sum is over all possible values of x_{n-1}.

5.3 TIME-INVARIANT MARKOV CHAINS WITH FINITE STATE SPACE

First we will look at Markov chains that have a finite state space. Let us suppose there are states numbered from $1, \ldots, K$. In this case the transition probabilities can be put in a matrix.

$$\mathbf{P}^{[n-1,n]} = \begin{bmatrix} p_{1,1}^{[n-1,n]} & \cdots & p_{1,K}^{[n-1,n]} \\ \vdots & \vdots & \vdots \\ p_{K,1}^{[n-1,n]} & \cdots & p_{K,K}^{[n-1,n]} \end{bmatrix}$$

where $p_{i,j}^{[n-1,n]} = P(X^{(n)} = j | X^{(n-1)} = i)$ is the transition probability of going from state i at time $n-1$ to state j at time n. We note that each row of $\mathbf{P}^{[n-1,n]}$ forms a probability distribution because $\sum_j p_{i,j}^{[n-1,n]} = 1$.

We will restrict ourselves to *time invariant* Markov chains where the transition probabilities only depend on the states, not the time n. These are also called *homogeneous* Markov chains. In this case, we can leave out the time index and the transition probability matrix of the Markov chain is given by

$$\mathbf{P} = \begin{bmatrix} p_{1,1} & \cdots & p_{1,K} \\ \vdots & \vdots & \vdots \\ p_{K,1} & \cdots & p_{K,K} \end{bmatrix} \tag{5.1}$$

where $p_{i,j} = P(X^{(n+1)} = j|X^{(n)} = i)$ is the probability of going to state j from state i in one step for all times $n = 1, \ldots$. This is the matrix on one-step transition probabilities. We should note transition probabilities are conditional probabilities given the initial state even though we are not writing them as such.

Higher Transition Probability Matrices

The matrix of two-step transition probabilities is given by

$$\mathbf{P}^{[2]} = \begin{bmatrix} p^{[2]}_{1,1} & \cdots & p^{[2]}_{1,K} \\ \vdots & \vdots & \vdots \\ p^{[2]}_{K,1} & \cdots & p^{[2]}_{K,K} \end{bmatrix}$$

where $p^{[2]}_{i,j} = P(X^{(n+2)} = j|X^{(n)} = i)$ is the probability of going from state i to state j in two steps, for all n. The superscript [2] here means in 2 steps. We can partition the event

$$(X^{(2)} = j|X^{(0)} = i) = \bigcup_k (X^{(2)} = j, X^{(1)} = k|X^{(0)} = i)$$

into disjoint events in terms of the state at time 1. Since the probability of a union of disjoint events is the sum of the probabilities, the transition probability of going from state i to state j in two steps is

$$\begin{aligned} P(X^{(2)} = j|X^{(0)} = i) &= \sum_{k=1}^{K} P(X^{(2)} = j, X^{(1)} = k|X^{(0)} = i) \\ &= \sum_{k=1}^{K} P(X^{(2)} = j|X^{(1)} = k, X^{(0)} = i) \\ &\quad \times P(X^{(0)} = i)\,. \end{aligned}$$

Since the Markov chain satisfies the Markov property

$$P(X^{(2)} = j|X^{(1)} = k, X^{(0)} = i) = P(X^{(2)} = j|X^{(1)} = k)\,.$$

Hence the two-step transition probability is given by

$$p^{[2]}_{i,j} = \sum_{k=1}^{K} p_{i,k} \times p_{k,j}\,. \tag{5.2}$$

We recognize this is the formula for matrix multiplication. Hence the matrix of two-step transition probabilities is given by

$$\begin{aligned} \mathbf{P}^{[2]} &= \mathbf{P} \times \mathbf{P} \\ &= \mathbf{P}^2 \end{aligned} \tag{5.3}$$

which is the square of the one-step transition probability matrix. This can be generalized. Partition the conditional event that we are in state j at time $m+n$ given we started in state i at time 0 on the basis of the state at time n. We get

$$(X^{(m+n)} = j|X^{(0)} = i) = \bigcup_k (X^{(m+n)} = j, X^{(n)} = k|X^{(0)} = i)$$

in terms of the state at time n. The probability of a union of disjoint events is the sum of the probabilities of each of the individual events making up the union. Hence

$$P(X^{(m+n)} = j|X^{(0)} = i) = \sum_{k=1}^{K} P(X^{(m+n)} = j, X^{(n)} = k|X^{(0)} = i)\,.$$

We find the joint probability in terms of a conditional probability times a marginal probability (both are conditional on $X^{(0)} = i$) to get

$$\begin{aligned} P(X^{(m+n)} = j|X^{(n)} = i) &= \sum_k P(X^{(m+n)} = j|x^{(n)} = k, X^{(0)} = i) \\ &\quad \times P(X^{(0)} = i)\,. \end{aligned}$$

By the Markov property,

$$P(X^{(m+n)} = j|x^{(n)} = k, X^{(0)} = i) = P(X^{(m+n)} = j|X^{(n)} = k)\,.$$

Hence the joint probability is

$$P(X^{(m+n)} = j|X^{(0)} = i) = \sum_k P(X^{(m+n)} = j|X^{(n)} = k) \times P(X^{(n)} = k|x^{(0)} = i)\,.$$

It follows that

$$p_{i,j}^{[m+n]} = \sum_{k=1}^{K} p_{i,k}^{[m]} \times p_{k,j}^{[n]}\,. \tag{5.4}$$

Since the state space is finite, we can write this in matrix form as

$$\mathbf{P}^{[m+n]} = \mathbf{P}^{[m]} \times \mathbf{P}^{[n]}\,. \tag{5.5}$$

This is called the Chapman-Kolmogorov equation. It follows that the matrix of n-step transition probabilities is the matrix of one-step transition probabilities to the power n.

$$\mathbf{P}^{[n]} = \mathbf{P}^{n}\,. \tag{5.6}$$

The occupation probability distribution. Suppose at time 0, the occupation probability distribution is given by

$$\boldsymbol{\alpha}^{(0)} = (\alpha_1^{(0)}, \ldots, \alpha_m^{(0)})$$

where $\alpha_i^{(0)} = P(X^{(n)} = i)$ is the probability that the chain is in state i at time 0. We can find the occupation probability distribution at time 1 by partitioning the event

$$(X^{(1)} = j) = \bigcup_i (X^{(1)} = j, X^{(0)} = i)\,.$$

The probability of a disjoint union is the sum of the probabilities, so

$$P(X^{(1)} = j) = \sum_{i=1}^{K} P(X^{(1)} = j, X^{(0)} = i)\,.$$

The joint probability is the marginal probability times the conditional probability

$$P(X^{(1)} = j) = \sum_{i=1}^{K} P(X^{(0)} = i) \times P(X^{(1)} = j | X^{(0)} = i)\,.$$

In other words

$$\alpha_j^{(1)} = \sum_{i=1} \alpha_i^{(0)} \times p_{i,j}\,. \tag{5.7}$$

This is the $\boldsymbol{\alpha}_0$ times the jth column of the transition probability matrix. We can write this in matrix form in the finite state space case as

$$\boldsymbol{\alpha}^{(1)} = \boldsymbol{\alpha}^{(0)} \times \mathbf{P}\,. \tag{5.8}$$

We can continue this process, linking each occupation probability distribution to the previous occupation probability distribution by multiplication by the one-step matrix of transition probabilities. Repeating this process back to the initial state we get

$$\boldsymbol{\alpha}^{(n)} = \boldsymbol{\alpha}^{(0)} \times \mathbf{P}^n\,. \tag{5.9}$$

Note, the ith row of the K by k identity matrix $\mathbf{I}$ would be a probability distribution where all the probability is in state i. Since $\mathbf{P}^n = \mathbf{I} \times \mathbf{P}^n$ where $\mathbf{I}$ is the K by K identity matrix. Hence the ith row of $\mathbf{P}^n$ is the occupation probability distribution at time n given that we started in state i at time 0.

Long-Run Distribution of the Chain and the Steady State Equation

If a Markov chain possesses a limiting distribution that is independent of the initial state (or initial distribution of states), it is called the *long-run distribution* of the chain.

$$\boldsymbol{\pi} = \lim_{n\to\infty} \boldsymbol{\alpha}^{(n)}\,.$$

Note that not all chains will have long-run distribution. However, when a chain does possess a limiting distribution,

$$\begin{aligned} \lim_{n\to\infty} \mathbf{I} \times \mathbf{P}^n &= \lim_{n\to\infty} \mathbf{P}^n \\ &= \begin{bmatrix} \pi_1 & \dots & \pi_K \\ \vdots & \vdots & \vdots \\ \pi_1 & \dots & \pi_K \end{bmatrix}. \end{aligned} \tag{5.10}$$

Hence all the rows of $\mathbf{P}^n$ approach the long-run distribution of the chain. Let the chain run one more step and we get

$$\mathbf{P}^{[n+1]} = \mathbf{P}^{[n]} \times \mathbf{P}\,.$$

If we take the limit of both sides of this equation where we let $n \to \infty$ both $\mathbf{P}^{[n+1]}$ and $\mathbf{P}^{[n]}$ will approach the same limit. Hence

$$\begin{bmatrix} \pi_1 & \dots & \pi_K \\ \vdots & \vdots & \vdots \\ \pi_1 & \dots & \pi_K \end{bmatrix} = \begin{bmatrix} \pi_1 & \dots & \pi_K \\ \vdots & \vdots & \vdots \\ \pi_1 & \dots & \pi_K \end{bmatrix} \times \mathbf{P}\,.$$

Each row gives the same equation

$$\boldsymbol{\pi} = \boldsymbol{\pi} \times \mathbf{P}\,. \tag{5.11}$$

We call this the steady state equation. Thus the long-run distribution, if it exists, satisfies the steady state equation. The steady state probability of entering state j from state i equals $\pi_i p_{i,j}$. The steady state probability of being in state j equals the weighted sum of probabilities of entering state j in one step from all the states, where each state is weighted by its steady state probability. Thus the steady state equation says the probability flowing into each state at the steady state equals its steady state probability.

Note, however, that it is possible to have a solution of the steady state equation, but not have a long-run distribution.

Example 5 *Sometimes a chain only visits some states only at times n where n is divisible evenly by some integer greater than one. For example, the Markov chain with transition matrix given by*

$$\mathbf{P} = \begin{bmatrix} 0 & 1 \\ 1 & 0 \end{bmatrix}.$$

alternates between two states. We note that

$$\begin{pmatrix} .5 & .5 \end{pmatrix} \begin{bmatrix} 0 & 1 \\ 1 & 0 \end{bmatrix} = \begin{pmatrix} .5 & .5 \end{pmatrix}$$

so $\begin{pmatrix} .5 & .5 \end{pmatrix}$ is a solution of the steady state equation for the Markov chain having that transition matrix. However

$$\mathbf{P}^{(n)} = \begin{bmatrix} 0 & 1 \\ 1 & 0 \end{bmatrix}$$

when n is odd, and

$$\mathbf{P}^{(n)} = \begin{bmatrix} 1 & 0 \\ 0 & 1 \end{bmatrix}$$

when n is even. Thus this chain does not have a long-run distribution as it just cycles back and forth between the two states.

A chain that only visits states on some multiple are called *periodic* Markov chains. We will restrict ourselves to *aperiodic* Markov chains. In such a chain, after the chain has run long enough, it could be in any of the states at any time after that. We should note that many chains have the same long-run distribution.

Countably Infinite Discrete State Space

Suppose the state space consists of all integers $-\infty < k < \infty$. When the state space is countably infinite instead of being finite, we cannot represent the one-step transition probabilities in a finite matrix, nor can we represent the occupation probability distribution as a finite vector. However, the equations for the elements of these matrices and vectors corresponding to Equations 5.2, 5.4, 5.7 are still the same except that now the summation goes from $k = -\infty$ to ∞. Similarly, we can't write the steady state equation in matrix terms, but the equation for the elements of the steady state equation would be similar.

$$\pi_j = \sum_{i=-\infty}^{\infty} \pi_i p_{i,j} \tag{5.12}$$

for all $j = -\infty, \ldots, \infty$.

5.4 CLASSIFICATION OF STATES OF A MARKOV CHAIN

In this section we want to classify the states of a Markov chain according to their long-run behavior. Some questions we will answer are:

- If we are in state i are we certain to eventually reach state j, or is there a possibility we will never reach j?
- If we are certain to reach state j, what is the waiting time distribution until we first reach state j? What is the expected value of this waiting time distribution?
- If we are in state i are we certain to return to state i, or is there a possibility that we never return to i?
- If we are certain to return to state i, what is the waiting time distribution until our first return to state i? What is the expected value of this waiting time distribution?

First Passage and First Return Probabilities

Partition the event $(X^{(n)} = j|X^{(0)} = i)$ according to m, the the first time state j was reached.

$$(X^{(n)} = j|X^{(0)} = i) =$$

$$\bigcup_{m=1}^{n} (X^{(n)} = j, X^{(m)} = j, X^{(m-1)} \neq j, \ldots, X^{(1)} \neq j | X^{(0)} = i) .$$

Hence

$$\begin{aligned} p_{i,j}^{[n]} &= \sum_{m=1}^{n} P(X^{(n)} = j, X^{(m)} = j, X^{(m-1)} \neq j, \ldots, X^{(1)} \neq j | X^{(0)} = i) \\ &= \sum_{m=1}^{n} P(X^{(n)} = j | X^{(m)} = j, X^{(m-1)} \neq j, \ldots, X^{(1)} \neq j, X^{(0)} = i) \\ &\quad \times P(X^{(m)} = j, X^{(m-1)} \neq j, \ldots, X^{(1)} \neq j | X^{(0)} = i) . \end{aligned}$$

Let the probability the first passage is in step m be

$$f_{i,j}^{[m]} = P(X^{(m)} = j, X^{(m-1)} \neq j, \ldots, X^{(1)} \neq j | X^{(0)} = i) .$$

Hence

$$p_{i,j}^{[n]} = \sum_{m=1}^{n} p_{j,j}^{[n-m]} \times f_{i,j}^{[m]} . \tag{5.13}$$

Solving for $f_{i,j}^{[n]}$ gives the recursion formula

$$f_{i,j}^{[n]} = p_{i,j}^{[n]} - \sum_{m=1}^{n-1} f_{i,j}^{[m]} p_{j,j}^{[n-m]} . \tag{5.14}$$

This enables the first passage probabilities to be found recursively from n-step transition probabilities and previous first passage probabilities. Similarly, if we start in state i and look at m, the first time we return to state i then the first return probabilities can be calculated recursively by

$$f_{i,i}^{[n]} = p_{i,i}^{[n]} - \sum_{m=1}^{n-1} f_{i,i}^{[m]} p_{i,i}^{[n-m]} . \tag{5.15}$$

Mean first passage time. Let $N_{i,j}$ be the number of steps until the chain reaches state j given it is in state i at time 0. $P(N_{i,j} = n) = f_{i,j}^{[n]}$. Assuming that for a given i and j, the first passage probabilities form a probability distribution (their sum over all n equals 1), then $N_{i,j}$ is a random variable and its mean is given by

$$\begin{aligned} m_{i,j} &= E(N_{i,j}) \\ &= \sum_{n=1}^{\infty} n \times f_{i,j}^{[n]} . \end{aligned}$$

Note: if the first passage probabilities sum to some number less than 1, they form a degenerate probability distribution, and there is a finite probability that starting from state i the chain never reaches state j. In that case, the formula gives the expected number of steps until the chain reaches state j given that it does reach state j.

A simpler way of finding the mean first passage time $m_{i,j}$ is to condition on the state after one step. Either we reach state j on the first step and $N_{i,j} = 1$, or we have used up a step and arrived in another state k and so a total of $1 + N_{k,j}$ steps will be needed to reach state j. Hence

$$\begin{aligned} N_{i,j} &= 1 \times p_{i,j} + \sum_{k \neq j} (1 + N_{k,j}) \times p_{i,k} \\ &= 1 \times \sum_{k} p_{i,k} + \sum_{k \neq j} p_{i,k} \times N_{k,j} . \end{aligned}$$

Taking expectations of both sides of this we get the system of linear equations

$$m_{i,j} = 1 + \sum_{k \neq j} p_{i,k} \times m_{k,j}$$

that the mean first passage times must satisfy for all possible values of i and j in the state space.

Classification of States of a Markov Chain

We can classify the states of a Markov chain by the first return probabilities $f_{i,i}^{[n]}$. When the first return probabilities $f_{i,i}^{[n]}$ sum to a number less than one, there is a finite probability that the chain will not return to state i. In that case we say that state i is a *transient* state. The chain may return to transient state a finite number of times, but eventually it will leave and never return.

When the first return probabilities for state i sum to one, the chain is certain to return to state i. In fact, because of the Markov property, given it has returned to state i, it is certain to return to state i again. Hence the chain is certain to return to state i infinitely often. In that case we say that the state i is a *recurrent* state. These are sometimes called *persistent* states. Recurrent states can be further classified by the mean first return time. When the mean first return time is finite, we say that the state is *positive recurrent*. The sequence of first return probabilities $f_{i,i}^{[n]}$ is going to zero at such a rate that $\sum_n n \times f_{i,i}^{[n]} j$ converges to a finite number. That means that $f_{i,i}^{[n]}$ must go to zero at a faster rate than n^{-2}. When the mean first return time is infinite we say the state is *null recurrent*. The chain is certain to return to a null recurrent state, but the first return probabilities are going to zero at a much slower rate so that $\sum_n n \times f_{i,i}^{[n]} j$ diverges. That means that $f_{i,i}^{[n]}$ must go to zero at a slower rate than n^{-2}. These classifications are summarized in Table 5.1.

Generating function of first return probabilities. The generating function of a sequence of numbers a_n for $n = 0, \ldots, \infty$ is defined to be

$$A(s) = \sum_{n=0}^{\infty} a_n s^n .$$

Table 5.1 Classification of states using first return probabilities $f_{i,i}^{[n]}$

Type	Sum	Mean
Transient	$\sum f_{i,i}^{[n]} < 1$	
Null recurrent	$\sum f_{i,i}^{[n]} = 1$	$\sum n\, f_{i,i}^{[n]} = \infty$
Positive recurrent	$\sum f_{i,i}^{[n]} = 1$	$\sum n\, f_{i,i}^{[n]} < \infty$

Note: the sum of the sequence $\sum a_n = A(1)$, and each term of the sequence can be found by taking the appropriate derivative of the generating function and substituting $s = 0$. Hence, the generating function of first return probabilities is given by

$$F_{i,i}(s) = \sum_{n=1}^{\infty} f_{i,i}^{[n]}\, s^n\,.$$

Note: $F_{i,i}(1) = \sum f_{i,i}^{[n]}$. Hence $F_{i,i}(1) = 1$ for a recurrent state (either null recurrent or positive recurrent) and $F_{i,i}(1) < 1$ for a transient state. The first derivative of the generating function is

$$F'_{i,i}(s) = \sum_{n=1}^{\infty} n f_{i,i}^{[n]}\, s^{n-1}\,.$$

Note: $F'_{i,i}(1) = \sum n f_{i,i}^{[n]}$, which is the mean return time for state i when that state is recurrent. When we multiply both sides of the Equation 5.13 by s^n and sum over $n = 1, \ldots$ we get

$$\sum_{n=1}^{\infty} p_{i,i}^{[n]} s^n = \sum_{n=1}^{\infty} \sum_{k=1}^{n} f_{i,i}^{[k]}\, s^k\, p_{i,i}^{[n-k]}\, s^{n-k}\,.$$

The left side is the generating function $P_{i,i}(s) - p_{i,i}^{(0)}$ and the right side is $F_{i,i}(s) \times P_{i,i}(s)$. Since the chain is in state i at time $n = 0$, $p_{i,i}^{(0)} = 1$,

$$P_{i,i}(s) - 1 = F_{i,i}(s) \times P_{i,i}(s)\,.$$

Solving this gives

$$P_{i,i}(s) = \frac{1}{1 - F_{i,i}(s)}\,.$$

Hence for transient states $P_{i,i}(1) = 1/(1 - F_{i,i}(1)) < \infty$ while for recurrent states. $P_{i,i}(1) = 1/(1 - F_{i,i}(1)) = \infty$. That means that a transient state will be returned to only a finite number of times, and after that will not be visited again, while the number of times a recurrent state will be returned to is without bound. Clearly the mean recurrence time for state i is $m_{i,i} = 1/\pi_i$. The limit of $p_{i,i}^{[n]}$ will be $1/m_{i,i}$,

Table 5.2 Classification of states using n-step transition probabilities $p_{i,i}^{[n]}$

Type	Sum	Limit as $n \to \infty$
Transient	$\sum p_{i,i}^{[n]} < \infty$	$\lim p_{i,i}^{[n]} = 0$
Null recurrent	$\sum p_{i,i}^{[n]} = \infty$	$\lim p_{i,i}^{[n]} = 1/\pi_i = 0$
Positive recurrent	$\sum p_{i,i}^{[n]} = \infty$	$\lim p_{i,i}^{[n]} = 1/\pi_i > 0$

so for a positive recurrent state the probability of being in the state i at some time n far in the future is approaching some finite number. On the other hand, the mean recurrence time for a null recurrent state is infinite, so for a null recurrent state the limit of $p_{i,i}^{[n]}$ is zero, and the probability of being in state i at some time n far in the future is going to zero. Thus we can classify the states of a Markov chain according to the sequence of n-step probabilities $\{p_{i,i}^{[n]}\}$. These are shown in Table 5.2.

Irreducible Markov Chains

State j is reachable from state i if $p_{i,j}^{[n]} > 0$ for some n. When state j is reachable from state i, and state i is reachable from state j, then the pair of states are said to communicate with each other. A set of states such that every pair of states inside the class communicate with each other, and no state outside the set is reachable from inside the set, is called a communicating class of states. It is a minimal closed set of states. A Markov chain consisting of a single communicating class is called an *irreducible* chain. Otherwise, it is called a reducible chain.

Theorem 2 *All states in an irreducible chain are the same type.*
Proof: Let i and j be arbitrary states in an irreducible chain. Since j is reachable from i then $c_1 = p_{i,j}^{[r]} > 0$ for some r, and since i is reachable from j then $c_2 = p_{j,i}^{[s]} > 0$ for some s. Hence

$$\begin{aligned} p_{i,i}^{[r+n+s]} &\geq p_{i,j}^{[r]} \times p_{j,j}^{[n]} \times p_{j,i}^{[s]} \\ &\geq c_1 c_2 \times p_{j,j}^{[n]} . \end{aligned}$$

If $p_{i,i}^{[n]} \to 0$ then so must $p_{j,j}^{[n]} \to 0$. If $\sum p_{j,j}^{[n]} = \infty$ then so must $\sum p_{i,i}^{[n]}$. Reversing the roles of i and j we see that $p_{i,i}^{[n]}$ and $p_{j,j}^{[n]}$ must have the same properties.

We will consider time-invariant Markov chains that are irreducible and aperiodic and where all states are positive recurrent. Chains having these properties are called *ergodic*. This type of chain is important as there are theorems which show that for this type of chain, the time average of a single realization approach the average of all possible realizations of the same Markov chain (called the *ensemble*) at some particular time point. This means that we can estimate long-run probabilities for this type chain by taking the time average of a single realization of the chain. The ergodic theorem for irreducible aperiodic Markov chains is

Theorem 3 *In an ergodic Markov chain the limits*

$$u_j = \lim_{n\to\infty} p_{i,j}^{[n]} > 0$$

exist and are independent of the initial state i. Also

$$\sum_j u_j = 1$$

and

$$u_j = \sum_i u_i p_{i,j} \,. \tag{5.16}$$

Conversely, if the chain has $u_j > 0$ satisfying the conditions above, then the chain is ergodic and $u_j = 1/m_j$, the reciprocal of the mean return time. A proof of this theorem is given in Feller (1968).

Equation 5.16 says that the u_j are a solution of the steady state equation. Thus the theorem says that if a unique non-zero solution of the steady state equation exists the chain is ergodic, and vice-versa. A consequence of this theorem is that time averages from an ergodic Markov chain will approach the steady state probabilities of the chain. Note, however, for an aperiodic irreducible Markov chain that has all null recurrent states, the mean recurrence times are infinite, and hence $u_j = 0$ for such a chain. The only solution to the steady state equation for such a chain is $u_j = 0$ for all j. It is very important that we make sure all chains that we use are ergodic and contain only positive recurrent states! Note also that the theorem does not say anything about the rate the time averages converges to the steady state probabilities.

5.5 SAMPLING FROM A MARKOV CHAIN

We now look at how estimates from a sample from an ergodic time-invariant Markov chain compares to the the long-run distribution of the chain. The occupation probability distribution for a Markov chain at time n actually represents the average (at time n) over the *ensemble*, which is the set of all possible Markov chains with those transition probabilities. The long-run distribution of a Markov chain describes the probability that the chain is in each particular state after the chain has been running a long time. The long-run distribution for an ergodic chain can be thought of the average over the ensemble at some time n a long time in the future. We are going to take a sample from a single realization of the Markov chain after we have let it run a period of time called the *burn-in* time, which we consider long enough for the chain to have reached the long-run distribution. Then we will take a time average over our sample. The question we wish to cover is how well does this time average over a single realization represent the ensemble average.

Example 6 *First, we will notice that there are many Markov chains that will have the same long-run distribution. We have two transition probability matrices* $\mathbf{P}_1$ *and* $\mathbf{P}_2$ *that describe the movement through a finite state-space with five elements. They are:*

$$\boldsymbol{P_1} = \begin{bmatrix} 0.35 & 0.35 & 0.10 & 0.10 & 0.10 \\ 0.15 & 0.55 & 0.10 & 0.10 & 0.10 \\ 0.15 & 0.15 & 0.10 & 0.20 & 0.40 \\ 0.15 & 0.15 & 0.40 & 0.10 & 0.20 \\ 0.15 & 0.15 & 0.20 & 0.40 & 0.10 \end{bmatrix}$$

and

$$\boldsymbol{P_2} = \begin{bmatrix} 0.499850 & 0.499850 & 0.000100 & 0.000100 & 0.000100 \\ 0.299850 & 0.699850 & 0.000100 & 0.000100 & 0.000100 \\ 0.000150 & 0.000150 & 0.199900 & 0.299900 & 0.499900 \\ 0.000150 & 0.000150 & 0.499900 & 0.199900 & 0.299900 \\ 0.000150 & 0.000150 & 0.299900 & 0.499900 & 0.199900 \end{bmatrix}.$$

We look at $\mathbf{P}_1^{[n]}$ *and* $\mathbf{P}_2^{[n]}$ *where* $n = 2^5$. *They are:*

$$\boldsymbol{P_1}^{[2^5]} = \begin{bmatrix} 0.187500 & 0.312500 & 0.166667 & 0.166667 & 0.166667 \\ 0.187500 & 0.312500 & 0.166667 & 0.166667 & 0.166667 \\ 0.187500 & 0.312500 & 0.166667 & 0.166667 & 0.166667 \\ 0.187500 & 0.312500 & 0.166667 & 0.166667 & 0.166667 \\ 0.187500 & 0.312500 & 0.166667 & 0.166667 & 0.166667 \end{bmatrix}$$

and

$$\boldsymbol{P_2}^{(2^5)} = \begin{bmatrix} 0.373162 & 0.622060 & 0.001593 & 0.001593 & 0.001593 \\ 0.373162 & 0.622060 & 0.001593 & 0.001593 & 0.001593 \\ 0.001838 & 0.002940 & 0.331741 & 0.331741 & 0.331741 \\ 0.001838 & 0.002940 & 0.331741 & 0.331741 & 0.331741 \\ 0.001838 & 0.002940 & 0.331741 & 0.331741 & 0.331741 \end{bmatrix}.$$

Row i *represents the occupation probability distribution at time* n *given a start in state* i. *We see that the first chain has converged its long-run distribution (to within 6 significant digits). We see the second chain is far from convergence since these occupation probabilities are very different for the different rows. We let the second chain run further to* $n = 2^{10}$, *and* $\mathbf{P}_2^{[n]}$ *now equals*

$$\boldsymbol{P_2}^{[2^{10}]} = \begin{bmatrix} 0.325360 & 0.542358 & 0.044094 & 0.044094 & 0.044094 \\ 0.325360 & 0.542358 & 0.044094 & 0.044094 & 0.044094 \\ 0.049640 & 0.082642 & 0.289239 & 0.289239 & 0.289239 \\ 0.049640 & 0.082642 & 0.289239 & 0.289239 & 0.289239 \\ 0.049640 & 0.082642 & 0.289239 & 0.289239 & 0.289239 \end{bmatrix}.$$

We see that the chain still has not converged to a long-run distribution since the rows are clearly not the same. We let the second chain run further to $n = 2^{16}$ *and* $\mathbf{P}_2^{[n]}$

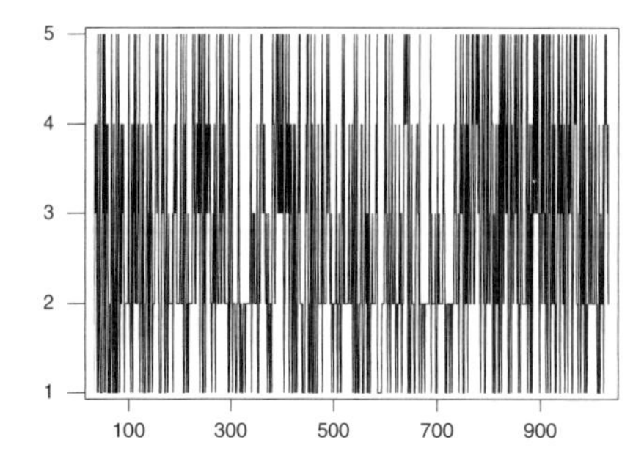

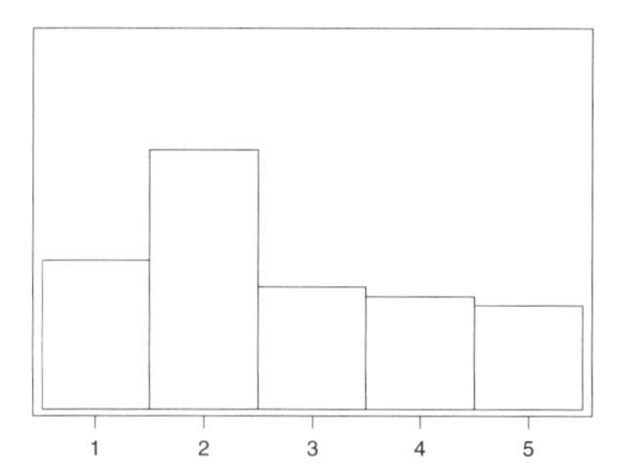

Figure 5.1 State history of the first Markov chain for 1000 steps after burn-in of $n = 2^5$, and histogram of the values after burn-in.

now equals

$$\boldsymbol{P_2}^{[2^{16}]} = \begin{bmatrix} 0.187500 & 0.312500 & 0.166667 & 0.166667 & 0.166667 \\ 0.187500 & 0.312500 & 0.166667 & 0.166667 & 0.166667 \\ 0.187500 & 0.312500 & 0.166667 & 0.166667 & 0.166667 \\ 0.187500 & 0.312500 & 0.166667 & 0.166667 & 0.166667 \\ 0.187500 & 0.312500 & 0.166667 & 0.166667 & 0.166667 \end{bmatrix}.$$

We see that the second chain has now converged to the same long-run distribution as the first chain.

However, the first chain converged in 32 *steps, while the second required around* 65000 *steps. This shows that many chains will have the same long-run distribution. There is no way that we can set a burn-in time that will work for all chains. A chain that has the same long-run distribution but requires several million steps to converge could be easily found. The traceplot shows the step-by-step history of a Markov chain as it moves through the states. The traceplot of the first chain for 1000 steps after its burn in of* $n = 2^5$ *is shown in Figure 5.1 together with a histogram of the states at those steps. We see that the chain is moving among all states fairly rapidly. The histogram of sample values shows that the proportion of time spent in each of the states is not too far from the long-run distribution.*

The traceplot of the second chain for 1000 steps after its burn-in of $n = 2^{16}$ *is shown in Figure 5.2, together with the the histogram of the states at those 1000 steps. This chain has got stuck in a group of states for the whole time period looked at, and has completely missed another group of states. We say that the first chain has much better mixing properties than the second chain which only changes between the groups of states very infrequently. Clearly, the second chain would have to be run a very long time before the histogram would approach the shape of the long-run distribution.*

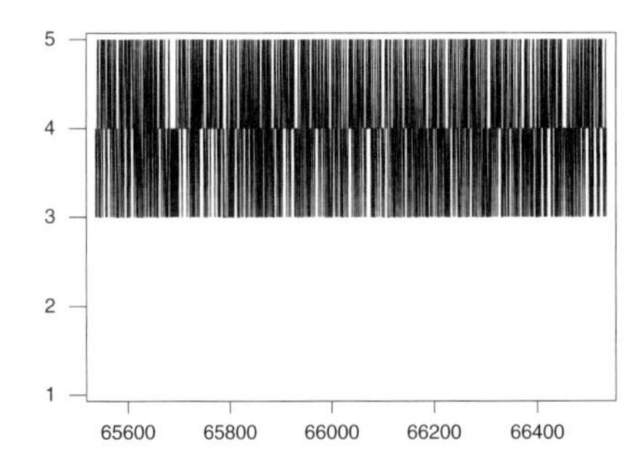

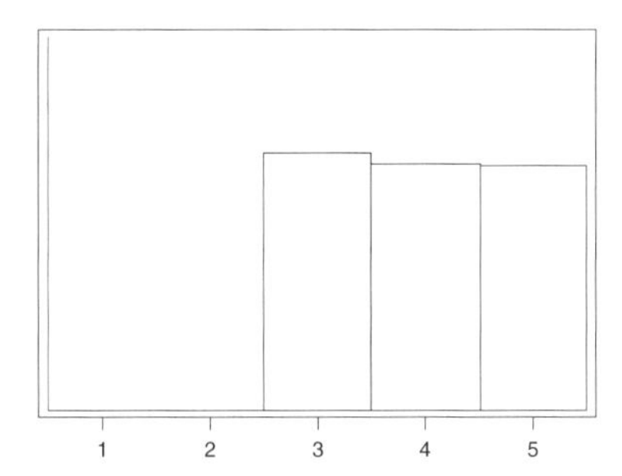

Figure 5.2 State history of the second Markov chain for 1000 steps after burn-in of $n = 2^{16}$, and histogram of the values after burn-in.

5.6 TIME-REVERSIBLE MARKOV CHAINS AND DETAILED BALANCE

So far, we have considered that we knew the one-step transition probabilities $p_{i,j}$ for all values of i and j, and found the long-run distribution π_i for all values of i from that. In this section we look at this the other way around. We start with the long-run distribution and want to find a Markov chain with that distribution. That means we find the set of one-step transition probabilities that will have that long run distribution. First we note that summing across rows

$$\sum_j p_{i,j} = 1$$

for all i, since the chain must go to some state on the next step. When a chain is at steady state, the total amount of probability flowing out of state j must balance the total amount of probability flowing into state j. Note that $\sum_i \pi_i p_{i,j}$ is the total flow of probability going into state j when we are at steady state. The total probability going out of state j at steady state is $\sum_i \pi_j p_{j,i}$. This total balance must be true for all states j.

Time-reversible Markov chains. If we look at the states of a Markov chain in the reverse time order they also form a Markov chain called the backwards chain. Let the transition probabilities for the backwards chain be

$$\begin{aligned} q_{i,j} &= P(X^{(n)} = j | X^{(n+1)} = i) \\ &= \frac{P(X^{(n)} = j, X^{(n+1)} = i)}{P(X^{(n+1)} = i)} \\ &= \frac{P(X^{(n)} = j) P(X^{(n+1)} = i | X^{(n)} = j)}{P(X^{(n+1)} = i)} . \end{aligned}$$

When the chain is at steady state

$$q_{i,j} = \frac{\pi_j p_{j,i}}{\pi_i} .$$

The Markov chain is said to be time reversible when the backwards Markov chain and the forward Markov chain have the same transition probabilities. In other words $q_{i,j} = p_{i,j}$ for all states i and j. Then it follows that the transition probabilities satisfy

$$\pi_i p_{i,j} = \pi_j p_{j,i}$$

for all states i and j. This is called "detailed balance." We note that $\pi_i p_{i,j}$ is the amount of probability of going from state i to state j when we are at steady state. Similarly $\pi_j p_{j,i}$ is the amount of probability going into state i from state j when we are at steady state. The detailed balance condition says that for every pair of states the flows between the two states balance each other.

Theorem 4 *A set of transition probabilities satisfying the detailed balance condition will have steady state distribution* $\boldsymbol{\pi}$.
Proof:

$$\begin{aligned} \sum_i \pi_i p_{i,j} &= \sum_i \pi_j p_{j,i} \\ &= \pi_j \sum_i p_{j,i} \\ &= \pi_j \sum_j p_{i,j} \\ &= \pi_j \end{aligned}$$

which is the steady state probability of state j. *This holds for all states* i *and* j. *Thus the Markov chain with transition matrix that satisfies the detailed balance condition has the steady state distribution* $\boldsymbol{\pi}$.

Thus, to find a Markov chain that has the desired steady state probabilities, we have to find the transition probabilities of a Markov chain that satisfies the detailed balance condition. The detailed balance condition won't be satisfied by the transition probabilities of most Markov chains.

The Metropolis Algorithm

Metropolis et al. (1953) discovered an algorithm that finds transition probabilities for a Markov chain that will yield the desired steady state distribution. They show how starting with an arbitrary set of transition probabilities a new set of transition probabilities can be found that satisfy the detailed balance condition. This is done by only accepting some of the transitions. First note, the transition from any state to itself always satisfies the detailed balance since clearly $\pi_i p_{i,i} = \pi_i p_{i,i}$. This means that we can change the probability $p_{i,i}$ at will, and not destroy detailed balance. Let $p_{i,j}$ for all i and j be the transition probabilities for a Markov chain. If $\pi_i p_{i,j} < \pi_j p_{j,i}$ then we want to accept all the transitions from i to j and only some of the transitions from j to i to bring these back into balance. Similarly if $\pi_i p_{i,j} > \pi_j p_{j,i}$ we bring them back into balance by only accepting some of the transitions from i to j and all of the transitions from j to i. We can add extra probability to the transition from i to itself to compensate for the unaccepted transitions. The details of the algorithm are:

- Start from the transition probabilities $p_{i,j}$ and the desired steady state probabilities π_i.
- For each pair of states i and j, define the acceptance probability

$$\alpha_{i,j} = \min\left[\frac{\pi_j p_{j,i}}{\pi_i p_{i,j}}; 1\right] .$$

- Then for each i and $j \neq i$ let $p'_{i,j} = \alpha_{i,j} p_{i,j}$.
- Let

$$p'_{i,i} = 1 - \sum_{j \neq i} p'_{i,j} .$$

  Then π is the steady state distribution for the Markov chain with transition probabilities given by $p'_{i,j}$.

Theorem 5 *The Markov chain having transition probabilities given by $p'_{i,j}$ satisfies the detailed balance condition, and thus it has the desired steady state distribution.*
Proof:
First, we note that these are transition probabilities since $\sum_j p'_{i,j} = 1$ for all i. Then we note that

$$\begin{aligned}
\pi_i p'_{i,j} &= \pi_i \alpha_{i,j} p_{i,j} \\
&= \pi_i \times \min\left[\frac{\pi_j p_{j,i}}{\pi_i p_{i,j}}; 1\right] \times p_{i,j} \\
&= \min\left[\pi_j p_{j,i}, \pi_i p_{i,j}\right]
\end{aligned}$$

and that

$$\begin{aligned}
\pi_j p'_{j,i} &= \pi_j \alpha_{j,i} p_{j,i} \\
&= \pi_j \times \min\left[\frac{\pi_i p_{i,j}}{\pi_j p_{j,i}}; 1\right] \times p_{j,i} \\
&= \min\left[\pi_j p_{j,i}, \pi_i p_{i,j}\right] .
\end{aligned}$$

Thus it satisfies the detailed balance condition, so the Markov chain with transition probabilities given by $p'_{i,j}$ has the desired steady state distribution $\boldsymbol{\pi}$.

Example 7 *Suppose we have a transition matrix given by*

$$\mathbf{P} = \begin{bmatrix} .1 & .2 & .3 & .4 \\ .1 & .2 & .3 & .4 \\ .3 & .4 & .2 & .1 \\ .3 & .4 & .2 & .1 \end{bmatrix}$$

and we want to achieve the steady state distribution $\boldsymbol{\pi} = (.2, .3, .4, .1)$. *Then*

$$\alpha_{1,2} = \min\left[\frac{.3 \times .1}{.2 \times .2}, 1\right] = \frac{3}{4} \quad \alpha_{1,3} = \min\left[\tfrac{.4\times.3}{.2\times.3}, 1\right] = 1 \quad \alpha_{1,4} = \min\left[\frac{.1 \times .3}{.2 \times .4}, 1\right] = \frac{3}{8}$$

$$\alpha_{2,1} = \min\left[\frac{.2 \times .2}{.3 \times .1}, 1\right] = 1 \quad \alpha_{2,3} = \min\left[\tfrac{.4\times.4}{.3\times.2}, 1\right] = 1 \quad \alpha_{2,4} = \min\left[\frac{.1 \times .4}{.3 \times .4}, 1\right] = \frac{1}{3}$$

$$\alpha_{3,1} = \min\left[\frac{.2 \times .3}{.4 \times .3}, 1\right] = \frac{1}{2} \quad \alpha_{3,2} = \min\left[\tfrac{.3\times.3}{.4\times.4}, 1\right] = \tfrac{9}{16} \quad \alpha_{3,4} = \min\left[\frac{.1 \times .2}{.4 \times .1}, 1\right] = \frac{1}{2}$$

$$\alpha_{4,1} = \min\left[\frac{.2 \times .4}{.1 \times .3}, 1\right] = 1 \quad \alpha_{4,2} = \min\left[\tfrac{.3\times.4}{.1\times.4}, 1\right] = 1 \quad \alpha_{4,3} = \min\left[\frac{.4 \times .1}{.1 \times .2}, 1\right] = 1$$

Then for each $j \neq i$ *let* $p'_{i,j} = \alpha_{i,j} \times p_{i,j}$, *and let* $p'_{i,i} = 1 - \sum_{j \neq i} p'_{i,j}$. *This gives*

$$\mathbf{P}' = \begin{bmatrix} .400000 & .150000 & .300000 & .150000 \\ .100000 & .466667 & .300000 & .133333 \\ .150000 & .225000 & .575000 & .050000 \\ .300000 & .400000 & .200000 & .100000 \end{bmatrix}.$$

It is easily seen that the steady state distribution of the Markov chain having transition matrix $\mathbf{P}'$ *is given by* $\boldsymbol{\pi} = (.2, .3, .4, .1)$.

5.7 MARKOV CHAINS WITH CONTINUOUS STATE SPACE

Sometimes the state space of a Markov chain consists of all possible values in an interval. In that case, we would say the Markov chain has a one-dimensional continuous state space. In other Markov chains, the state space consists of all possible values in a rectangular region of dimension p and we would say the Markov chain has a p-dimensional continuous state space. In both of these cases, there are an uncountably infinite number of possible values in the state space. This is far too many to have a transition probability function associated with each pair of values. The probability of a transition between all but a countable number of pairs of possible values must be zero. Otherwise the sum of the transition probabilities would be infinite. A state to state transition probability function won't work for all pairs of states. Instead we define the transition probabilities from each possible state x to each possible measurable set of states A.[2] We call

$$P(x_0, A) = P(X^{(n+1)} \in A | X^{(n)} = x_0)$$

the *transition kernel* of the Markov chain for all values of x and all possible measurable sets A. Since we are restricting ourselves to time invariant Markov chains, this transition kernel does not depend on n.

[2]When the state space has a single dimension, measurable sets are all sets that can be created from sets $X \leq x$ for all values of x in a finite number of steps by using unions, intersections, and complements. When the state space has p continuous dimensions the measurable sets are all sets that can be created from regions of the form $(X_1 \leq x_1, \ldots, X_p \leq x_p)$ for all possible values of $x_1, \ldots, x_p$ in an finite number of steps by using unions, intersections, and complements. Thus the probabilities of all the measurable sets can be built up from the cumulative distribution function *CDF* of the transition or the joint *CDF* of the transition using the rules of probability for single dimensional and p-dimensional state spaces, respectively.

The main results that hold for discrete Markov chains continue to hold for Markov chains with a continuous state space with some modifications. In the single dimensional case, the probability of a measurable set A can be found from the probability of the transition *CDF* which is

$$F(v|x) = P(X^{(n+1)} \leq v | X^{(n)} = x) \, .$$

When it is absolutely continuous we can take the derivative

$$f(v|x) = \frac{\partial P(X^{(n+1)} \leq v | X^{(n)} = x)}{\partial v} \tag{5.17}$$

which gives us the probability density function of the one-step transition.[3] Note, this conditional probability density given the initial state corresponds to the one-step transition probabilities $p_{i,j}$ of a discrete Markov chain. The probability density of two-step transitions would be given by

$$f^{[2]}(v|x) = \int_{\infty}^{\infty} f(w|x) f(v|w) dw$$

which is analogous to Equation 5.3. The general Chapman-Kolmogorov equation for Markov chains with a one-dimensional continuous state space would be

$$f^{[m+n]}(v|x) = \int_{\infty}^{\infty} f^{[m]}(w|x) f^{[n]}(v|w) dw \, . \tag{5.18}$$

This is analogous to Equation 5.5. If the Markov chain possesses a limiting transition density independent of the initial state,

$$\lim_{n \to \infty} f^{[n]}(v|x) = g(v)$$

which does not depend on initial state x, $g(v)$ is called the steady state density of the Markov chain and is the solution of the steady state equation

$$g(v) = \int_{-\infty}^{\infty} g(w) f(v|w) dw \, . \tag{5.19}$$

This is analogous to Equation 5.11. When the continuous state space has dimension p, the integrals are multiple integrals over p dimensions, and the joint density function is found by taking p partial derivatives.

[3]More generally, the transition *CDF* may be a combination of an absolutely continuous part and a discrete part. Then we would have the transition density given in Equation 5.16 for the absolute continuous part and a transition probability function $p_{i,j}$ for the discrete part. The number of points in the discrete part can be either finite, or countably infinite, otherwise the sum of the discrete probabilities would be infinite. Random variables having this type distributions are discussed in Mood et al. (1974). In practice, we will be dealing with either absolutely continuous transition *CDF*, or a mixed absolutely continuous and discrete transition *CDF*, where the only discrete part is $p_{i,i}$, the probability of remaining at the same state, which may be needed for the Metropolis algorithm.

In general, the probability of passing to state j from state i must be zero at all but at most a countably infinite number of states, or else they would not have a finite sum. This means, that we can't classify states by their first return probabilities and return probabilities as given in Tables 5.1 and 5.2, because these would equal 0 for almost all states. Redefining the definitions for recurrence is beyond the scope of this book, and Gamerman (1997) outlines the changes required. What is important to us is that under the required modifications, the main results we found for discrete Markov chains continue to hold for Markov chains with continuous state space. These are:

1. All states of an irreducible Markov chain are the same type: transient, null recurrent, or positive recurrent.

2. All states in an ergodic (irreducible and aperiodic) Markov chain are positive recurrent if and only if there is a unique non-zero solution to the steady state Equation 5.18.

3. The time average of a single realization of an ergodic Markov chain approaches the steady state distribution that is the average of the ensemble (all possible realizations of the chain) at some fixed time point far in the future.

4. Time reversible Markov chains satisfy the detailed balance condition. Let $g(x)$ be the steady state density and $f(v|x)$ be the density function of the one-step transitions
$$g(x)f(v|x) = g(v)f(x|v)$$
for all states x and v. Any chain satisfying the detailed balance condition will have steady state density $g(x)$.

Thus, in the continuous case, we will use procedures that are analogous to those for the discrete case.

Main Points

- A stochastic process moves through a set of possible states in a probabilistic way. We model a stochastic process by a sequence of random variables indexed by time.

- A Markov chain is a special type of stochastic process where the future development of the process, given the present state and the past states, only depends on the present state. This is known as the Markov property.

- If the probability of transition from state i to state j is the same for all times, the Markov chain is called time-invariant or homogeneous.

- The one-step transition probabilities for a time-invariant Markov chain with a finite state space can be put in a matrix $\mathbf{P}$, where the $p_{i,j}$ is the probability of a transition from state i to state j in one-step. This is the conditional probability
$$p_{i,j} = P(X^{(n+1)} = j|X^{(n)} = i)$$

which is the same for all times n.

- Higher transition probabilities can be built up from the one-step transition probabilities using the Chapman-Kolmogorov equation, which takes the form $\mathbf{P}^{[n+m]} = \mathbf{P}^{[n]} \times \mathbf{P}^{[m]}$ for Markov chains with a finite state space.

- If a Markov chain has a limiting distribution that is independent of the initial state, it is called the long-run distribution of the Markov chain and is a solution of the steady state equation

$$\pi = \pi \mathbf{P}$$

for Markov chains with a finite state space.

- The Chapman-Kolmogorov equation and the steady state equation takes the form of infinite sums when the Markov chain has an infinite number of discrete states.

- The states of a Markov chain with a discrete state space can be classified using either the first return probabilities $f_{i,i}^{[n]}$ or the n-step transition probabilities $p_{i,i}^{[n]}$.

- There is a finite probability that the chain will never return to a transient state. It will be visited at most a finite number of times, eventually leaving and never returning. The limiting probability of $p_{i,i}^{[n]}$ is zero.

- A positive recurrent state is certain to recur infinitely often. The mean recurrence time is finite, and the limiting probability of $p_{i,i}^{[n]}$ is positive.

- A null recurrent state is certain to recur infinitely often. However, the mean recurrence time is infinite, and the limiting probability of $p_{i,i}^{[n]}$ is zero.

- A Markov chain is irreducible if every state j can be reached from every other state i.

- All states in an irreducible chain are the same type.

- An aperiodic irreducible Markov chain with positive recurrent states has a unique non-zero solution to the steady state equation, and vice-versa. These are known as ergodic Markov chains.

- For an irreducible Markov chain with null recurrent states, the only solution to the steady state equation is identically equal to zero.

- The time average of a single realization of an ergodic Markov chain approaches the steady state probabilities, which are the average of the ensemble (all possible realizations of the Markov chain) at a single time point far in the future.

- A Markov chain is time reversible if the one-step transition probabilities for the backwards chain where the states are taken in reverse time order are the same as the one-step probabilities of the (forwards) Markov chain.

- A time-reversible Markov chain has transition probabilities satisfying the detailed balance condition.

$$\pi_i p_{i,j} = \pi_j p_{j,i} \quad \text{for all states} \quad i, j$$

Thus the probability flows between each pair of states balance.

- If a Markov chain has transition probabilities that satisfy the detailed balance condition, then π is the steady state distribution for the Markov chain.

- The Metropolis algorithm starts with transition probabilities and a steady state distribution, and finds a new set of transition probabilities that satisfy the detailed balance condition by only accepting some of the transitions.

- For Markov chains with a continuous state space, there are too many states for us to use a transition probability function. Instead we define a transition kernel which measures the probability of going from each individual state to every measurable set of states.

- For a Markov chain with continuous state space, if the transition kernel is absolutely continuous, the Chapman-Kolmogorov and the steady state equations are written as integral equations involving the transition density function.

Exercises

5.1 Let the matrix of one-step transition probabilities for a Markov chain be

$$\mathbf{P} = \begin{bmatrix} .5 & .3 & .2 \\ .2 & .3 & .5 \\ .6 & .3 & .1 \end{bmatrix}.$$

(a) Find the steady state probability distribution $\boldsymbol{\pi}$.

(b) i. Find the matrix of two-step transition probabilities $\mathbf{P}^{[2]}$.

ii. Find the matrix of three-step transition probabilities $\mathbf{P}^{[3]}$.

iii. Find the matrix of three-step transition probabilities $\mathbf{P}^{[4]}$.

(c) Calculate the first return probabilities $f_{1,1}^{[1]}$, $f_{1,1}^{[2]}$, $f_{1,1}^{[3]}$, and $f_{1,1}^{[4]}$.

5.2 Let the matrix of one-step transition probabilities for a Markov chain be

$$\mathbf{P} = \begin{bmatrix} .4 & .3 & .2 & 1 \\ .2 & .3 & .2 & .3 \\ ..3 & .3 & .2 & .2 \end{bmatrix}.$$

(a) Find the steady state probability distribution $\boldsymbol{\pi}$.

(b) i. Find the matrix of two-step transition probabilities $\mathbf{P}^{[2]}$.

ii. Find the matrix of three-step transition probabilities $\mathbf{P}^{[3]}$.

iii. Find the matrix of three-step transition probabilities $\mathbf{P}^{[4]}$.

(c) Calculate the first return probabilities $f_{3,3}^{[1]}$, $f_{3,3}^{[2]}$, $f_{3,3}^{[3]}$, and $f_{3,3}^{[4]}$.

5.3 Let the matrix of one-step transition probabilities for a Markov chain be

$$\mathbf{P} = \begin{bmatrix} .4 & .3 & .3 \\ .2 & .5 & .3 \\ .1 & .6 & .2 \end{bmatrix}.$$

Suppose we need to have the long-run distribution be $\pi = (.5, .3, .2)$. Use the Metropolis algorithm to find the matrix of transition probabilities that will have the desired long-run distribution.

5.4 Let the matrix of one-step transition probabilities for a Markov chain be

$$\mathbf{P} = \begin{bmatrix} .9 & .0 & .1 \\ .1 & .8 & .1 \\ .3 & .4 & .3 \end{bmatrix}.$$

Suppose we need to have the long-run distribution be $\pi = (.2, .3, .5)$. Use the Metropolis algorithm to find the matrix of transition probabilities that will have the desired long-run distribution.

6

Markov Chain Monte Carlo Sampling from Posterior

When there are many parameters or if the prior is very diffuse compared to the likelihood, direct Monte Carlo sampling from the posterior using the methods developed in Chapter 2 is very inefficient, because a huge prior sample size is necessary to get a reasonable sized sample from the posterior. In this chapter we will look at Markov chain Monte Carlo (MCMC) methods for generating samples from the posterior distribution. Here we don't draw our sample from the posterior distribution directly. Rather we set up a Markov chain that has the posterior distribution as its limiting distribution. The Metropolis-Hastings (M-H) algorithm, Gibbs sampler, and the substitution sampler (data augmentation algorithm) are methods of doing this. We let the Markov chain run a long time until it (hopefully) has approached the limiting distribution. Any value taken after that initial run-in time approximates a random draw from the posterior distribution. Values taken from the Markov chain at time points close to each other are highly correlated, however, values at widely separated points in time are approximately independent. An approximately random sample from the posterior distribution can be found by taking values from a single run of the Markov chain at widely spaced time points after the initial run-in time. Alternatively, it can be found by taking values from independent runs of the Markov chain that have each been run long enough to approach stationarity. Gelman et al. (2004) and Gilks et al. (1996) give a good coverage of MCMC methods.

Surprisingly MCMC methods are more efficient than the direct procedures for drawing samples from the posterior when we have a large number of parameters. We run the chain long enough so that it approaches the long-run distribution. A value taken after that is a draw from the long-run distribution.

We want to find a Markov chain that has the posterior distribution of the parameters given the data as its long-run distribution. Thus the parameter space will be the state space of the Markov chain. We investigate how to find a Markov chain that satisfies this requirement. We know that the long-run distribution of a ergodic Markov chain is a solution of the steady state equation. That means that the long-run distribution π of a finite ergodic Markov chain with one-step transition matrix P satisfies the equation

$$\pi = P\pi \, .$$

This says the long-run probability of a state equals the weighted sum of one-step probabilities of entering that state from all states each weighted by its long-run probability. The comparable steady state equation that $\pi(\theta)$, the long-run distribution of a Markov chain with a continuous state space, satisfies is given by

$$\int_A \pi(\theta)d\theta = \int \pi(\theta)P(\theta, A)d\theta \tag{6.1}$$

for all A (measureable subsets of parameter space) where $P(\theta, A)$ is the transition kernel of the chain.[1] The left side of Equation 6.1 is the total steady state probability of set A while the right side of the equation is the steady state flow of probability into set A.

The transition kernels we will encounter consist of two parts. The first part takes points in the parameter space to other points in the parameter space in a continuous way, similar to a probability density function. The second part takes each point in the parameter space to itself, with finite probability.

Equation 6.1 that gives the long-run distribution of the Markov chain says probability of a set of states A, equals the weighted integral of one-step probabilities of entering that set of states from all states each weighted by its long-run probability density. Thus, we will set the long-run distribution $\pi(\theta)$ for the Markov chain equal to the posterior density $g(\theta|y)$. Generally we will only know the unscaled posterior density $g(\theta|y) \propto g(\theta) \times f(y|\theta)$, not the exact posterior. Fortunately, we will see that the unscaled posterior is all we need to know to find a Markov chain that has the exact posterior as its long-run distribution. After the chain has run for a large number of steps (called the *burn-in* time), a draw from the chain can be considered to be a random draw from the posterior. However, a sequence of draws after this time will be a dependent sample not a random sample.

In Section 6.1 we show how the Metropolis-Hastings algorithm can be used to find a Markov chain that has the posterior as its long-run distribution in the case of a single parameter. There are two kinds of candidate densities we can use, random-walk candidate densities, or independent candidate densities. We see how the chain

[1] A transition kernel is a function from a point in the p-dimensional parameter space to measureable sets in the parameter space. Measurable sets consist of all sets that can be constructed from joint intervals of the form $(\theta_1 \leq x_1, \ldots, \theta_p \leq x_p)$ for using the operations of union, intersection, and complement. Thus the probabilities of the measureable set can be found from the joint cumulative distribution functions of the parameters using the laws of probability.

moves around the parameter space using each type. In Section 6.2 we show how the Metropolis-Hastings algorithm can be used when we have multiple parameters. Again, we can use random-walk candidate densities for the multivariate parameter or independent candidate densities for the multivariate parameter. In Section 6.3 we show how, in the multiple parameter case, the blockwise Metropolis-Hastings algorithm can be used. In Section 6.4 we show how the Gibbs sampling algorithm is a special case of the blockwise Metropolis-Hastings algorithm.

Finding a Markov Chain That Has the Posterior as Its Long-Run Distribution

We need to find a probability transition kernel $P(\theta, A)$ that satisfies

$$\int g(\theta|y)P(\theta, A)d\theta = \int_A g(\theta|y)d\theta \qquad \text{for all } A$$

where we know $g(\theta|y)$ except for the scale factor needed to make it an exact density. There are many possible transition kernels that have the posterior as their long-run distribution! It turns out to be remarkably easy find one! In Chapter 5 we showed for a finite ergodic Markov chain that finding transition probabilities that balance the steady state flow between each pair of states (detailed balance) was a sufficient condition for the long-run distribution of the Markov chain having that transition probability matrix to be π. This will hold for an ergodic Markov chain with a continuous state space as well. If a transition kernel $P(\theta|A)$ is found that balances the steady state flow between every possible pair of states, then the long-run distribution for the ergodic Markov chain with that transition kernel is $g(\theta|y)$. This is shown in the following theorem.

Theorem 6 *Let $q(\theta, \theta')$ be a candidate distribution that generates a candidate θ' given starting value θ. If for all θ, θ' the candidate distribution $q(\theta', \theta)$ satisfies the reversibility condition*

$$g(\theta|y) \times q(\theta, \theta') = g(\theta'|y) \times q(\theta', \theta) \text{ for all } \theta, \theta' \tag{6.2}$$

then $g(\theta|y)$ is the long-run distribution for the Markov Chain with probability kernel

$$P(\theta, A) = \int_A q(\theta, \theta')d\theta' + r(\theta)\delta_A(\theta)$$

where $r(\theta) = 1 - \int q(\theta, \theta')d\theta'$ is the probability the chain remains at θ, and where $\delta_A(\theta)$ is the indicator function of set A.

$$\delta_A(\theta) = \begin{cases} 1 & \text{if} \quad \theta \in A \\ 0 & \text{if} \quad \theta \notin A \end{cases} .$$

Proof:

$$\begin{aligned}
\int g(\theta|y)P(\theta,A)d\theta &= \int\int_A g(\theta|y)q(\theta,\theta')d\theta' d\theta + \int g(\theta|y)r(\theta)\delta_A(\theta)d\theta \\
&= \int_A\int g(\theta|y)q(\theta,\theta')d\theta d\theta' + \int_A g(\theta|y)r(\theta)d\theta \\
&= \int_A\int g(\theta'|y)q(\theta',\theta)d\theta d\theta' + \int_A g(\theta|y)r(\theta)d\theta \\
&= \int_A g(\theta'|y)(1-r(\theta'))d\theta' + \int_A g(\theta|y)r(\theta)d\theta \\
&= \int_A g(\theta|y)d(\theta)
\end{aligned}$$

qed.

Where the first line is by substitution, the second follows by reversing the order of integration, the third follows by the reversibility condition, the fourth follows by substitution, and the fifth follows since both θ and θ' are dummies of integration.

6.1 METROPOLIS-HASTINGS ALGORITHM FOR A SINGLE PARAMETER

Unfortunately, most candidate distributions don't satisfy the reversibility condition. For some θ and θ'

$$g(\theta|y)\,q(\theta,\theta') \neq g(\theta'|y)\,q(\theta',\theta)$$

the probability of moving from θ to θ' is not the same as the probability of moving in the reverse direction. Metropolis et al. (1953) supplied the solution. They restored the balance by introducing a probability of moving

$$\alpha(\theta,\theta') = min\left[1, \frac{g(\theta'|y)\,q(\theta',\theta)}{g(\theta|y)\,q(\theta,\theta')}\right].$$

We do not need to know the exact posterior. If we multiply the posterior $g(\theta|y)$ by a constant k, both the factor k occurs in both the numerator and the denominator so it cancels out. The algorithm only requires that we know the unscaled posterior! Similarly, we can multiply the candidate density by a constant, and since it occurs in both numerator and denominator it will also cancel out. All we need is the part that gives the shape of the candidate density. Hastings (1970) made some significant improvements and extended the algorithm so now it is known as the Metropolis-Hastings algorithm.

The revised candidate distribution $\alpha(\theta,\theta')\times q(\theta,\theta')$ satisfies the reversibility condition so $g(\theta|y)$ is the long-run distribution for the Markov chain having probability kernel

$$P(\theta,A) = \int_A \alpha(\theta,\theta')q(\theta,\theta')d\theta' + r(\theta)\delta_A(\theta)$$

where $r(\theta) = 1 - \int \alpha(\theta, \theta') \times q(\theta, \theta')d\theta'$ is the probability the chain remains at θ, and $\delta_A(\theta)$ is the indicator function of A.

Steps of Metropolis-Hastings Algorithm

1. Start at an initial value $\theta^{(0)}$.
2. Do for $n = 1, \ldots, n$.
   (a) Draw θ' from $q(\theta^{(n-1)}, \theta')$.
   (b) Calculate the probability $\alpha(\theta^{(n-1)}, \theta')$.
   (c) Draw u from $U(0, 1)$.
   (d) if $u < \alpha(\theta^{(n-1)}, \theta')$ then let $\theta^{(n)} = \theta'$, else let $\theta^{(n)} = \theta^{(n-1)}$.

We should note that having the candidate density $q(\theta, \theta')$ close to the target $g(\theta|y)$ leads to more candidates being accepted. In fact, when the candidate density is exactly the same shape as the target

$$q(\theta, \theta') = k \times g(\theta'|y)$$

the acceptance probability

$$\begin{aligned} \alpha(\theta, \theta') &= min\left[1, \frac{g(\theta'|y)\, q(\theta', \theta)}{g(\theta|y)\, q(\theta, \theta')}\right] \\ &= min\left[1, \frac{g(\theta'|y)g(\theta|y)}{g(\theta|y)g(\theta'|y)}\right] \\ &= 1\,. \end{aligned}$$

Thus, in that case, all candidates will be accepted.

Single Parameter with a Random-Walk Candidate Density

Metropolis et al. (1953) considered Markov chains with a random-walk candidate distribution. Suppose we look at the case where there is a single parameter θ. For a random-walk candidate generating distribution the candidate is drawn from a symmetric distribution centered at the current value. Thus the candidate density is given by

$$q(\theta, \theta') = q_1(\theta' - \theta)$$

where $q_1()$ is a function symmetric about 0. Because of the symmetry $q_1(\theta' - \theta) = q_1(\theta - \theta')$, so for a random-walk candidate density, the acceptance probability simplifies to be

$$\begin{aligned} \alpha(\theta, \theta') &= min\left[1, \frac{g(\theta'|y)\, q(\theta', \theta)}{g(\theta|y)\, q(\theta, \theta')}\right] \\ &= min\left[1, \frac{g(\theta'|y)}{g(\theta|y)}\right]. \end{aligned}$$

This means that a candidate θ' that has a higher value of the target density than target density of the current value θ will always be accepted. The chain will always move "uphill." On the other hand, a candidate with a lower target value will only be accepted with a probability equal to the proportion of target density value to current density value. There is a certain probability that the chain will move "downhill." This allows a chain with a random-walk candidate density to move around the whole parameter space over time. However a chain with a random-walk candidate density will generally have many accepted candidates, but most of the moves will be a short distance. Thus, it might take a long time to move around the whole parameter space.

Example 8 *Suppose we have a unscaled target density given by*

$$g(\theta|y) = .8 \times e^{-\frac{1}{2}\theta^2} + .2 \times \frac{1}{2} e^{-\frac{1}{2\times 2^2}(\theta-3)^2} .$$

This is a mixture of a normal$(0, 1^2)$ *and a normal*$(3, 2^2)$. *However, we only need to know the unscaled target since multiplying by a constant would multiply both the numerator and denominator by the constant which would cancel out. Let us use the normal candidate density with variance* $\sigma^2 = 1$ *centered around the current value as our random-walk candidate density distribution. Its shape is given by*

$$q(\theta, \theta') = e^{-\frac{1}{2}(\theta'-\theta)^2} .$$

Let the starting value be $\theta = 2$. *Figure 6.1 shows six consecutive draws from a random-walk Metropolis-Hastings chain together with the unscaled target and the candidate density. In a Metropolis-Hastings chain with a random-walk candidate density, the candidate density centered around the starting value, so it changes every time a candidate is accepted. Since the random-walk candidate density is symmetric about the current value, the acceptance probability*

$$\begin{aligned} \alpha &= \min\left(\frac{g(\theta'|y)}{g(\theta|y)} \times \frac{q(\theta', \theta)}{q(\theta, \theta')}, 1\right) \\ &= \min\left(\frac{g(\theta'|y)}{g(\theta|y)}, 1\right) . \end{aligned}$$

We see it only depends on the target density. It always moves "uphill," and moves "downhill" with some probability. The candidate is accepted if a random draw from a uniform$(0, 1)$ *is less than* α. *Table 6.1 gives a summary of the first six draws of this chain. Figure 6.2 shows the traceplot and a histogram for 1000 draws from the Metropolis-Hastings chain using a random-walk candidate density with standard deviation equal to 1. We see that the chain is moving through the space satisfactorily. We note that when it occasionally goes into the tail region, it does not stay there for long because the tendency to go "uphill" moves it back towards the central region that has higher values of the probability density. We see that the histogram has a way to go before it is close to the true posterior. Figure 6.3 shows the histograms for 5000 and 20000 draws from the Metropolis-Hastings chain. We see the chain is getting closer to the true posterior as the number of draws increases.*

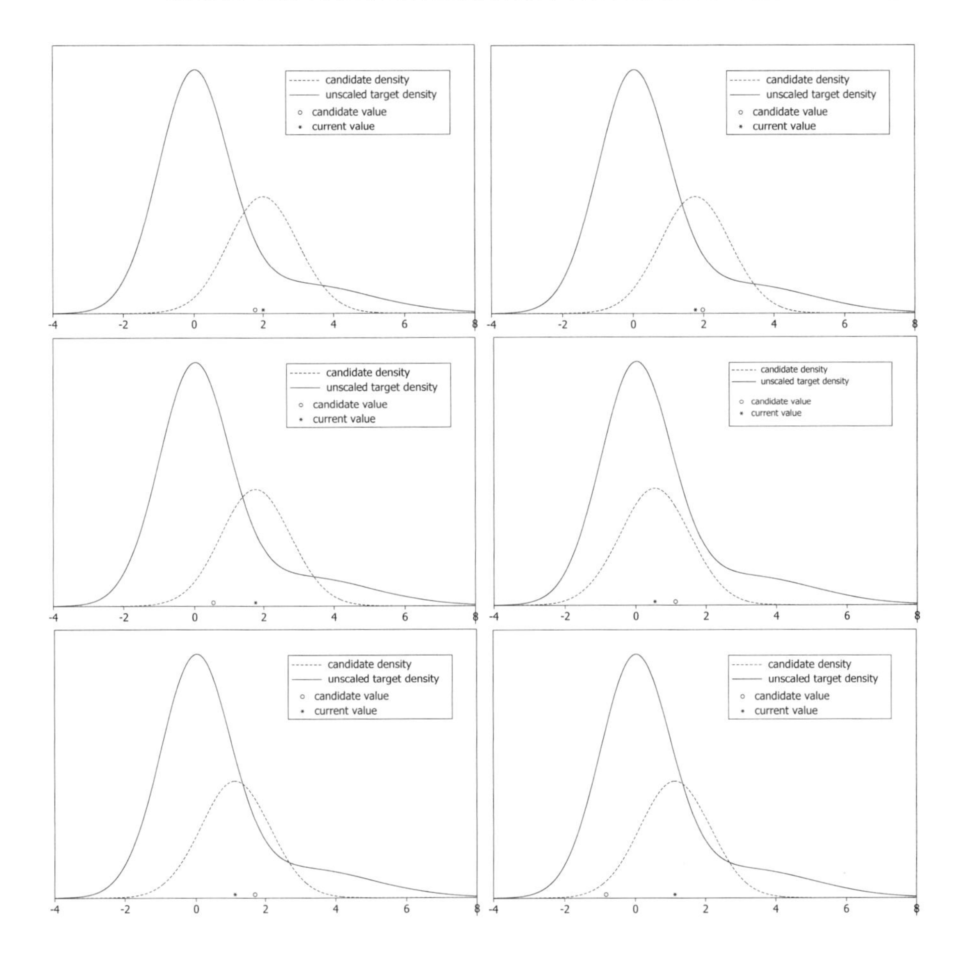

Figure 6.1 Six consecutive draws from a Metropolis-Hastings chain with a random-walk candidate density. Note: the candidate density is centered around the current value.

With random-walk candidate density function, the Markov chain will accept a large proportion of candidates. However, the accepted candidate will be close to the previous current value. The Markov chain will move through the state space (parameter space), however it may not be very fast.

Single Parameter with an Independent Candidate Density

Hastings (1970) introduced Markov chains with candidate generating density that did not depend on the current value of the chain. These are called *independent* candidate distribution

$$q(\theta, \theta') = q_2(\theta')$$

Table 6.1 Summary of first six draws of the chain using the random-walk candidate density

Draw	Current value	Candidate	α	u	Accept
1	2.000	1.767	1.000	.773	yes
2	1.767	1.975	.804	.933	no
3	1.767	.547	1.000	.720	yes
4	.547	1.134	.659	.240	yes
5	1.134	1.704	.553	.633	no
6	1.134	-.836	1.000	.748	yes

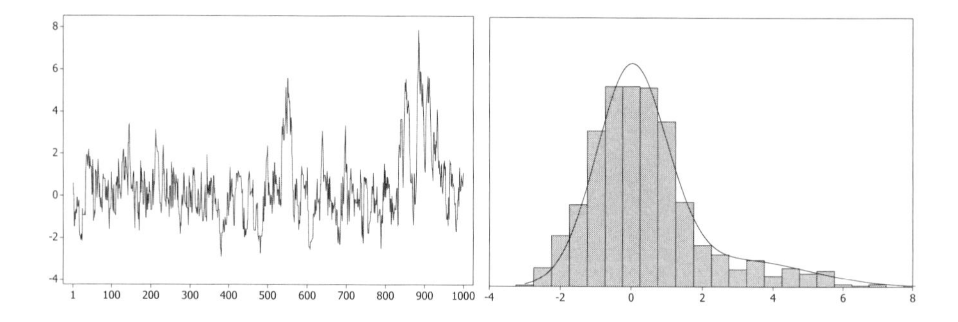

Figure 6.2 Trace plot and histogram of 1000 Metropolis-Hastings values using the random-walk candidate density with standard deviation 1.

for some function $q_2(\theta)$ that must dominate the target in the tails. This is similar to the requirement that the candidate density for acceptance-rejection-sampling must dominate the target in the tails we noted in Chapter 2. We can make sure this requirement is met by graphing logarithms of the target and the candidate density. It is preferable that the candidate density has the same mode as the target as this leads to a higher proportion of accepted candidates. For an independent candidate density, the acceptance probability simplifies to be

$$\begin{aligned} \alpha(\theta, \theta') &= min \left[1, \frac{g(\theta'|y)\, q(\theta', \theta)}{g(\theta|y)\, q(\theta, \theta')} \right] \\ &= min \left[1, \frac{g(\theta'|y)}{g(\theta|y)} \times \frac{q_2(\theta)}{q_2(\theta')} \right] . \end{aligned}$$

The acceptance probability is a product of two ratios that give the chain two opposite tendencies. The first ratio $\frac{g(\theta'|y)}{g(\theta|y)}$ gives the chain a tendency to move "uphill" with respect to the target. The second ratio $\frac{q_2(\theta)}{q_2(\theta')}$ gives the chain a tendency to move towards the tails of the candidate density. The acceptance probability depends on the balance of these two opposing tendencies.

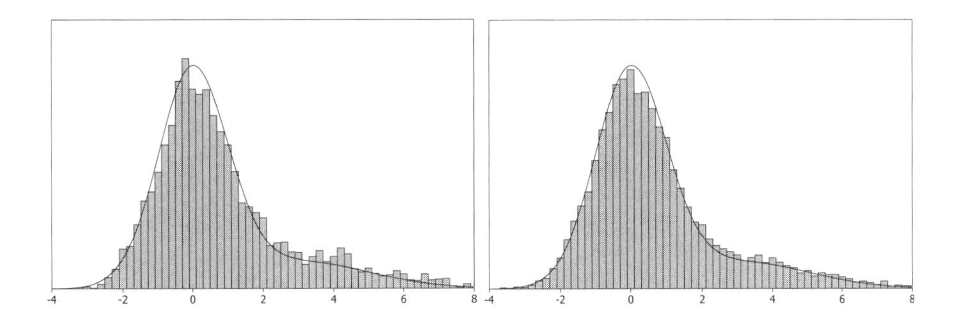

Figure 6.3 Histograms for 5000 and 20000 draws from the Metropolis-Hastings chain using random-walk candidate generating density with standard deviation 1

Table 6.2 Summary of first six draws of the chain using the independent candidate density

Draw	Current value	Candidate	α	u	Accept
1	.4448	-2.7350	.0402	.5871	no
2	.4448	2.3553	.2537	.0987	yes
3	2.3553	.1611	1.000	.7896	yes
5	.1611	-1.5480	.3437	.6357	no
4	.1611	-1.3118	.4630	.1951	yes
6	-1.3118	-2.6299	.1039	.4752	no

Example 8 (continued) *Suppose we use the same unscaled target density given by*

$$g(\theta|y) = .8 \times e^{-\frac{1}{2}\theta^2} + .2 \times \frac{1}{2} e^{-\frac{1}{2\times 2^2}(\theta-3)^2} .$$

This is a mixture of a normal$(0, 1^2)$ *and a normal*$(3, 2^2)$. *Let us use the normal*$(0, 3^2)$ *candidate density as the independent candidate density density. Its shape is given by*

$$q_2(\theta') = e^{-\frac{1}{2\times 3^2}(\theta')^2} .$$

Let the starting value be $\theta = 2$.

Figure 6.4 shows six consecutive draws from the Metropolis-Hastings chain using the independent candidate distribution along with the unscaled target and the candidate density. Table 6.2 gives a summary of the first 6 draws of this chain. Figure 6.5 shows the traceplot and a histogram for 1000 draws from the Metropolis-Hastings chain using the independent candidate density with mean equal to 0 and standard deviation equal to 3. The independent candidate density allows for large jumps. There may be fewer acceptances than with the random-walk candidate density, however, they will be larger, and we see that the chain is moving through the space very satisfactorily. We see that the histogram has a way to go before it is close to the

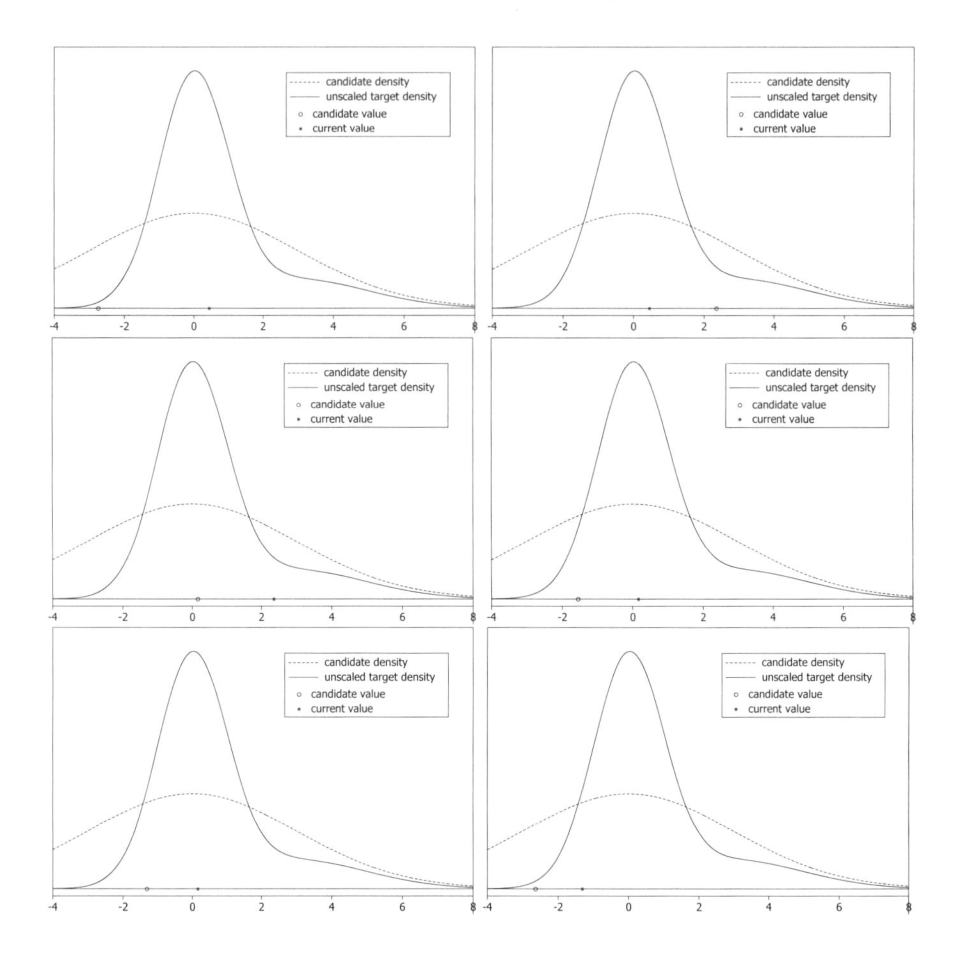

Figure 6.4 Six consecutive draws from a Metropolis-Hastings chain with an independent candidate density. Note: the candidate density remains the same, regardless of the current value.

true posterior. Figure 6.6 shows the histograms for 5000 and 20000 draws from the Metropolis-Hastings chain using the independent candidate density with mean and standard deviation equal to 0 and 3, respectively. We see the chain is getting closer to the true posterior as the number of draws increases.

Note we want to reach the long-run distribution as quickly as possible. There is a trade off between the number of candidates accepted and the distance moved. Generally a chain with a random-walk candidate density will have more candidates accepted, but each move will be a shorter distance. The chain with an independent candidate density will have fewer moves accepted, but the individual moves can be very large. We can see this in comparing the traceplot of the random-walk chain in Figure 6.2 with the trace plot of the independent chain Figure 6.5. The random-walk

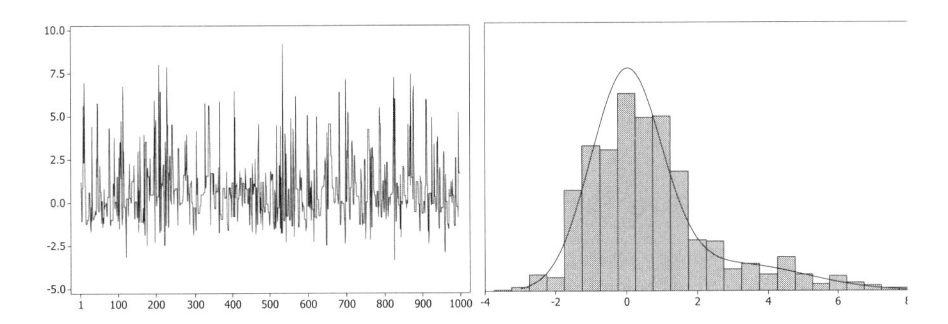

Figure 6.5 Trace plot and histogram of 1000 Metropolis-Hastings values using the independent candidate density with mean 0 and standard deviation 3.

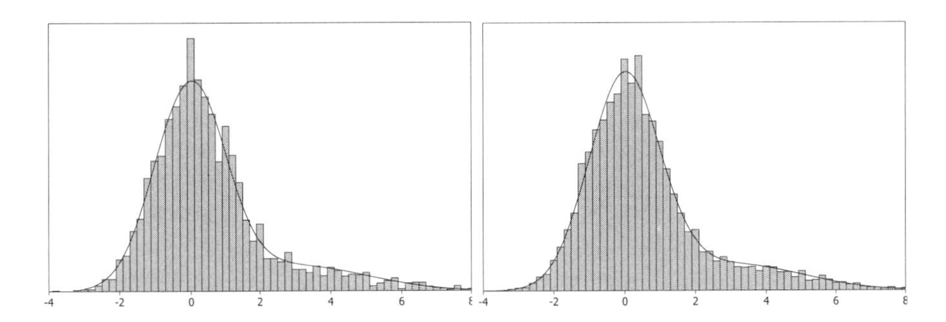

Figure 6.6 Histograms for 5000 and 20000 draws from the Metropolis-Hastings chain using independent candidate generating density with mean 0 and standard deviation 3.

chain is moving more slowly through the whole space, despite more candidates being accepted.

6.2 METROPOLIS-HASTINGS ALGORITHM FOR MULTIPLE PARAMETERS

Suppose we have p parameters $\theta_1, \ldots, \theta_p$. Let the parameter vector be

$$\boldsymbol{\theta} = (\theta_1, \ldots, \theta_p)\,.$$

Let $q(\boldsymbol{\theta}', \boldsymbol{\theta})$ be the candidate density when the chain is at $\boldsymbol{\theta}$ and let $g(\boldsymbol{\theta}|y)$ be the posterior density. The reversibility condition will be

$$g(\boldsymbol{\theta}|y)q(\boldsymbol{\theta}, \boldsymbol{\theta}') = g(\boldsymbol{\theta}'|y)q(\boldsymbol{\theta}', \boldsymbol{\theta})$$

for all states. Most chains won't satisfy the reversibility condition. There will be some values $\boldsymbol{\theta}$ and $\boldsymbol{\theta}'$ where it does not hold. The balance can be restored by introducing the probability of moving which in this case is given by

$$\alpha(\boldsymbol{\theta}, \boldsymbol{\theta}') = \min\left[1, \frac{g(\boldsymbol{\theta}'|y)\, q(\boldsymbol{\theta}', \boldsymbol{\theta})}{g(\boldsymbol{\theta}|y)\, q(\boldsymbol{\theta}, \boldsymbol{\theta}')}\right].$$

Multiple Parameters with a Random-Walk Candidate Density

For a random-walk candidate distribution the candidate is drawn from a symmetric distribution centered at the current value. Suppose we have p parameters $\theta_1, \ldots, \theta_p$ represented by the parameter vector $\boldsymbol{\theta}$. Since we are using a random-walk candidate density it given by

$$q(\boldsymbol{\theta}, \boldsymbol{\theta}') = q_1(\theta_1' - \theta_1, \ldots, \theta_p' - \theta_p)$$

where the function $q_1(\,, \ldots, \,)$ is symmetric about 0 for each of its arguments. Thus we can write the candidate density as

$$q(\boldsymbol{\theta}, \boldsymbol{\theta}') = q_1(\boldsymbol{\theta}' - \boldsymbol{\theta})$$

where $q_1(\,)$ is a vector function symmetric about the vector $\mathbf{0}$. (It is symmetric in each of its arguments.) Because of the symmetry $q_1(\boldsymbol{\theta}' - \boldsymbol{\theta}) = q_1(\boldsymbol{\theta} - \boldsymbol{\theta}')$, so for a random-walk candidate density, the acceptance probability simplifies to be

$$\begin{aligned}\alpha(\boldsymbol{\theta}, \boldsymbol{\theta}') &= min\left[1, \frac{g(\boldsymbol{\theta}'|y)\, q(\boldsymbol{\theta}', \boldsymbol{\theta})}{g(\boldsymbol{\theta}|y)\, q(\boldsymbol{\theta}, \boldsymbol{\theta}')}\right] \\ &= min\left[1, \frac{g(\boldsymbol{\theta}'|y)}{g(\boldsymbol{\theta}|y)}\right].\end{aligned}$$

This means that a candidate $\boldsymbol{\theta}'$ that has a higher value of the target density than target density of the current value $\boldsymbol{\theta}$ will always be accepted. The chain will always move "uphill." On the other hand, a candidate with a lower target value will only be accepted with a probability equal to the proportion of target density value to current density value. There is a certain probability that the chain will move "downhill." This allows a chain with a random-walk candidate density to move around the whole parameter space.

Example 9 *Suppose there are two parameters, θ_1 and θ_2. It is useful to try a target density that we know and could approach analytically, so we know what a random sample from the target should look like. We will use a bivariate normal$(\boldsymbol{\mu}, \mathbf{V})$ distribution with mean vector and covariance matrix equal to*

$$\boldsymbol{\mu} = \begin{pmatrix} 0 \\ 0 \end{pmatrix} \quad \text{and} \quad \mathbf{V} = \begin{bmatrix} 1 & \rho \\ \rho & 1 \end{bmatrix}.$$

Suppose we let $\rho = .9$. Then the unscaled target (posterior) density has formula

$$g(\theta_1, \theta_2) \propto e^{-\frac{1}{2(1-.9^2)}(\theta_1^2 + 2\times.9\times\theta_1\theta_2 + \theta_2^2)}.$$

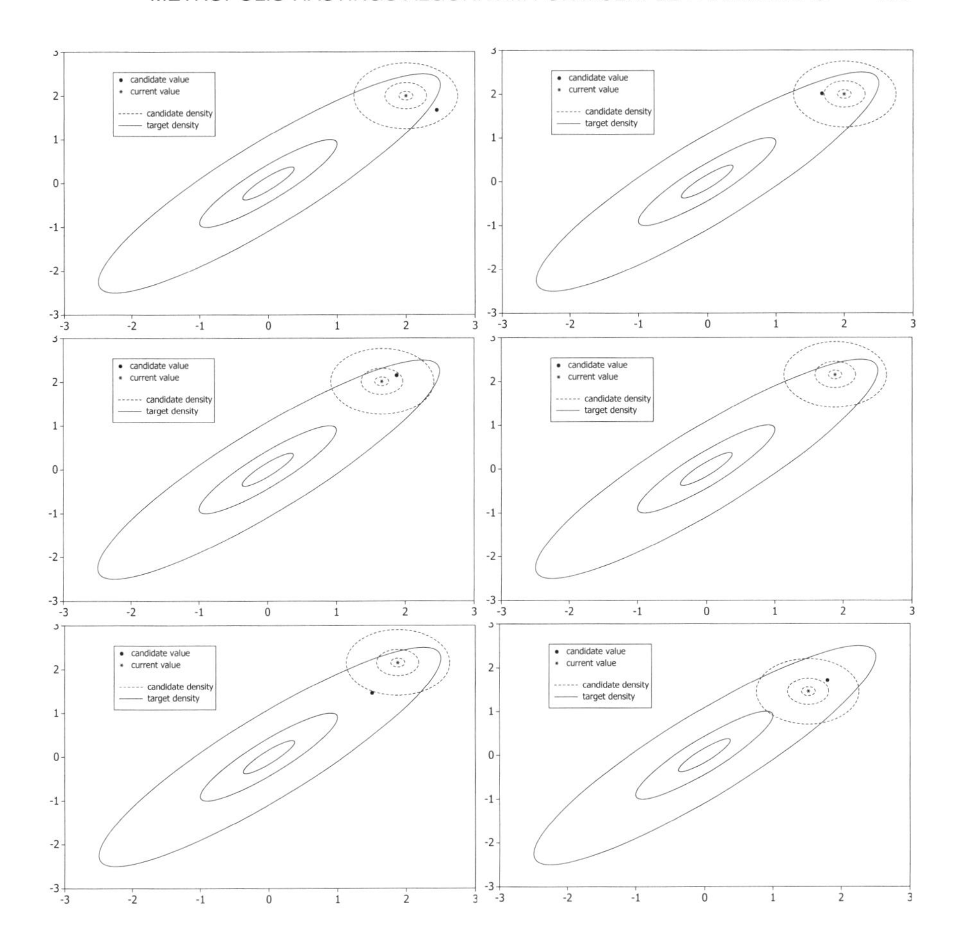

Figure 6.7 Six consecutive draws from a Metropolis-Hastings chain with a random-walk candidate density.

We will use a bivariate normal density centered around the current value with both standard deviations equal to .4, and correlation equal to 0. The random-walk candidate density in its unscaled form is given by

$$q_1(\theta_1' - \theta_1, \theta_2' - \theta_2) \quad \propto \quad e^{-\frac{1}{2\times .4^2}((\theta_1'-\theta_1)^2+(\theta_2'-\theta_2)^2)} .$$

Since this is a random-walk chain, we we will always accept a candidate value that is uphill from the current value. A candidate value that is downhill from the current value will be accepted with probability $\alpha = \frac{g(\theta_1',\theta_2'|y)}{g(\theta_1,\theta_2|y)}$. *We start the chain at initial value* $(\theta_2, \theta_2) = (1.0000, 1.0000)$. *Figure 6.7 shows six consecutive draws from the chain. Table 6.3 gives a summary of the first six draws of this chain. Note that the second and fifth candidates are uphill of their current values so will be accepted, while the first, third, fourth, and sixth candidate values were downhill from their*

Table 6.3 Summary of first six draws of the chain using the random-walk candidate density

Draw	Current value (θ_1, θ_2)	Candidate value (θ_1, θ_2)	α	u	Accept
1	(2.0000, 2.0000)	(2.4493, 1.6772)	.1968	.4905	no
2	(2.0000, 2.0000)	(1.6657, 2.0148)	1.0000	.3304	yes
3	(1.6657, 2.0148)	(1.8776, 2.1542)	.7842	.3571	yes
4	(1.8776, 2.1542)	(1.7016, 3.0999)	.0037	.1535	no
5	(1.8776, 2.1542)	(1.5099, 1.4566)	1.0000	.7961	yes
6	(1.5099, 1.4566)	(1.7918, 1.7138)	.6268	.9632	no

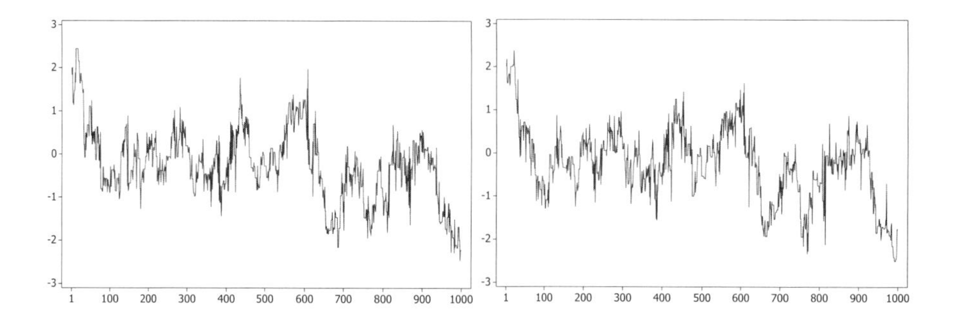

Figure 6.8 Trace plots of θ_1 and θ_2 for 1000 steps of the Metropolis-Hastings chain with the random-walk candidate density.

respective current values. Of these downhill candidates, only the third was accepted since only its randomly drawn u is less than its calculated acceptance probability α. Figure 6.8 shows traceplots for 1000 draws from the chain for the two parameters θ_1 and θ_2. We see that the chain is moving through the space satisfactorily, but at a slow rate. After it leaves a tail region, it generally takes a long time before it returns. Figure 6.9 shows a scatterplot of θ_2 versus θ_1 for the first 1000 draws of the random-walk chain. This distribution of the joint posterior sample shows the same shape as shown by the level curves of the joint target as shown in Figure 6.7. Figure 6.10 shows histograms for θ_1 and θ_2 for 5000 and 20000 steps of the Metropolis-Hastings algorithm. These can be considered samples from the marginal posterior densities of θ_1 and θ_2 respectively. However they are not random samples as the draws are serially correlated, not independent.

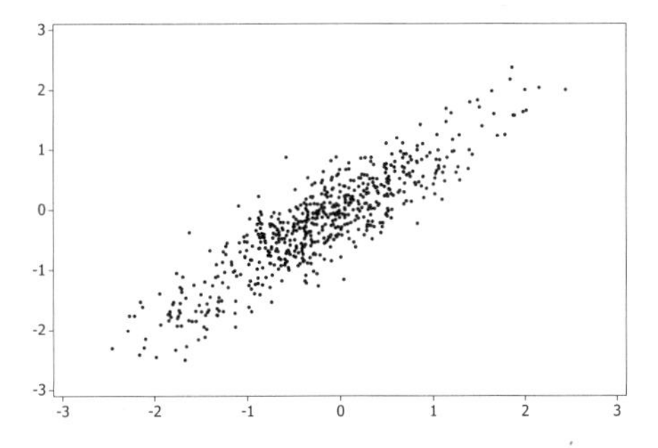

Figure 6.9 Scatterplot of θ_2 versus θ_1 for 1000 steps of the Metropolis-Hastings chain using the random-walk candidate density.

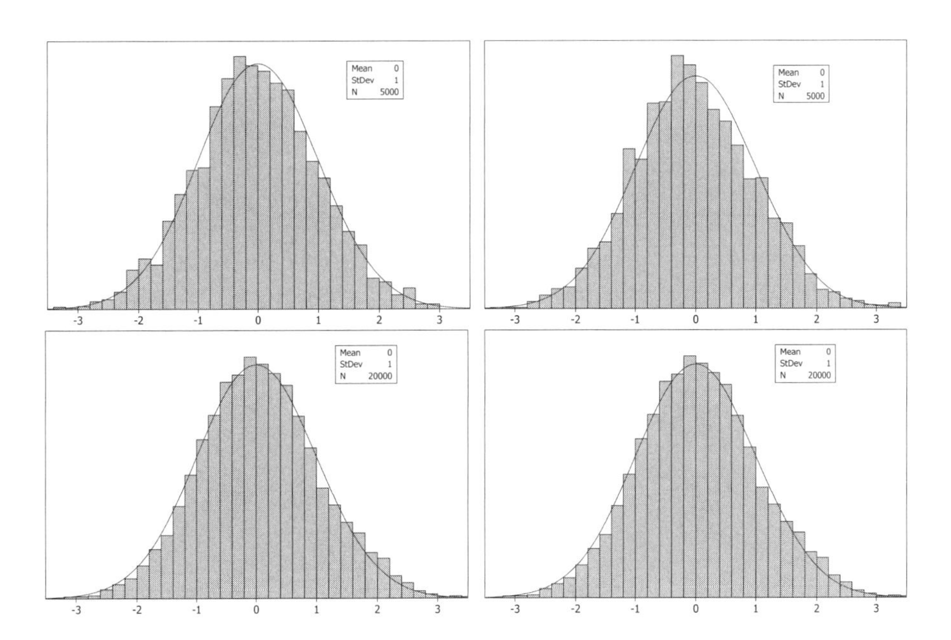

Figure 6.10 Histograms of θ_1 and θ_2 for 5000 and 20000 steps of the Metropolis-Hastings chain using the random-walk candidate density.

Multiple Parameters with an Independent Candidate Density

The density the candidate is drawn from does not depend on the current value when an independent candidate density is used. Hence for multiple parameters

$$q(\boldsymbol{\theta}, \boldsymbol{\theta}') \quad = \quad q_2(\boldsymbol{\theta}') \, .$$

Table 6.4 Summary of first six draws of the chain using the independent candidate density

Draw	Current value (θ_1, θ_2)	Candidate value (θ_1, θ_2)	α	u	Accept
1	(2.000, 1.500)	(.0211, .4976)	1.0000	.2960	yes
2	(.0211, .4976)	(−.4646, .2987)	.7696	.9506	no
3	(.0211, .4976)	(−.2923, .6001)	.6519	.1581	yes
4	(−.2923, .6001)	(.0692, .4722)	1.0000	.9093	yes
5	(.0692, .4722)	(−1.0615, −1.3032)	.8750	.9038	no
6	(.0692, .4722)	(.6357, 1.1574)	.8177	.1745	yes

The acceptance probability for a chain that uses an independent candidate density simplifies to

$$\begin{aligned}\alpha(\boldsymbol{\theta}, \boldsymbol{\theta}') &= \min\left[1, \frac{g(\boldsymbol{\theta}'|y)\, q(\boldsymbol{\theta}', \boldsymbol{\theta})}{g(\boldsymbol{\theta}|y)\, q(\boldsymbol{\theta}, \boldsymbol{\theta}')}\right] \\ &= \min\left[1, \frac{g(\boldsymbol{\theta}'|y)}{g(\boldsymbol{\theta}|y)} \times \frac{q_2(\boldsymbol{\theta})}{q_2(\boldsymbol{\theta}')}\right].\end{aligned}$$

Example 9 (continued) *Suppose we use the same bivariate target density as before with density given by*

$$g(\theta_1, \theta_2) \propto e^{-\frac{1}{2(1-.9^2)}(\theta_1^2 + 2\times .9 \times \theta_1\theta_2 + \theta_2^2)}.$$

We will use a bivariate independent candidate density given by

$$q_2(\theta_1', \theta_2') \propto e^{-\frac{1}{2\times 1.2^2}((\theta_1')^2 + 2\times .9\times \theta_1'\theta_2' + (\theta_2')^2)}.$$

This candidate density matches the correlation structure of the target. This is shown by having the same shape level curves as the target, only the ellipses are slightly larger. This candidate density will dominate the target in all directions. Most candidates will be accepted. The acceptance probability will be given by

$$\begin{aligned}\alpha(\boldsymbol{\theta}, \boldsymbol{\theta}') &= min\left[1, \frac{g(\boldsymbol{\theta}'|y)\, q(\boldsymbol{\theta}', \boldsymbol{\theta})}{g(\boldsymbol{\theta}|y)\, q(\boldsymbol{\theta}, \boldsymbol{\theta}')}\right] \\ &= \min\left[1; \frac{g(\boldsymbol{\theta}'|y)}{g(\boldsymbol{\theta}|y)} \times \frac{q_2(\boldsymbol{\theta})}{q_2(\boldsymbol{\theta}')}\right].\end{aligned}$$

The first six draws from the chain are shown in Figure 6.11 along with the level curves of the target. Table 6.4 gives a summary of the first six draws. We see the first, third, fourth, and sixth candidates are accepted, and the others are rejected. Figure 6.12 shows the traceplots for 1000 draws from the chain. With this independent

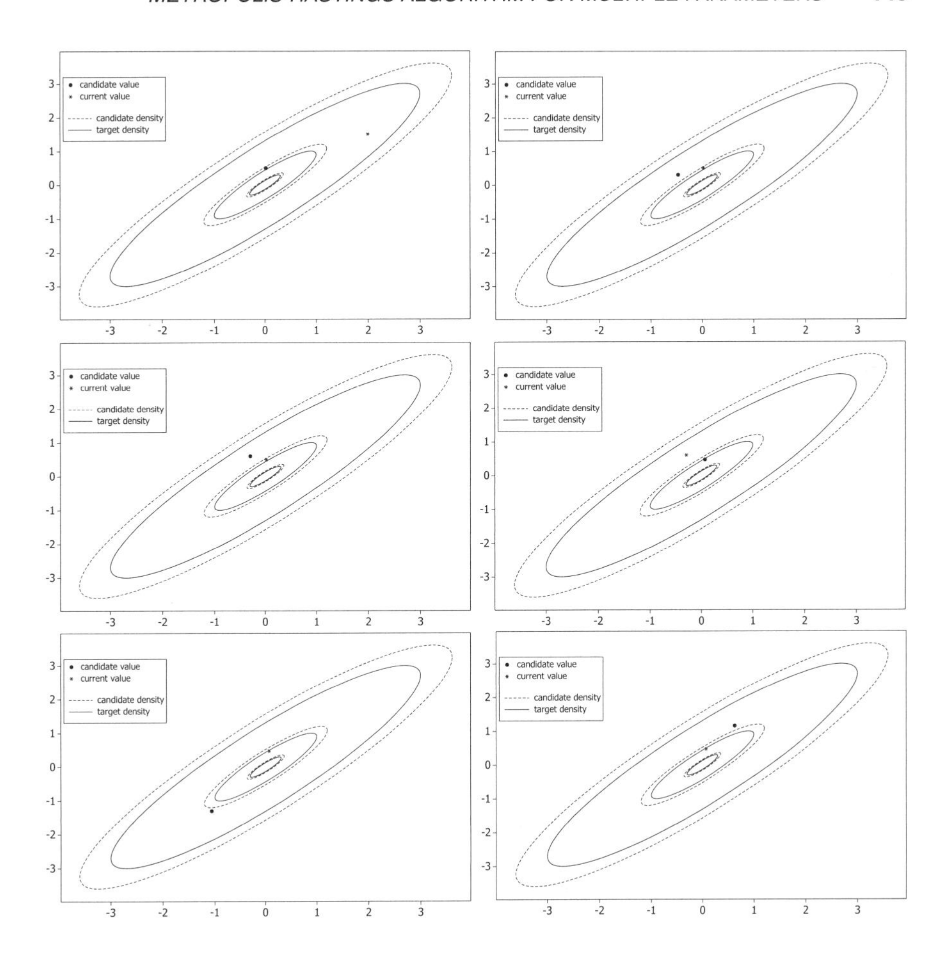

Figure 6.11 Six consecutive draws from a Metropolis-Hastings chain with an independent candidate density, together with the level curves of the target density and the independent candidate density.

candidate distribution that matches the correlation structure of the target, we expect many candidates to be accepted. The moves can be clear across the parameter space. Thus, the chain is moving around the parameter space very satisfactorily. Figure 6.13 shows the scatterplot of θ_2 versus θ_1 for 1000 consecutive draws from the independent chain. We see they show a similar shape to the level curves of the target. Figure 6.14 shows the histograms of θ_1 and θ_2 for 5000 and 20000 steps together with the correct marginal distributions which are both normal$(0, 1^2)$ distributions. We see these histograms are approaching the true marginal distributions.

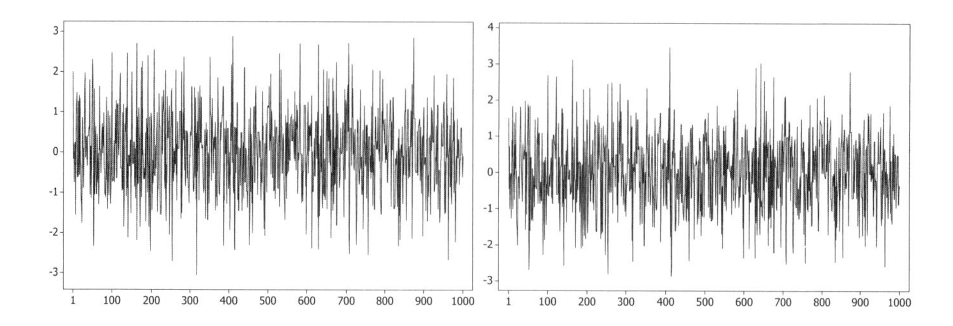

Figure 6.12 Trace plots of θ_1 and θ_2 for 1000 steps of the Metropolis-Hastings chain using the independent candidate density.

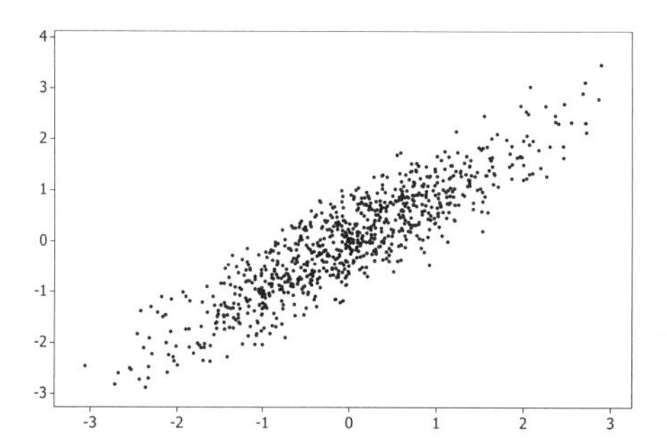

Figure 6.13 Scatterplot of θ_2 versus θ_1 for 1000 steps of the Metropolis-Hastings chain using the independent candidate density.

6.3 BLOCKWISE METROPOLIS-HASTINGS ALGORITHM

Let the parameter vector be partitioned into blocks

$$\boldsymbol{\theta} = \boldsymbol{\theta_1}, \boldsymbol{\theta_2}, \ldots, \boldsymbol{\theta_J},$$

where $\boldsymbol{\theta_j}$ is a block of parameters. Let $\boldsymbol{\theta_{-j}}$ be all the other parameters not in block j. It may be easier to find the conditional kernel for each block of parameters that converges to its respective conditional posterior density than to find the single overall kernel that converges to the joint posterior density. Hastings (1970) suggested that, instead of applying the Metropolis-Hastings algorithm to the whole parameter vector $\boldsymbol{\theta}$ all at once, that the algorithm be applied sequentially to each block of parameters $\boldsymbol{\theta_j}$ in turn, conditional on knowing the values of all other parameters not in that block. Let $P_j(\boldsymbol{\theta_j}, A_j)|\boldsymbol{\theta_{-j}})$ be the transition kernel for the M-H algorithm applied to parameter block θ_j holding all the other parameter blocks fixed. He proved the following theorem

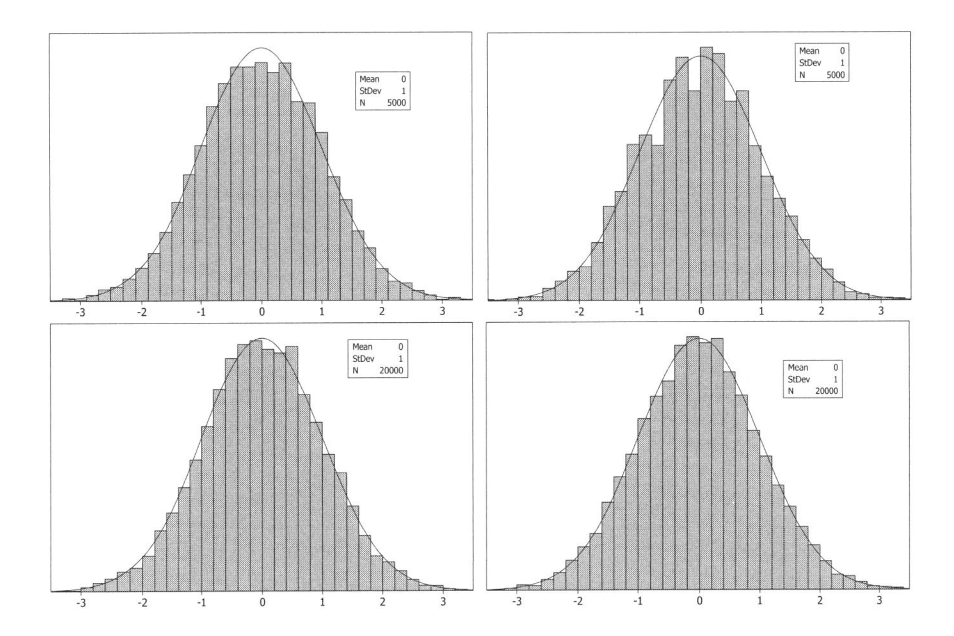

Figure 6.14 Histograms of θ_1 and θ_2 for 5000 and 20000 draws of the Metropolis-Hastings chain with the independent candidate density.

Theorem 7 *The transition kernel*

$$P(\theta, A) = \prod_{j=1}^{J} P_j(\theta_j, A_j|\theta_{-j})$$

has the posterior density $g(\theta|y)$ as its long-run distribution when we cycle through the parameters, holding the rest at their most recent values.

The practical significance of this *product of kernels principle* is that we do not have to run each subchain to convergence before we move on the the next. If we draw a block of parameters from each subchain in turn, given the most recent values of the other parameters, this process will converge to the posterior distribution of the whole parameter vector.

Steps of Blockwise Metropolis-Hastings

1. Start at point in parameter space $\boldsymbol{\theta}_1^{(0)}, \ldots, \boldsymbol{\theta}_j^{(0)}$.

2. For $n = 1, \ldots, N$

    - For $j = 1, \ldots, J$

– draw candidate from

$$q(\boldsymbol{\theta}_j^{(n-1)}, \boldsymbol{\theta}_j' | \boldsymbol{\theta}_1^{(n)}, \ldots, \boldsymbol{\theta}_{j-1}^{(n)}, \boldsymbol{\theta}_{j+1}^{(n-1)}, \ldots, \boldsymbol{\theta}_J^{(n-1)}) .$$

– Calculate the acceptance probability

$$\alpha(\boldsymbol{\theta}_j^{(n-1)}, \boldsymbol{\theta}_j' | \boldsymbol{\theta}_1^{(n)}, \ldots, \boldsymbol{\theta}_{j-1}^{(n)}, \boldsymbol{\theta}_{j+1}^{(n-1)}, \ldots, \boldsymbol{\theta}_J^{(n-1)}) .$$

– Draw u from $U(0,1)$.
– if $u < \alpha(\boldsymbol{\theta}_j^{(n-1)}, \boldsymbol{\theta}_j')$ then let $\boldsymbol{\theta}_j^{(n)} = \boldsymbol{\theta}_j'$, else let $\boldsymbol{\theta}_j^{(n)} = \boldsymbol{\theta}_j^{(n-1)}$.

Example 9 (continued) *Let us continue with the same bivariate posterior density as the target as before. The unscaled target (posterior) density has formula*

$$g(\theta_1, \theta_2) \quad \propto \quad e^{-\frac{1}{2(1-.9^2)}(\theta_1^2 + 2\times.9\times\theta_1\theta_2 3 + \theta_2^2)} .$$

We will let each of the parameters θ_1 and θ_2 be in separate blocks, and will perform the blockwise Metropolis-Hastings algorithm. The conditional density of θ_1 given the value of θ_2 is normal(m_1, s_1^2) where

$$m_1 = \rho\theta_2 \quad \text{and} \quad s_1^2 = (1-\rho^2) .$$

We will use the candidate density of θ_1 given the value of θ_2 to be normal$(m_1, .75^2)$. Similarly, the conditional density of θ_2 given the value of θ_1 is normal(m_2, s_2^2) where

$$m_2 = \rho\theta_1 \quad \text{and} \quad s_2^2 = (1-\rho^2)$$

and we will use the candidate density of θ_2 given the value of θ_1 to be normal$(m_2, .75^2)$. We alternate between drawing a candidate θ_1 given the current value of θ_2 then either accepting it or not, then drawing a candidate θ_2 given the current value of θ_1 and either accepting it or not. The first three steps from the are shown in Figure 6.15. The summary of the first three steps are shown in Table 6.5. We see on the first step, the candidate value of θ_1 is found, and is accepted. Then the candidate value of θ_2 is found, but is not accepted. On the second step, the candidate value for θ_1 is found and accepted, then the candidate value for θ_2 is found and accepted. On the third step, the candidate value for θ_1 is selected but not accepted, then the candidate value for θ_2 is found and is accepted. Figure 6.16 shows the traceplots of θ_1 and θ_2 for the first 1000 steps of the blockwise Metropolis-Hastings chain. We see the chain is moving through the whole parameter space, but it only moves slowly.

Figure 6.17 shows the scatterplot of θ_2 versus θ_1 for the first 1000 steps of the blockwise Metropolis-Hastings chain. We see that the shape of the sample is approaching the shape of the target. Figure 6.18 shows the histograms for the sample of θ_1 and θ_2, together with their exact marginal posteriors for 5000 and 20000 draws. We see that the histograms of the samples are approaching the exact marginal posteriors, but fairly slowly.

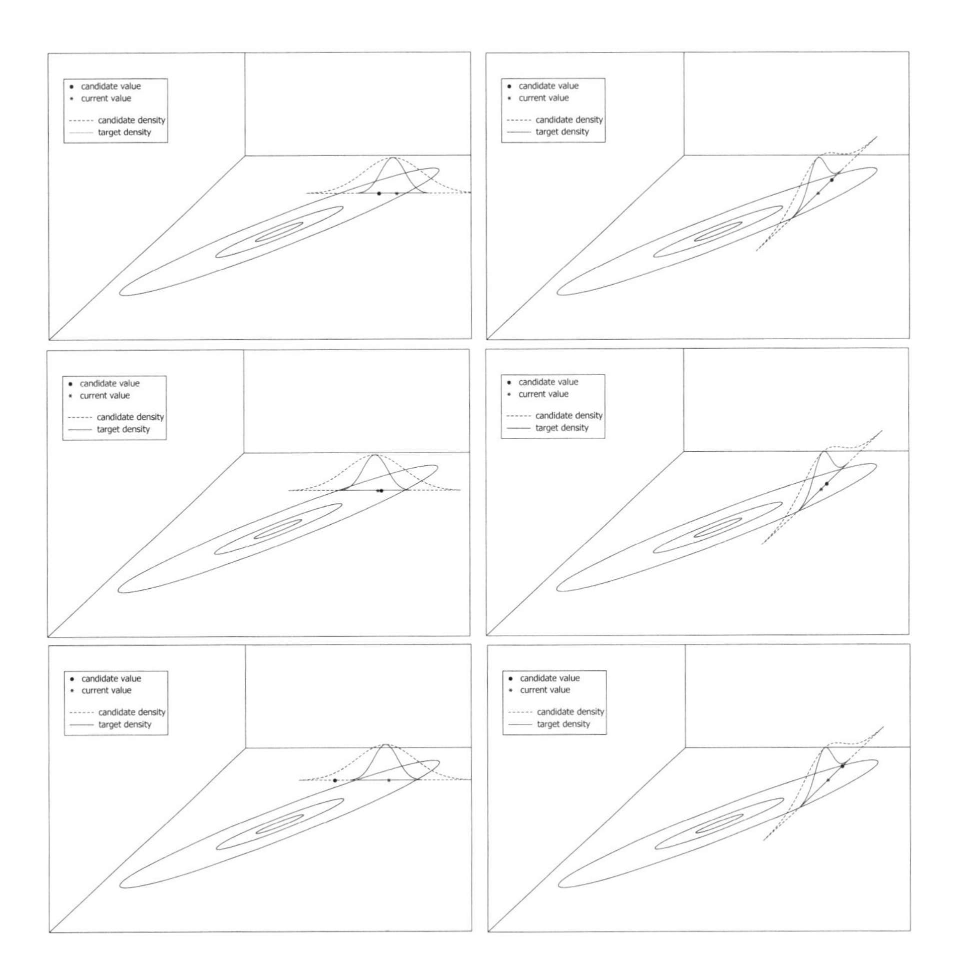

Figure 6.15 Three consecutive draws from a Blockwise Metropolis-Hastings chain.

Table 6.5 Summary of first three draws of the blockwise chain

Draw	Current value (θ_1, θ_2)	Candidate value (θ_1, θ_2)	α	u	Accept
1A	(2.0000, 1.5000)	(1.5405, 1.5000)	.6033	.1780	yes
1B	(1.5405, 1.5000)	(1.5405, 2.0141)	.2782	.6066	no
2A	(1.5405, 1.5000)	(1.6292, 1.5000)	.9128	.0139	yes
2B	(1.6292, 1.5000)	(1.6292, 1.7115)	.9012	.2842	yes
3A	(1.6292, 1.7115)	(.2254, 1.7115)	.0006	.7505	no
3B	(1.62922, 1.7115)	(1.6292, 2.2534)	.1357	.1327	yes

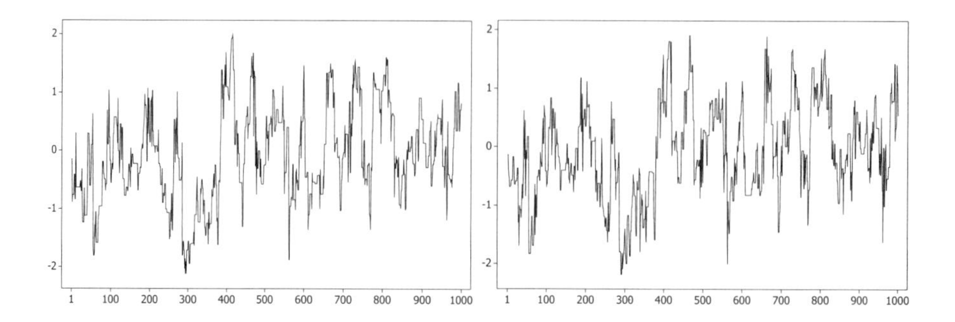

Figure 6.16 Trace plots of θ_1 and θ_2 for 1000 steps of the blockwise Metropolis-Hastings chain.

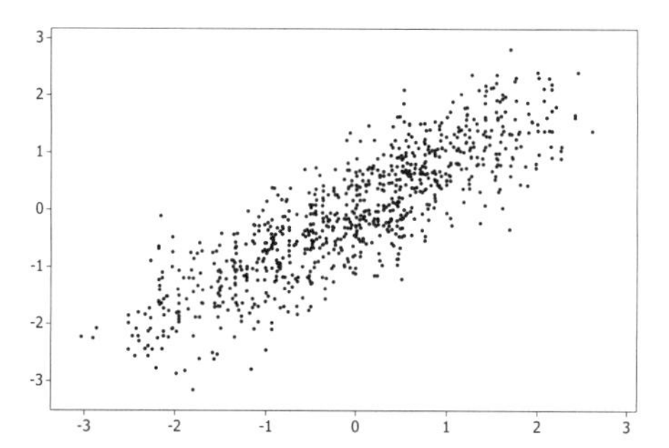

Figure 6.17 Scatterplot of θ_2 versus θ_1 for 1000 steps of the blockwise Metropolis-Hastings chain.

In the blockwise Metropolis-Hastings algorithm the candidate density for the block of parameters $\boldsymbol{\theta}_j$ given all the other parameters $\boldsymbol{\theta}_{-j}$ and the data $\mathbf{y}$ must dominate the true conditional density in the tails. That is

$$q(\boldsymbol{\theta}_j, \boldsymbol{\theta}'_j|\boldsymbol{\theta}_{-j}) \quad > \quad g(\boldsymbol{\theta}_j|\boldsymbol{\theta}_{-j}, \mathbf{y}) \,.$$

At each step for each block in turn, we draw the candidate $\boldsymbol{\theta}_j$ from the candidate density, calculate the acceptance probability $\alpha(\boldsymbol{\theta}_j, \boldsymbol{\theta}'_j)$, and either move that block of parameters to that candidate $\boldsymbol{\theta}'_j$, or keep that block at the current value $\boldsymbol{\theta}_j$, depending on whether or not a random draw from a *uniform*$(0,1)$ random variable is less than the acceptance probability.

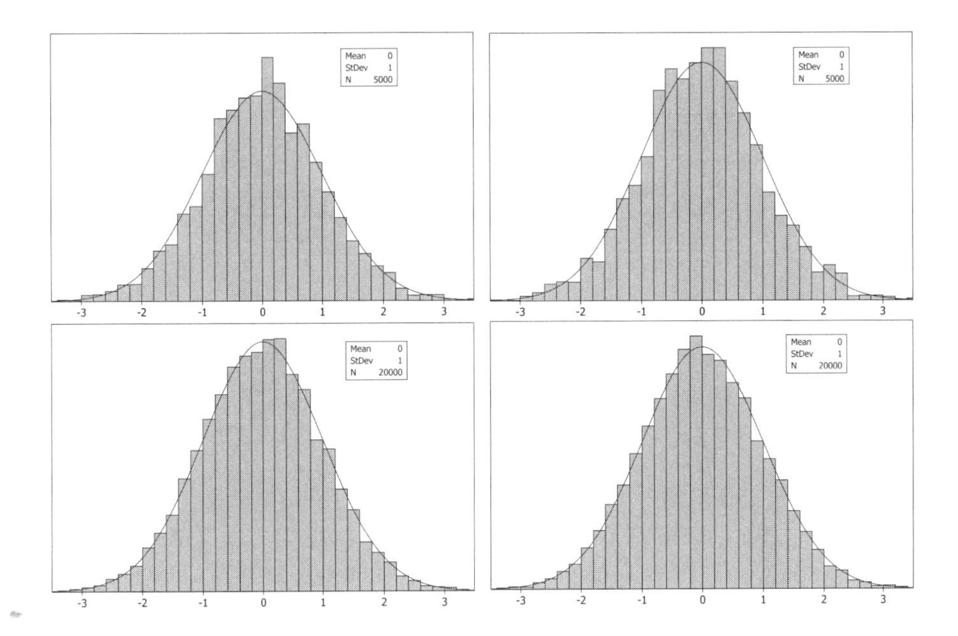

Figure 6.18 Histograms of θ_1 and θ_2 for 5000 and 20000 draws of the blockwise Metropolis-Hastings chain.

6.4 GIBBS SAMPLING

Suppose we decide to use the true conditional density as the candidate density at each step for every block of parameters given the others. In that case

$$q(\boldsymbol{\theta}_j, \boldsymbol{\theta}'_j|\boldsymbol{\theta}_{-j}) \quad = \quad g(\boldsymbol{\theta}_j|\boldsymbol{\theta}_{-j}, \mathbf{y}),$$

so at step n for block $\boldsymbol{\theta}_j$ the acceptance probability will be

$$\begin{aligned} \alpha(\boldsymbol{\theta}_j^{(n-1)}, \boldsymbol{\theta}'_j|\boldsymbol{\theta}_1^{(n)}, \ldots, \boldsymbol{\theta}_{j-1}^{(n)}, \boldsymbol{\theta}_{j+1}^{(n-1)}, \ldots, \boldsymbol{\theta}_J^{(n-1)}) \\ = \min\left[1, \frac{g(\boldsymbol{\theta}_j{}'|\boldsymbol{\theta}_{-j}\mathbf{y})\, q(\boldsymbol{\theta}'_j, \boldsymbol{\theta}_j|\boldsymbol{\theta}_{-j})}{g(\boldsymbol{\theta}_j|\boldsymbol{\theta}_{-j}\mathbf{y})\, q(\boldsymbol{\theta}_j, \boldsymbol{\theta}_{-j}|\boldsymbol{\theta}_{-j})}\right] \\ = 1\,. \end{aligned}$$

so the candidate will be accepted at each step. The case where we draw each candidate block from its true conditional density given all the other blocks at their most recently drawn values is known as Gibbs sampling. This algorithm was developed by Geman and Geman (1984) as a method for recreating images from a noisy signal. They named it after Josiah Willard Gibbs who had determined a similar algorithm could be used to determine the energy states of gasses at equilibrium. He would cycle through the particles, drawing each one conditional on the energy levels of all other particles. His algorithm became the basis for the field of statistical mechanics. As we see, the

Gibbs sampling algorithm is just a special case of the blockwise Metropolis-Hastings algorithm. In Chapter 10 we will look at the class of hierarchical models where the Gibbs sampling algorithm is particularly well suited.

Example 9 (continued) *Suppose we look at the same bivariate target density as before. The target density is given by*

$$g(\theta_1, \theta_2) \quad \propto \quad e^{-\frac{1}{2(1-.9^2)}(\theta_1^2 + 2\times .9 \times \theta_1 \theta_2 3 + \theta_2^2)} .$$

The conditional density of θ_1 *given* θ_2 *is normal*(m_1, s_1^2) *where*

$$m_1 = \rho\theta_2 \quad \text{and} \quad s_1^2 = (1 - \rho^2) .$$

Similarly, the conditional density of θ_2 *given* θ_1 *is normal*(m_2, s_2^2) *where*

$$m_2 = \rho\theta_2 \quad \text{and} \quad s_2^2 = (1 - \rho^2) .$$

We will alternate back and forth, first drawing θ_1 *from its density given the most recently drawn value of* θ_2*, then drawing* θ_2 *from its density given the most recently drawn value of* θ_1*. We don't have to calculate the acceptance probability since we know it will always be 1. Figure 6.19 shows the first three steps of the algorithm. We see the candidate density is the same as the true conditional density, so all candidates are accepted. Figure 6.20 shows traceplots for the first 1000 steps of the Gibbs sampling chain. Figure 6.21 shows the scatterplot of* θ_2 *versus* θ_1 *for the first 1000 draws from the Gibbs sampling chain. Figure 6.22 shows histograms of* θ_1 *and* θ_2 *together with their exact marginal posteriors for 5000 and 2000 steps of the Gibbs sampler.*

6.5 SUMMARY

In this chapter we have shown several ways to find a Markov chain that has the posterior distribution of the parameters given the data as its long-run distribution. These are generally based on the Metropolis-Hastings algorithm which balances the flow between each pair of states at equilibrium, by only accepting some of the candidates. There are different types based on how the candidates are drawn. These include random-walk chains, where the candidate is drawn from a distribution symmetric about the current value; independent chains, where the the distribution the candidate is drawn from does not depend on the current value; and blockwise chains, where candidates for each block of parameters are drawn and either accepted or not in turn. The Gibbs sampling algorithm was developed from a statistical mechanics perspective, where we cycling through the parameter blocks, drawing each candidate conditional on the most recently drawn values of the other parameter blocks. However, we saw, that the Gibbs sampling algorithm can really be considered a special case of the blockwise Metropolis-Hastings algorithm.

We can use any of these methods to draw a sample from the posterior. However, the samples won't be random. Draws from a Markov chain are serially correlated.

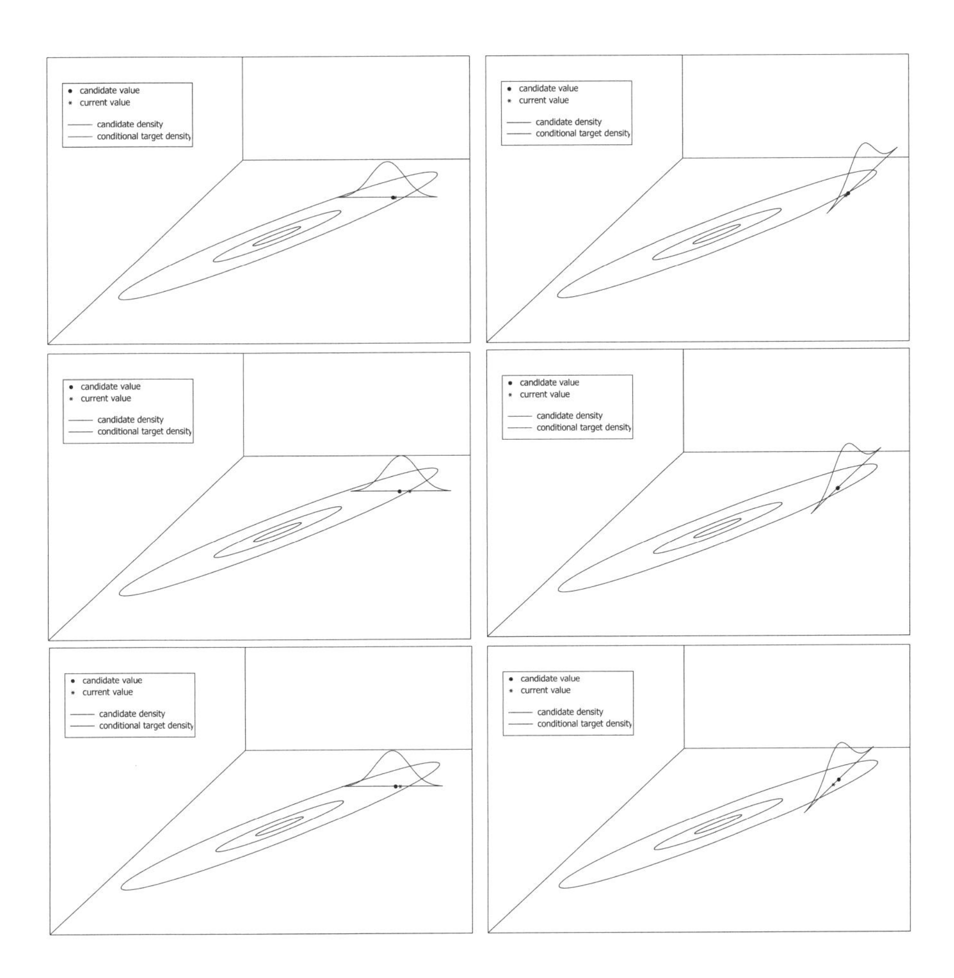

Figure 6.19 Three consecutive draws from the Gibbs sampling chain.

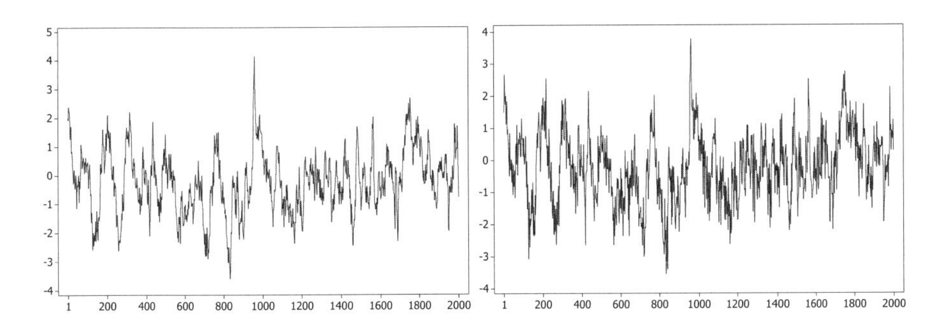

Figure 6.20 Trace plots of θ_1 and θ_2 for 1000 steps of the Gibbs sampling chain.

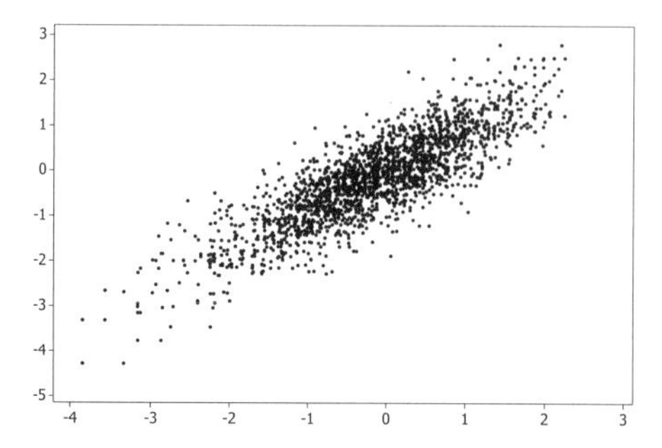

Figure 6.21 Scatterplot of θ_2 versus θ_1 for 1000 steps of the Gibbs sampling chain.

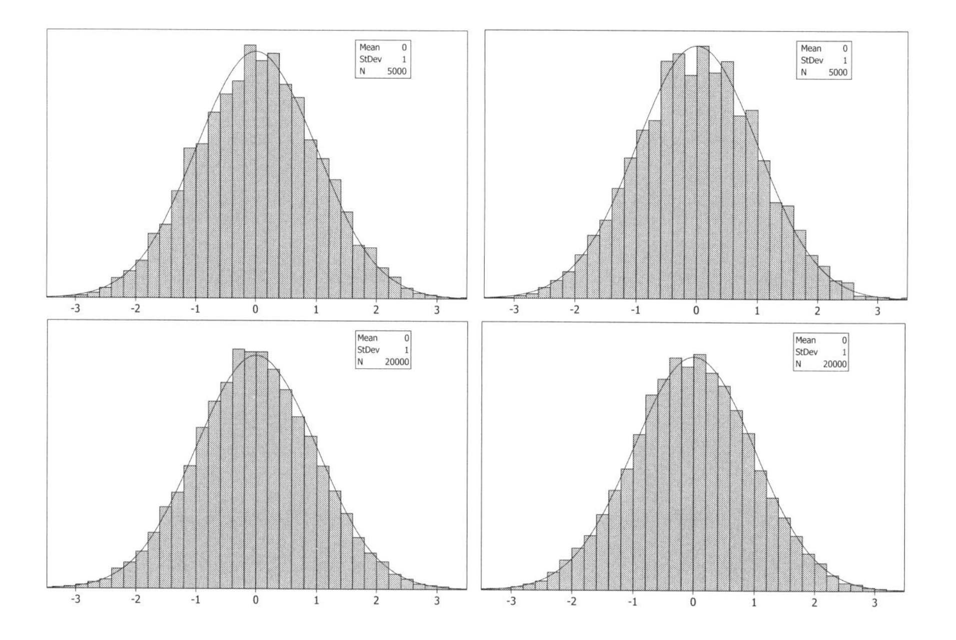

Figure 6.22 Histograms of θ_1 and θ_2 for 5000 and 20000 draws of the Gibbs sampling chain.

We would like the samples to be random, at least approximately, in order to do our inferences. In Chapter 7 we will look at how we can obtain an approximately random sample from the nonrandom sample we have obtained from the Markov chain.

There are many candidate distributions we can choose from. Each of them would have its own mixing properties. We would like to choose a chain that has good mixing properties. In other words, a chain that moves through the parameter space quickly. The trace plots give an indication of the mixing properties.

Main Points

- The detailed balance condition says that the steady state flow of probability between each pair of states is in balance. In the case of a continuous parameter, when we have a candidate density that satisfies the detailed balance condition

$$g(\theta|y)q(\theta,\theta') = g(\theta'|y)q(\theta',\theta)$$

for all θ, θ' then the posterior $g(\theta|y)$ is the long-run distribution of the Markov chain with probability kernel

$$P(\theta, A) = \int_A q(\theta,\theta')d\theta' + r(\theta)\delta_A(\theta)$$

where $r(\theta) = 1 - \int q(\theta,\theta')d\theta'$ is the probability the chain remains at θ, and where $\delta_A(\theta)$ is the indicator function of set A.

$$\delta_A(\theta) = \begin{cases} 1 & \text{if} \quad \theta \in A \\ 0 & \text{if} \quad \theta \notin A \end{cases} .$$

Unfortunately most candidate densities won't satisfy the detailed balance condition.

- The Metropolis-Hastings algorithm restores the balance by introducing an acceptance probability that depends on the current and candidate values and then only accepting some of the transitions.

- A random-walk candidate density is symmetric about the current value.

- An independent candidate density is the same regardless of the current value. The independent candidate density must have heavier tails than the posterior density.

- The trace plot shows how the chain is moving through the parameter space.

- The blockwise Metropolis-Hastings algorithm draws each block of parameters in turn from a candidate distribution conditional on all the other parameters not in that block. Then the we either move that block of parameters to that candidate, or we stay at the curent values for that block of parameters according to our acceptance probability which is calculated using the candidate density and the conditional density for that block given all the other parameters and the data. The algorithm cycles through all the blocks of parameters in turn.

- The Gibbs sampling algorithm is a special case of blockwise Metropolis-Hastings where the candidate density for each block of parameters is its correct conditional distribution given all other parameters not in its block and the observed data. The acceptance probability is always 1, so every candidate is accepted.

Exercises

6.1 Let the unscaled posterior have shape given by

$$g(\theta|\mathbf{y}) \propto .6 \times e^{-\frac{1}{2}(\theta^2)} + .4 \times \frac{1}{2} e^{-\frac{1}{2\,2^2}(\theta-3)^2}.$$

This is a mixture of two normal distributions.

(a) Use the Minitab macro *NormMixMHRW.mac* or the equivalent R-function `NormMixMH` to draw a Markov chain Monte Carlo sample of size 5000 from the posterior using the Metropolis-Hastings algorithm with *normal*$(0, 1^2)$ random-walk candidate density.

(b) Use the Minitab macro *NormMixMHRW.mac* or the equivalent R-function `NormMixMH` to draw a Markov chain Monte Carlo sample of size 5000 from the posterior using the Metropolis-Hastings algorithm with *normal*$(0, .5^2)$ random-walk candidate density.

(c) Comment on how the two chains are moving through the parameter space.

(d) Form a histogram for each of the two samples. Do they resemble the shape of the posterior?

6.2 Let the unscaled posterior have the same shape as in the previous exercise.

(a) Show that a *normal*$(0, 3^2)$ independent candidate density will dominate the target.

(b) Use the Minitab macro *NormMixMHInd.mac* or the equivalent R-function `NormMixMH` to draw a Markov chain Monte Carlo sample of size 5000 from the posterior using the Metropolis-Hastings algorithm with *normal*$(0, 3^2)$ independent candidate density.

(c) Comment on how the chain is moving through the parameter space.

(d) Form a histogram of the sample. Does it resemble the shape of the posterior? Explain why or why not.

(e) Use the Minitab macro *NormMixMHInd.mac* or the equivalent R-function `NormMixMH` to draw a Markov chain Monte Carlo sample of size 5000 from the posterior using the Metropolis-Hastings algorithm with *normal*$(0, 2^2)$ independent candidate density.

(f) Comment on how the chain is moving through the parameter space.

(g) Form a histogram of the sample. Does it resemble the shape of the posterior? Explain why or why not.

6.3 Let the unscaled posterior have shape given by

$$g(\theta|\mathbf{y}) \propto .5 \times e^{-\frac{1}{2}(\theta+2)^2} + .5 \times \frac{1}{2} e^{-\frac{1}{2\,2^2}(\theta-4)^2}.$$

This is a mixture of two normal distributions.

(a) Use the Minitab macro *NormMixMHRW.mac* or the equivalent R-function `NormMixMH` to draw a Markov chain Monte Carlo sample of size 5000 from the posterior using the Metropolis-Hastings algorithm with *normal*$(0, 1^2)$ random-walk candidate density.

(b) Use the Minitab macro *NormMixMHRW.mac* or the equivalent R-function `NormMixMH` to draw another Markov chain Monte Carlo sample of size 5000 from the posterior using the Metropolis-Hastings algorithm with *normal*$(0, .5^2)$ random-walk candidate density.

(c) Comment on how each of the chains is moving through the parameter space.

(d) Form a histogram of the two sample. Do they resemble the shape of the posterior?

6.4 Let the unscaled posterior have the same shape as in the previous exercise.

(a) Show that a *normal*$(0, 3^2)$ independent candidate density will dominate the target.

(b) Use the Minitab macro *NormMixMHInd.mac* or the equivalent R-function `NormMixMH` to draw a Markov chain Monte Carlo sample of size 5000 from the posterior using the Metropolis-Hastings algorithm with *normal*$(0, 3^2)$ independent candidate density.

(c) Comment on how the chain is moving through the parameter space.

(d) Form a histogram of the sample. Does it resemble the shape of the posterior? Explain why or why not.

(e) Use the Minitab macro *NormMixMHInd.mac* or the equivalent R-function `NormMixMH` to draw a Markov chain Monte Carlo sample of size 5000 from the posterior using the Metropolis-Hastings algorithm with *normal*$(0, 2^2)$ random-walk candidate density.

(f) Comment on how the chain is moving through the parameter space.

(g) Form a histogram of the sample. Does it resemble the shape of the posterior? Explain why or why not.

6.5 Let the unscaled posterior have shape given by

$$g(\theta|\mathbf{y}) \propto .5 \times e^{-\frac{1}{2}(\theta+3)^2} + .5 \times e^{-\frac{1}{2}(\theta-3)^2} .$$

This is a mixture of two normal distributions.

(a) Use the Minitab macro *NormMixMHRW.mac* or the equivalent R-function `NormMixMH` to draw a Markov chain Monte Carlo sample of size 5000 from the posterior using the Metropolis-Hastings algorithm with *normal*$(0, 1^2)$ random-walk candidate density.

(b) Use the Minitab macro *NormMixMHRW.mac* or the equivalent R-function `NormMixMH` to draw another Markov chain Monte Carlo sample of size 5000 from the posterior using the Metropolis-Hastings algorithm with *normal*$(0, .5^2)$ random-walk candidate density.

(c) Comment on how each of the chains is moving through the parameter space.

(d) Form a histogram of the two sample. Do they resemble the shape of the posterior?

6.6 Let the unscaled posterior have the same shape as in the previous exercise.

(a) Show that a *normal*$(0, 3^2)$ independent candidate density will dominate the target.

(b) Use the Minitab macro *NormMixMHInd.mac* or the equivalent R-function `NormMixMH` to draw a Markov chain Monte Carlo sample of size 5000 from the posterior using the Metropolis-Hastings algorithm with *normal*$(0, 3^2)$ independent candidate density.

(c) Comment on how the chain is moving through the parameter space.

(d) Form a histogram of the sample. Does it resemble the shape of the posterior? Explain why or why not.

(e) Use the Minitab macro *NormMixMHInd.mac* or the equivalent R-function `NormMixMH` to draw a Markov chain Monte Carlo sample of size 5000 from the posterior using the Metropolis-Hastings algorithm with *normal*$(0, 2^2)$ independent candidate density.

(f) Comment on how the chain is moving through the parameter space.

(g) Form a histogram of the sample. Does it resemble the shape of the posterior? Explain why or why not.

6.7 In this exercise, we compare the effectiveness of several ways to draw a sample from the Markov chain when the multiple parameters are strongly correlated.

(a) Use the Minitab macro *BivNormMHRW.mac* or the equivalent R-function `bivnormMH` to draw a sample of 1000 from a *bivariate normal* target distribution when the two parameters have the correlation coefficient $\rho = .9$. This macro draws candidates for both parameters in a single step using a random-walk candidate density.

(b) Use the Minitab macro *BivNormMHIND.mac* or the equivalent R-function `bivnormMH` to draw a sample of 1000 from a *bivariate normal* target distribution when the two parameters have the correlation coefficient $\rho = .9$. This macro draws candidates for both parameters in a single step using an independent candidate density.

(c) Use the Minitab macro *BivNormMHBL.mac* or the equivalent R-function `bivnormMH` to draw a sample of 1000 from a *bivariate normal* target distribution when the two parameters have the correlation coefficient $\rho = .9$. This macro draws a candidate for each parameter separately, conditional on the most recent value of the other parameter. Then it uses a Metropolis-Hastings step to either accept the candidate, or continue with the current value of that parameter.

(d) Use the Minitab macro *BivNormMHGibbs.mac* or the equivalent R-function `bivnormMH` to draw a sample of 1000 from a *bivariate normal* target distribution when the two parameters have the correlation coefficient $\rho = .9$. This macro draws a candidate for each parameter separately, conditional on the most recent value of the other parameter. The candidate density used is the correct conditional density, so all the candidates are accepted.

(e) Comment on how the different chains are moving through the parameter space. Which chain is moving through the parameter space more effectively?

7

Statistical Inference from a Markov Chain Monte Carlo Sample

In the previous chapter, we learned how to set up a Markov chain that has the posterior (target) as its long-run distribution. The method was based on the Metropolis-Hastings algorithm, which balances the steady state flow between each pair of states by only accepting some of the candidates. We found that there were several ways of implementing the algorithm. These methods differ in how they choose the candidates. The first was using a random-walk candidate density, where at each step, the candidate is drawn from a symmetric distribution centered around the current value. The second was using an independent candidate density, where the same candidate density is used, independent of the current value. We also saw that for multiple parameters we could either draw a multivariate candidate for all the parameters at once or we could run the algorithm blockwise. In the blockwise case, we would draw candidate from each block of parameters in sequence, given the most recent values of all the parameters in the other blocks. The acceptance probability is calculated from the candidate density for that block of parameters, and the true conditional density of that block of parameters given the other blocks of parameters at their most recent values. We either accept the candidate, and move to it, or stay at the current value for that block. Then we repeat the process for the next block. After we finish the last block, we go back to the first block and repeat the whole process. Although the Gibbs sampling algorithm was developed separately from a statistical mechanics perspective, it turns out to be a special case of blockwise Metropolis-Hastings algorithm where the candidate for each block is chosen from the true conditional density of that block given the other parameters. Thus all candidates are accepted when we use Gibbs sampling.

We have flexibility in choosing which method to use, and which candidate density we should use for that method. We would like to find a chain that could be at any point in the parameter space in a few steps from any starting point. A chain like that is said to have good mixing properties. So, we would like to choose a chain that has good mixing properties, a chain that moves through the parameter space quickly. In Section 7.1, we look at the mixing properties of the various types of chain. We will see that an independent chain where the candidate density has the same relationship structure as the target generally will have very good mixing properties. In Section 7.2 we will develop a method for determining a candidate density with the same relationship structure as the target, but with heavier tails. This method will be based on matching the curvature of the target at its mode.

After we have let the chain run a long time, the state the chain is in does not depend on the initial state of the chain. This length of time is called the *burn-in* period. A draw from the chain after the burn-in time is approximately a random draw from the posterior. However, the sequence of draws from the chain after that time is not a random sample from the posterior, rather it is a dependent sample. In Chapter 3, we saw how we could do inference on the parameters using a random sample from the posterior. In Section 7.3 we will continue with that approach to using the Markov chain Monte Carlo sample from the posterior. We will have to *thin* the sample so that we can consider it to be approximately a random sample. A chain with good mixing properties will require a shorter *burn-in* period and less thinning.

7.1 MIXING PROPERTIES OF THE CHAIN

We have seen that it is easy to find a Markov chain that will have the target (posterior) distribution as its long-run distribution by using the Metropolis-Hastings algorithm. The chain will be ergodic and aperiodic. However, this does not answer the question of how long the chain must be run until an outcome from the chain has the target distribution. If the long-run distribution of the Markov chain has more than one mode, and one of the modal regions has a very low probability of entering and leaving, it can take an exceedingly long time until convergence. Examples can be constructed where that time could exceed any fixed value. In the general case, we can't decide beforehand how many iterations will be needed. However, most of the target distributions we will look at are regression type models that do not have multiple modes. They will be well behaved.

We have a great deal of flexibility here in how we apply the Metropolis-Hastings algorithm. We can select the type of chain by deciding on the type of candidate density we will use. We can choose to do all the parameters at once, or to do the parameters in blocks. We can choose the specific candidate density of our selected type. We would like to find a chain that moves through the entire parameter space at a fast rate. In other words a chain that has good mixing properties. The mixing properties of the chain depend on our choices. The traceplots of the parameters drawn from the chain give an indication of the mixing properties of the chain. The different types of chains will have different patterns in their traceplots.

Random-Walk Chains

A random-walk chain will accept many candidates, but the moves will be small ones. We observed this in the traceplots shown in Figure 6.2 and Figure 6.8. There is a long time between the visits to each tail. Generally a random-walk chain will move through the whole parameter space, but it may take a very long time. And if the long-run distribution has multiple modes that are widely separated, it may take such a long time to move between modes, that we may fail to realize that there was a mode that had not been visited. This is similar to what we observed for discrete Markov chains in Example 7.

Independent Chains

An independent candidate chain may have better mixing properties. The candidate density should dominate the target in all directions. It is good to have a candidate density that has a shape close to that of the target so that more candidates will be accepted. This means that the candidate density should have the same shape covariance structure as the target. The independent candidate chain will not accept as many candidates as a random-walk chain, but the moves will be larger. We observed this in Figures 6.5 and 6.12. Thus an independent chain with a well chosen candidate density will move through the parameter space fairly quickly.

Blockwise Metropolis-Hastings Chains

When the Metropolis-Hastings algorithm is run blockwise and the parameters in different blocks are highly correlated, the candidate for a block won't be very far from the current value for that block. Because of this, the chain will move slowly around the parameter space very slowly. We observed this in Figure 6.15 and in the traceplots in Figure 6.16. The traceplots of the parameters of a blockwise Metropolis-Hastings chain look much more like those for a random-walk chain than those for an independent chain.

Gibbs Sampling Chain

Despite the fact that every candidate will be accepted, the Gibbs sampling chain moves through the parameter space similarly to the blockwise Metropolis-Hastings chain. This is shown in Figure 6.19 and the traceplots in Figure 6.20.

Recommendation

From the above discussion we recommend that the Metropolis-Hastings algorithm be used with an independent candidate generating distribution. This is particularly helpful when the parameters are correlated. The candidate distribution should have a similar shape to the target, but heavier tails in all directions. Because it is close to the target, many candidates will be accepted, and since it dominates the target in the tails large moves are possible. The chain will have good mixing properties. In the next section we will look at a way to find a suitable candidate density that has those properties.

7.2 FINDING A HEAVY-TAILED MATCHED CURVATURE CANDIDATE DENSITY

We want to find a candidate density that has a similar shape to the target density near its mode, but has heavier tails. This is not entirely straightforward when all we have is the unscaled target. This gives us the formula, but not the scale factor to make it an exact density. Without knowing the scale factor, we can't calculate measures of spread, nor tail probabilities directly. What we will do is find the mode of the posterior, and the curvature at the mode. The inverse of the curvature of the target at its mode gives us the variance of the normal density that has the same mode and curvature at the mode as the target. While this normal density is very similar shape to the target close to the mode, it may not be very similar as you get into the tails. The curvature of the target density at the mode is a local property of the target. So we are defining the global property of our candidate density (variance) from a local property (curvature at the mode) of the target. That is a dangerous assumption to use since the resulting density may have much lighter tails than the target. So instead of using the matched curvature normal, we will use the corresponding *Student's t* with low degrees of freedom. This candidate density will dominate the target in the tails. We will show how to do this for a single parameter. The procedure for a multivariate parameter is shown in the Appendix at the end of this chapter.

Procedure for a Single Parameter

A function and its logarithm both have their modes at the same value. Since it is easier to use logarithms, we let the logarithm of the target be

$$l(\theta|\mathbf{y}) = \log(g(\theta|\mathbf{y})) .$$

Let the starting value be θ_0. We find the mode by iterating through the following steps until convergence. Let $n = 1$.

1. At step n, we find the first two derivatives of $l(\theta|\mathbf{y})$ and evaluate them at θ_{n-1}.
2. The quadratic function $f(\theta) = a\theta^2 + b\theta + c$ that matches the values of the first two derivatives of $l(\theta|\mathbf{y})$ at the value θ_{n-1} is found by letting

$$a = \frac{1}{2}\left(\left. \frac{d^2 l(\theta|\mathbf{y})}{d\theta^2} \right|_{\theta=\theta_{n-1}} \right), \tag{7.1}$$

$$b = \left(\left. \frac{d\, l(\theta|\mathbf{y})}{d\theta} \right|_{\theta=\theta_{n-1}} \right) - \left(\left. \frac{d^2 l(\theta|\mathbf{y})}{d\theta^2} \right|_{\theta=\theta_{n-1}} \right) \times \theta_{n-1}, \tag{7.2}$$

and

$$c = l(\theta_{n-1}|\mathbf{y}) - a\,\theta_{n-1}^2 - b\,\theta_{n-1}\,. \tag{7.3}$$

Note, the logarithm of a *normal* density will give us a quadratic function.

3. We let θ_n be the value that is the mode of the quadratic. This value is given by

$$\begin{aligned} \theta_n &= \frac{-b}{2a} \qquad (7.4) \\ &= \theta_{n-1} - \frac{\left(\left.\frac{d\,l(\theta|\mathbf{y})}{d\theta}\right|_{\theta=\theta_{n-1}}\right)}{\left(\left.\frac{d^2 l(\theta|\mathbf{y})}{d\theta^2}\right|_{\theta=\theta_{n-1}}\right)}\,. \end{aligned}$$

4. If $\theta_n = \theta_{n-1}$, then we let the mode $\hat{\theta} = \theta_n$ and stop, else we let $n = n + 1$ and return to step 1.

This algorithm is known as the Gauss-Newton algorithm. It converges to the mode provided the initial starting value is close to the true value. The usual modifications required when we don't know whether or not the initial value is near the true value is to only take a small proportion of the suggested correction at each step. This is discussed in Jennrich (1995). The curvature of $l(\theta|\mathbf{y})$, the logarithm of the target density, is given by

$$\frac{\frac{d^2 l(\theta|\mathbf{y})}{d\theta^2}}{\left(1 + \left(\frac{d\,l(\theta|\mathbf{y})}{d\theta}\right)^2\right)^{\frac{3}{2}}}\,.$$

At the mode $\hat{\theta}$, the first derivative equals 0, so the curvature at the mode equals the second derivative evaluated at the mode. When we exponentiate the quadratic that matches the curvature of $l(\theta, \mathbf{y})$ at the mode $\hat{\theta}$, we get the *normal*(m, s^2) density with $m = \hat{\theta}$ and curvature that matches the curvature of the target at the mode $\hat{\theta}$. The curvature of a *normal*(m, s^2) density at the mode m equals $-\frac{1}{s^2}$, the negative reciprocal of the variance. Thus we have found the variance of the *normal* density that has mode equal to the mode of the target density, and matches its curvature at the mode.

In general, this matched curvature normal will not be a suitable candidate density, because we determined a global property of the density (the variance) from a local property (the curvature of the density at a single point). The spread may be much too small. However, it will provide us the basis for finding a suitable density.

Example 10 *Suppose the unscaled target density is given by*

$$g(\theta|\mathbf{y}) \propto \theta^{-4}\, e^{-\frac{5}{2\theta}}\,.$$

Table 7.1 Summary of first twenty steps of the Gauss-Newton algorithm.

n	θ_{n-1}	$l(\theta_{n-1})$	$\frac{dl(\theta\|\mathbf{y})}{d\theta}$	$\frac{d^2l(\theta\|\mathbf{y})}{d\theta^2}$	a	b	c	θ_n
1	.5000	–2.2274	2.0000	–24.000	–12	14	–6.2274	.5833
2	.5833	–2.1297	.4898	–13.434	–6.7172	8.3265	–4.7012	.6198
3	.6198	–2.1201	.0542	–10.588	–5.2939	6.6165	–4.1874	.6249
4	.6249	–2.1200	.0009	–10.246	–5.1228	6.4036	–4.1211	.6250
⋮	⋮	⋮	⋮	⋮	⋮	⋮	⋮	⋮
20	.6250	–2.1200	.0000	–10.240	–5.1200	6.4000	–4.1200	.6250

We decide to let the initial value $\theta_0 = .5$ and the step $n = 1$. We evaluate the log of the unscaled target, its derivative, and its second derivative at the current value θ_{n-1}. We find the quadratic $y = a\,\theta^2 + b\,\theta + c$ that is tangent to the unscaled target and matches its curvature at θ_0 using Equations 7.1–7.3 to solve for a, b, and c, respectively. The next value θ_1 will be where the quadratic achieves its maximum. It is found using Equation 7.4. Then we let $n = n + 1$ and repeat this process until we have converged. Figure 7.1 shows the first four steps of this process. The graphs on the left side are the log *$target$ and the quadratic that matches the first two derivatives of the* log *$target$ at the current value. The graphs on the right side are the $target$ and the* normal *found by exponentiating the quadratic. We see that the values are converging. Table 7.1 gives the intermediate calculations of the first four steps, and the value it converges to. Note, the matched curvature normal has mean given by*

$$\begin{aligned} m &= -\frac{b}{2a} \\ &= .625 \end{aligned}$$

and variance given by

$$\begin{aligned} s^2 &= -\left[\frac{d^2 l(\theta|y)}{d\theta^2}\right]^{-1} \\ &= .0976563\,. \end{aligned}$$

We see that the matched curvature normal is heavier on the lower tail, but it is much lighter on the upper tail. We shall see that this will make it a very bad candidate distribution.

Warning: the candidate distribution must dominate the target. It is very important that the independent candidate distribution dominate the target. That means it must have heavier tails than the target. The acceptance probability for the independent chain is given by

$$\alpha(\theta, \theta') = min\left[1, \frac{g(\theta'|y)}{g(\theta|y)} \times \frac{q_2(\theta)}{q_2(\theta')}\right].$$

It is made up from two opposing tendencies. The first term $\frac{g(\theta'|y)}{g(\theta|y)}$ makes the chain want to go uphill with respect to the target. The second term $\frac{q_2(\theta)}{q_2(\theta')}$ makes the chain want to go downhill with respect to the candidate density. (Alternatively, we could say that it does not want to go uphill with respect to the candidate density.) When the candidate density has shorter tails than the target it is very unlikely to draw a candidate that is downhill due to the short tail of the candidate density, that part of the target density will be very underrepresented in the sample.

Example 10 (continued) *Suppose we use the matched curvature normal candidate density. Figure 7.2 shows the traceplot and the histogram for the first 1000 draws from the chain, together with the target density. Although the chain appears to be moving through the parameter space fairly quickly, it is not in fact moving through the whole parameter space. We see that the upper tail of the target density is severely underrepresented in the sample. This shows the short-tailed matched curvature normal candidate density is a bad choice. (We could have seen the matched curvature normal does not dominate in the upper tail by looking at the left graph on the fourth row of Figure 7.1.) Any inferences done with a sample from a chain using this candidate density will be incorrect because the upper tail of the posterior will not be represented.*

Use Student's t candidate density that is similar to the matched curvature normal. The way forward is to find a candidate density that is similar shape to the matched curvature normal, but with heavier tails than the target. The *Student's t* with low degrees of freedom will work very well. The Student's t with 1 degree of freedom will dominate all target distributions.

Example 10 (continued) *Suppose we use the Student's t density with 1 degrees of freedom as the candidate density. The shape of the candidate density is*

$$q_2(\theta) \propto \left[1 + \left(\frac{\theta - m}{s}\right)^2\right]^{-1}$$

given by Lee (2004). Figure 7.3 shows the log of the candidate density and the log of the target on the left and the candidate and the target on the right. We see in the right panel that the candidate density dominates the target as its asymptote is moving above that for the target in the upper tail. Figure 7.4 shows the traceplot and the histogram for the first 1000 draws from the chain, together with the target density. We see that the chain is moving through the whole parameter space very quickly, and all regions of the target are represented.

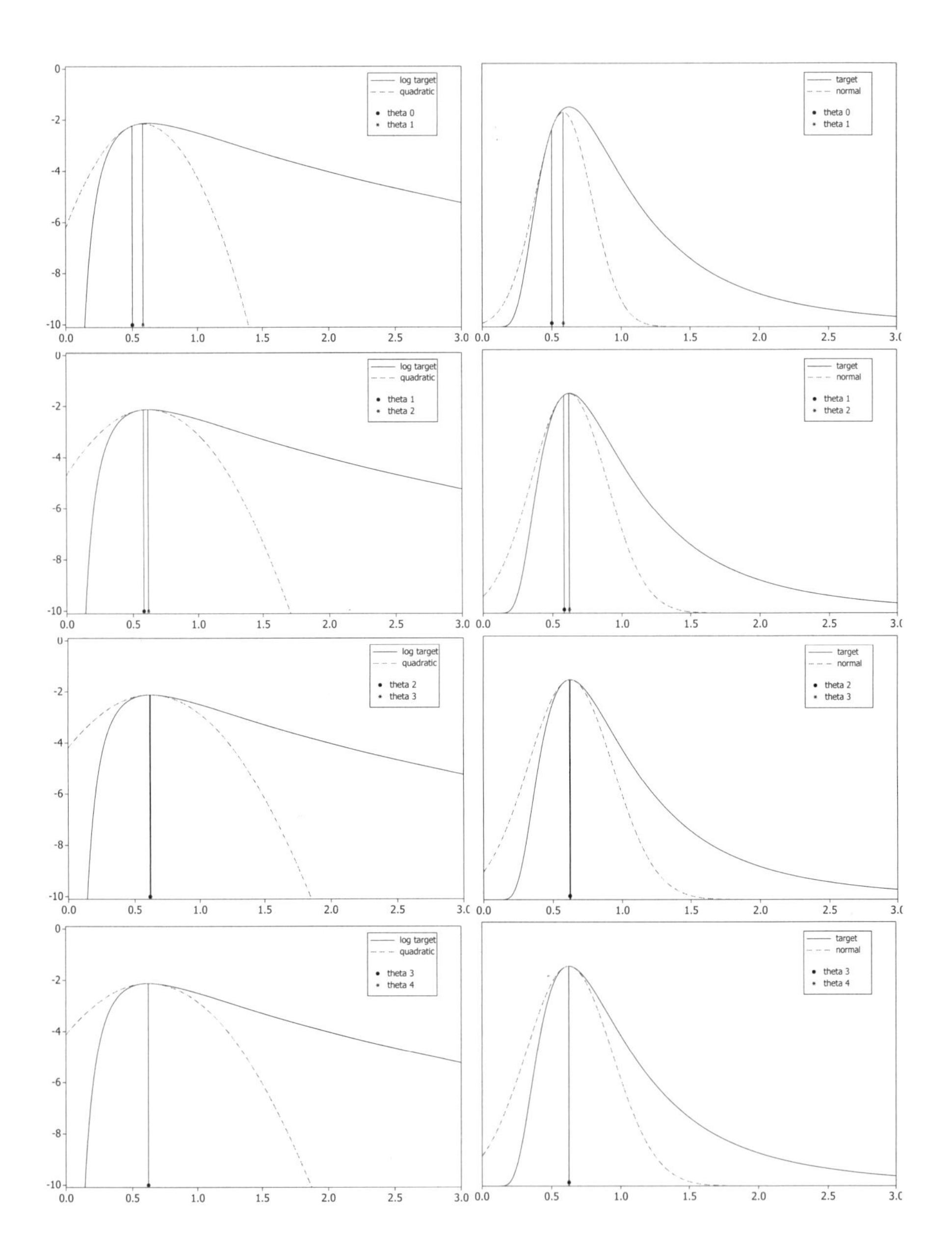

Figure 7.1 Four steps of the Gauss-Newton algorithm showing convergence to matched-curvature normal distribution. The left-hand panels are the logarithm of the target and the quadratic that matches the first two derivatives at the value θ_{n-1}, and the right-hand panels are the corresponding target and normal approximation.

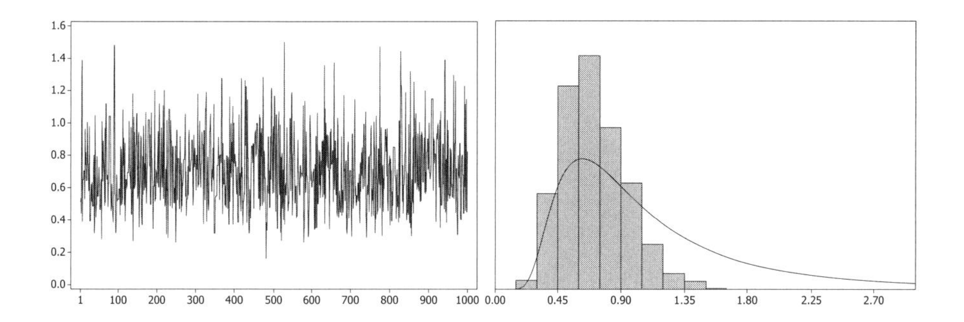

Figure 7.2 The traceplot and the histogram (with the target density) for the first 1000 draws from the chain using the matched curvature normal candidate distribution.

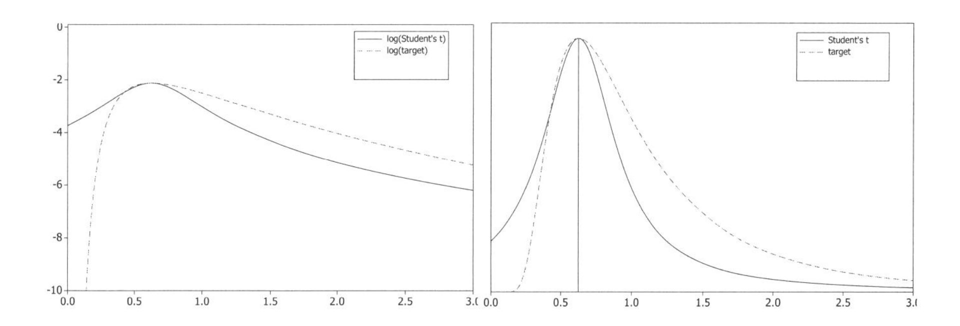

Figure 7.3 The log of the target with the log of the matched curvature *Student's t* with 1 degree of freedom in the left panel. The right panel contains the target density and the matched curvature *Student's t* density with 1 degrees of freedom.

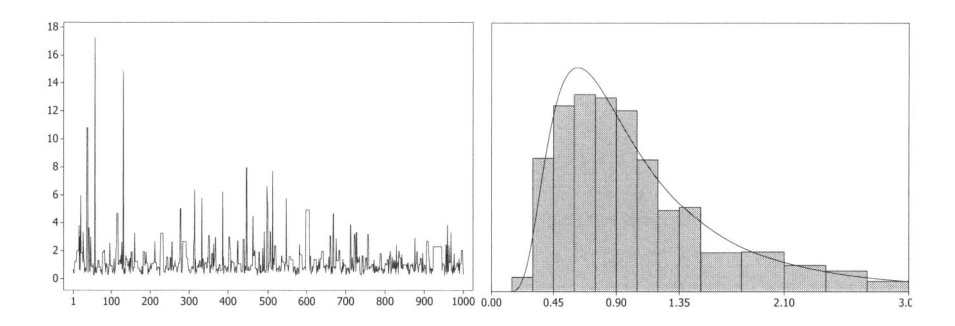

Figure 7.4 The traceplot and the histogram (with the target density) for the first 1000 draws from the chain using the matched curvature *Student's t* candidate distribution.

7.3 OBTAINING AN APPROXIMATE RANDOM SAMPLE FOR INFERENCE

Markov chain Monte Carlo samples are not independent random samples. This is unlike the case for samples drawn directly from the posterior by acceptance-rejection sampling. This means that it is more difficult to do inferences from the Markov chain Monte carlo sample. In this section we discus the differing points of view on this problem.

The burn-in time is the number of steps needed for the draw from the chain to be considered a draw from the long-run distribution. In other words, after the burn-in time the draw from the chain would be the same, no matter where the chain started from. Since we have set up the Markov chain to have the posterior as its long-run distribution, the draw is like a random draw from the posterior. However, consecutive draws from the sample continuing after the burn-in time will not be a random sample from the posterior. Subsequent values will be dependent on the first value drawn after burn-in.

Example 6 (continued) *We showed the second Markov chain, which has transition matrix*

$$\boldsymbol{P_2} = \begin{bmatrix} 0.499850 & 0.499850 & 0.000100 & 0.000100 & 0.000100 \\ 0.299850 & 0.699850 & 0.000100 & 0.000100 & 0.000100 \\ 0.000150 & 0.000150 & 0.199900 & 0.299900 & 0.499900 \\ 0.000150 & 0.000150 & 0.499900 & 0.199900 & 0.299900 \\ 0.000150 & 0.000150 & 0.299900 & 0.499900 & 0.199900 \end{bmatrix},$$

has a burn-in time of about $n = 2^{16}$ steps. Figure 5.2 shows the state history for 1000 steps of the chain after the burn-in time. We see this is extremely far from the true posterior. There is a group of states that has been missed completely. This demonstrates that the subsequent values of the chain are strongly dependent on the first draw after burn-in.

One school of thought considers that you should use all draws from your Markov chain Monte Carlo sample after the burn-in time. All the draws from the chain after that can be considered to be draws from the posterior, although they are not independent from each other. Using the ergodic theorems, the time-average of a single realization from the Markov chain approaches the average of all possible realizations of the chain, which is the mean of the long-run distribution. This school considers that to throw away any values from the Markov chain Monte Carlo sample is throwing away information. However, it is not clear how good this estimate would be. It is certainly not as accurate as you would get with a random sample from the posterior of the same sample size. Other inferences such as hypothesis tests or credible intervals found from this non-random sample would also be suspect since the tail proportions might not be as good estimates of tail probabilities of the true posterior as we would think they are, because of the dependency.

We advocate getting an approximately random sample from the posterior as the basis for inference. There are two ways this can be done. The first way is to run the Markov chain to the burn-in period. Then we draw that value which will be a random draw from the posterior. Then we run a separate run of the Markov chain again until the burn-in time, and then draw that value for the second random draw from the posterior. We repeat the process as many times as we need to get the size random sample we need for the posterior. The second way is to thin the Markov chain Monte Carlo sample to enough degree that the sample is approximately a random one. To do this, we have to thin enough so the next draw does not depend on the last one. This is essentially the same idea as the burn-in. Thus the gap until the next thinned value should be the same as the burn-in time. This way the thinned Markov chain Monte Carlo sample will be approximately a random sample from the posterior. Then we will do our inferences using the methods developed in Chapter 3 on the thinned, approximately random, sample.

Advocates of the first point of view would consider this to be very inefficient. We are discarding a number of draws equal to the burn-in time for each value into the random sample. They would say this is throwing away information. However, it is not data that is being discarded, rather it is computer-generated Markov chain Monte Carlo samples. If we want more, we can get them by running the Markov chain longer.

Estimates found using all draws after the burn-in would be much less precise than their sample size would suggest. This is due to the serial dependency of the draws from the Markov chain. The exact precision of the estimates would not be easily determined. To get estimates with precision similar to those from the approximate random sample would require the chain to be run almost as long as we would have to run the chain to get the approximately random sample. The values that were thinned out wouldn't be adding very much to the precision of the estimate. For all these reasons, we advocate getting an approximate random sample from the posterior to base our inference. Then the methods for putting error bounds on the inferences based on sample sizes will be justified.

Determining the Burn-in Time

Unfortunately, there is no exact way to determine how long a burn-in period is required from the output of the Markov chain. If there are some regions of high posterior probability, but widely separated by regions of low probability, the chain will generally spend a long time in each high-probability region before by chance, it moves into another high-probability region. Then it will generally be another long time before it gets back into the original region.

Examining sample autocorrelation function of the Markov chain Monte Carlo Sample. In Section 7.1, we saw that the traceplots of the Markov chain output are different for the different types of candidate distributions. Using a random-walk candidate distribution leads to a high proportion of accepted candidates, but relatively small moves at each step. This can lead to high autocorrelations at low lags. With an independent candidate distribution, there is a much smaller proportion of accepted candidates, but relatively large moves. Since many candidates are not accepted, the chain may stay at the same point for many steps, which also may lead to high autocorrelations at low lags. Examination of the sample autocorrelation function of the Markov chain output is a useful tool for helping to determine a suitable burn-in time. We look at it to see how many steps are needed for the autocorrelations to be statistically indistinguishable from zero. Bartlett (1946) showed that if the autocorrelations have gone to zero above lag q, the the variance of the autocorrelation at lag $k > q$ is approximately

$$Var(r_k) = \frac{1}{n}\left(1 + 2\sum_{i=1}^{q} \rho_k\right).$$

This is shown in Abraham and Ledolter (1983). Usually 95% bounds are shown on the sample autocorrelation function, and those within those bounds are assumed to be indistinguishable from zero. Occasional autocorrelations outside those bounds at high lags are not meaningful. The probability is .95 that any individual sample autocorrelation is within the bounds. This allows an occasional autocorrelation to be outside these bounds due to random chance alone. Even a random sample from an extremely heavy-tailed target density will show occasional extremely significant sample autocorrelations at high lags due to the number of steps that have occurred between extreme values just due to chance. These of course would be meaningless since they are just due to random chance.

Example 8 (continued) *Suppose we take a Markov chain Monte Carlo sample from the density given by*

$$g(\theta|y) = .8\, e^{-\frac{1}{2}\theta^2} + .2\, e^{-\frac{1}{2\times 2^2}(\theta-3)^2}.$$

using the normal$(0, 1^2)$ *random-walk candidate density. The autocorrelation function for the Markov chain Monte Carlo sample drawn from this chain is shown in the left graph of Figure 7.5 along with the 95% bounds. The sample autocorrelations have become indistinguishable from zero by lag 50. This indicates a burn-in period of 50, and thinning by including every 50*th *draw into the sample will give us an approximately random sample from the posterior.*

Suppose we decide to use the normal$(0, 3)$ *independent candidate density. The autocorrelation function for the Markov chain Monte Carlo sample drawn from this chain together with the 95% bounds are shown in the right graph of Figure 7.5b. We see the autocorrelations have become indistinguishable from zero by lag 10. This indicates a burn-in time of 10, and thinning by including every 10*th *draw into the final sample will give us an approximately random sample from the posterior. This again shows that a Metropolis-Hastings algorithm is more efficient when a well chosen independent candidate density is used than when a random-walk candidate density is used.*

Unfortunately, examination of the sample autocorrelation function is not a foolproof way for determining convergence to the posterior. If there are widely separated regions of high probability in the posterior the chain may have completely missed

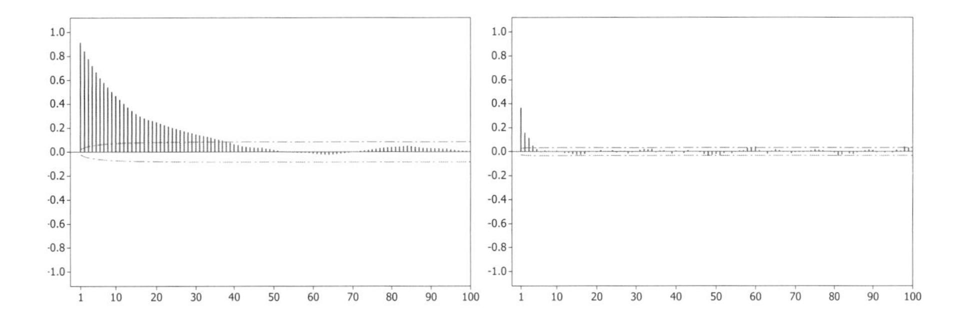

Figure 7.5 The sample autocorrelation function for the random-walk chain and the independent chain, respectively.

one of the regions. The sample autocorrelation function would look like it has gone to zero, but for a larger sample that did visit all the regions would have a sample autocorrelation function that takes a much longer time to become statistically indistinguishable from zero.

Gelman and Rubins potential improvement statistic. Gelman and Rubin (1992) developed a method for determining $\sqrt{\hat{R}}$, the potential improvement possible for a single parameter θ by increasing the burn-in time. This is an analysis of variance type approach using multiple chains started independently from an overdispersed starting distribution. If the chains have converged, the within-chain variance and the between-chain variance are estimating the same thing. It proceeds using the following steps.

1. Independently simulate m independent realizations of the Markov chain, each started independently from an overdispersed starting distribution. They suggest finding all the modes of the target distribution, and letting the starting distribution be approximated by a mixture of *Student's t* distributions, one centered at each node. Each realization of the chain should be $2n$ steps. We discard the first n steps of each chain as the proposed burn-in time.

2. Calculate the between-chain variance

$$\frac{B}{n} = \frac{\sum_{i=1}^{m} (\bar{x}_{i.} - \bar{x}_{..})^2}{m}$$

and the average of all the within-chain variances

$$W = \frac{\sum_{i=1}^{m} s_i^2}{m}$$

where s_i^2 is the variance from the last n steps of the ith chain.

3. The estimated variance equals

$$\hat{V} = \left(1 - \frac{1}{n}\right) \times W + \frac{1}{n} \times B\,.$$

4. The estimated scale reduction equals

$$\sqrt{\hat{R}} = \sqrt{\frac{\hat{V}}{W}}.$$

For values of n that are less than an adequate burn-in time, V should overestimate the variance of the target distribution since we start with overdispersed starting points. W will underestimate the variance.

The Gelman-Rubin statistic indicates the amount of improvement towards convergence that is possible if we allowed n to increase. Values of $\sqrt{\hat{R}}$ that are less than 1.10 show acceptable convergence. It should be noted that the value of the Gelman-Rubin statistics depends on the actual draws that occurred in the chains. It would give a different value if we repeated the process, so it is only indicative of what the burn-in time and thinning should be.

Example 8 (continued) *The target density is the mixture of a normal$(0, 1^2)$ and a normal$(3, 2^2)$ given by*

$$g(\theta|y) = .8 \times e^{-\frac{1}{2}\theta^2} + .2 \times \frac{1}{2} e^{-\frac{1}{2\times 2^2}(\theta-3)^2}.$$

We run 10 Metropolis-Hastings chains with random-walk normal$(0, 1^2)$ candidate density, each starting from a different point. We let $n = 50$ so run the chains 100 steps, and calculate the Gelman-Rubin statistic. The value is $\sqrt{\hat{R}} = 1.049$. This shows acceptable convergence after 50 draws.

We run 10 Metropolis-Hastings chains with independent normal$(0, 3^2)$ candidate densities starting from overdispersed starting points. We let $n = 10$ so run the chain for 20 steps. The value of the Gelman-Rubin statistic is $\sqrt{\hat{R}} = 1.014$, which indicates acceptable convergence after 10 draws. This shows the acceptable convergence after 10 steps. Again we see that the Metropolis-Hastings chain using a well-chosen independent candidate density converges faster than a Metropolis-Hastings chain with a random-walk candidate density.

Coupling with the past. Prop and Wilson (1996) and (1998) developed a method they refer to as coupling with the past, which can be used to determine an adequate burn-in time for the Markov chain. The chain will have converged to the target when the distribution at step n no longer depends on the state it started in at step 0. Prop and Wilson suggested starting several parallel chains at overdispersed starting points, and then giving them the same inputs and observing how many steps are required until they all converge. We would consider this to be a satisfactory burn-in time. That draw can be considered a random draw from the posterior. If the sample is found by repeating this process over and over the sample is called a perfect random sample from the posterior. When we are using parallel Metropolis-Hastings chain with random-walk candidate densities given the same inputs, the chains move in parallel when they accept on the same draw. They only get closer when some of the chains

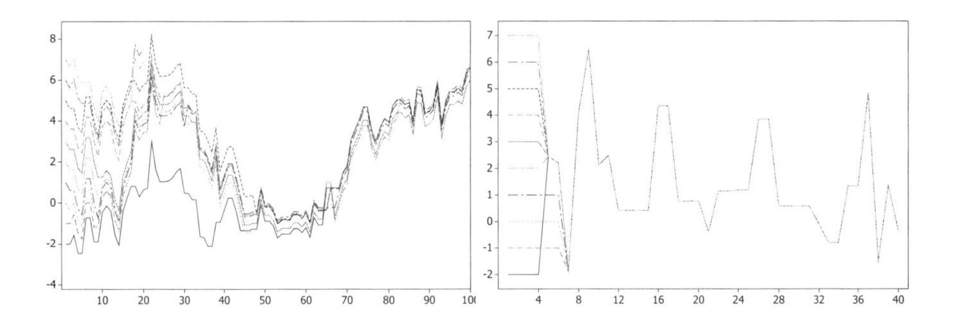

Figure 7.6 The outputs from multiple parallel Metropolis-Hastings chains with the same inputs for the random-walk chain and the independent chain, respectively.

accept values closer to the current values of the other chains, and the other chains don't accept and stay at their current values. This means the chains never converge to the same exact value, rather they get to within a suitable margin. On the other hand, when we are using parallel Metropolis-Hastings chains with an independent candidate densities given the same inputs, when the values are all accepted the chains have converged exactly to the same values.

Example 8 (continued) *The target density is the mixture of a normal*$(0, 1^2)$ *and a normal*$(3, 2^2)$ *given by*

$$g(\theta|y) = .8 \times e^{-\frac{1}{2}\theta^2} + .2 \times \frac{1}{2} e^{-\frac{1}{2\times 2^2}(\theta-3)^2} .$$

We run ten parallel Metropolis-Hastings chains from different starting positions, but with the same random inputs for 100 draws. First we used a random-walk chain, which chooses each normal candidate with standard deviation $\sigma = 1$ *centered around the current value. We observe how long it takes until all the parallel chains are converged to within a tolerance. The results are shown in the left side of Figure 7.6. We see that by around 50 steps, all have converged to a band. After that, they don't converge very much more. This indicates that a burn-in of 50 and thinning the chain by using every 50*th *draw would give approximately a random sample from the target.*

Next we used a Metropolis-Hastings chain with an independent normal$(0, 3^2)$ *candidate density. Again we run ten parallel chains from different starting positions, but with the same random inputs. The first 40 draws are shown in the right side of Figure 7.6. We see that all these chains have converged after 8 steps. This indicates that a burn-in of 10 and thinning the chain by using every* 10th *draw would give approximately a random sample from the target. Once again, the independent Metropolis-Hastings chain is shown to converge more quickly than the random-walk Metropolis-Hastings chain.*

All of the methods we have used to determine the burn-in time and hence the thinning required to get an approximate random sample are useful, however, none of them are foolproof. They all can give misleading results when there are widely separated multiple modes in the target density. Blindly following these methods can lead to finding a burn-in time that is actually much shorter than the real burn-in time that should be required. It is important if following these methods, that the modes should be found first, and the multiple starting points include some from a overdispersed distribution centered around each mode. Fortunately, many of the main statistical models including the logistic regression model, the Poisson regression model, the proportional hazards model, and the hierarchical mean model with covariates do not have multiple modes.

Main Points

- The appearance of the traceplot of a Metropolis-Hastings chain depends on the type of candidate density used.
- A random-walk Metropolis-Hastings chain will have many candidates accepted, but the moves will be small. This will lead to trace plot where visits to tails are infrequent.
- An independent Metropolis-Hastings chain will have fewer candidates accepted, but the moves may be large. This will lead to trace plot moving through the whole space more quickly.
- When the parameters are strongly correlated, a blockwise Metropolis-Hastings chain will move slowly through the parameter space. The trace plots will look similar to those from a random-walk chain.
- A Gibbs sampling chain is a special case of a blockwise Metropolis-Hastings chain. When the parameters are strongly correlated, the traceplots will look similar to those from a random-walk chain.
- An independent Metropolis-Hastings chain with a candidate distribution similar to the target but with heavier tails than the target will have the best mixing properties.
- A heavy-tailed matched curvature candidate density can be found by matching the mode and curvature at the target with a normal, then replacing the normal with a *Student's t* with degrees of freedom equal to 1.
- We need a random sample from the posterior to use for inference. We obtain this by discarding the first n draws, the burn-in time, then thinning the chain by taking every n^{th} draw.
- It is hard to determine the adequate burn-in time. Examination of the trace plots and examination of the sample autocorrelation functions can be helpful. However, when there are widely separated modal regions, the chain may have

completely missed some part of the parameter space, and this would not be apparent on either the trace plots or the sample autocorrelations at all.

- Gelman-Rubin statistic is also useful to determine the burn-in period, but it is also not infallible. It is an analysis of variance approach where several parallel chains are run, and the between-chain variation is compared to the within-chain variation. The Gelman-Rubin statistic shows the improvement possible by letting the chain run longer. When it is approximately 1, no further improvement is possible and the chain is approximately at the long-run distribution.

- Coupling with the past is another method that can be used to determine the burn-in time. It starts a series of parallel chains from overdispersed starting points and uses the same random input values until all chains have converged. The number of steps this takes is adequate burn-in time as it shows the state of the chain is independent of the starting point.

Exercises

7.1 Suppose the unscaled target density has shape given by

$$g(\theta|\mathbf{y}) \quad \propto \quad \theta^{-6} e^{-\frac{7}{2\theta}} \,.$$

Let the starting value be $\theta_0 = .5$ and go through six steps of the Gauss-Newton algorithm to find the mode of the target density and the matched curvature variance.

7.2 Let the unscaled posterior have shape given by

$$g(\theta|\mathbf{y}) \propto .6 \times e^{-\frac{1}{2}(\theta^2)} + .4 \times \frac{1}{2} e^{-\frac{1}{2\,2^2}(\theta-3)^2} \,.$$

This is a mixture of two normal distributions.

(a) Let $n = 25$. Run six Markov chains $2n$ steps each using the Minitab macro *NormMixMHRW.mac* or the R-function `NormMixMH`. Use a *normal*$(0, 1^2)$ random-walk candidate distribution. Then calculate the Gelman-Rubin statistic using the Minitab macro *GelmanRubin.mac* or the R-function `GelmanRubin`.

(b) Repeat where $n = 50$.

(c) Repeat where $n = 100$.

7.3 Let the unscaled posterior have the same shape as the previous problem.

(a) Let $n = 25$. Run six Markov chains $2n$ steps each using the Minitab macro *NormMixMHInd.mac* or the R-function `NormMixMH`. Use a *normal*$(0, 3^2)$ independent candidate distribution. Then calculate the Gelman-Rubin statistic using the Minitab macro *GelmanRubin.mac* or the R-function `GelmanRubin`.

(b) Repeat where $n = 50$.

(c) Repeat where $n = 100$.

7.4 Let the unscaled posterior have the same shape as the previous problem. Run six parallel Metropolis-Hastings chains starting from overdispersed starting points but with the same random input. Let each chain have a *normal*$(0, 1^2)$ random-walk candidate density. Run each chain 100 steps. Use the Minitab macro *NormMixMHRW.mac* or the R-function `NormMixMH`. Graph the traceplots of all the chains on the same graph to see how long it takes until they converge together.

7.5 Let the unscaled posterior have the same shape as the previous problem. Run six parallel Metropolis-Hastings chains starting from overdispersed starting points but with the same random input. Let each chain have a *normal*$(0, 3^2)$ independent candidate density. Run each chain 20 steps. Use the Minitab macro *NormMixMHInd.mac* or the R-function `NormMixMH`. Graph the traceplots of all the chains on the same graph to see how long it takes until they converge together.

7.6 What conclusions about Metropolis-Hastings chains with independent candidate density compared to Metropolis-Hastings chain with a random-walk candidate density can you draw from Exercises 7.2–7.5.

Appendix: Procedure for Finding the Matched Curvature Candidate Density for a Multivariate Parameter

Let the logarithm of the target density be

$$l(\boldsymbol{\theta}|\mathbf{y}) = \log(g(\boldsymbol{\theta}|\mathbf{y}))\,.$$

Let the starting value for the parameter be

$$\boldsymbol{\theta}_0 = \begin{pmatrix} \theta_{1\,0} \\ \vdots \\ \theta_{p\,0} \end{pmatrix}.$$

We will find the mode of the multivariate target by iterating through the following steps until convergence. Let $n = 1$.

1. At step n we evaluate all of the first two derivatives of $l(\boldsymbol{\theta}|\mathbf{y})$ at $\boldsymbol{\theta}_{n-1}$.

2. The second order equation that matches the values of the first two derivatives of $l(\boldsymbol{\theta}|\mathbf{y})$ at $\boldsymbol{\theta}_{n-1}$ is given by

$$f(\boldsymbol{\theta}) = \boldsymbol{\theta}'_{n-1}\mathbf{A}\boldsymbol{\theta}_{n-1} + \boldsymbol{\theta}'_{n-1}\mathbf{b} + c$$

where

$$2\,\mathbf{A} = \begin{bmatrix} \frac{\partial^2 l(\boldsymbol{\theta}|\mathbf{y})}{\partial\theta_1^2} & \cdots & \frac{\partial^2 l(\boldsymbol{\theta}|\mathbf{y})}{\partial\theta_1\theta_p} \\ \vdots & \vdots\vdots\vdots & \vdots \\ \frac{\partial^2 l(\boldsymbol{\theta}|\mathbf{y})}{\partial\theta_p\theta_1} & \cdots & \frac{\partial^2 l(\boldsymbol{\theta}|\mathbf{y})}{\partial\theta_1^2} \end{bmatrix}_{\boldsymbol{\theta}=\boldsymbol{\theta}_{n-1}},$$

$$\mathbf{b} = \begin{pmatrix} \frac{\partial l(\boldsymbol{\theta}|\mathbf{y})}{\partial\theta_1} \\ \vdots \\ \frac{\partial l(\boldsymbol{\theta}|\mathbf{y})}{\partial\theta_p} \end{pmatrix}_{\boldsymbol{\theta}=\boldsymbol{\theta}_{n-1}} - \mathbf{A}\boldsymbol{\theta}_{n-1} - \boldsymbol{\theta}'_{n-1}\mathbf{A}\,,$$

and

$$c = l(\boldsymbol{\theta}_{n-1}|\mathbf{y}) - \boldsymbol{\theta}'_{n-1}\mathbf{A}\boldsymbol{\theta}_{n-1} - \boldsymbol{\theta}'_{n-1}\mathbf{b}\,.$$

Note: the logarithm of a *multivariate normal* density will give us a second order equation.

3. We let $\boldsymbol{\theta}_n$ be the mode of the second order equation. It is given by

$$\boldsymbol{\theta}_n = -\frac{\mathbf{A}^{-1}\mathbf{b}}{2} \tag{A.1}$$

$$= \boldsymbol{\theta}_{n-1} - \begin{bmatrix} \frac{\partial^2 l(\boldsymbol{\theta}|\mathbf{y})}{\partial\theta_1^2} & \cdots & \frac{\partial^2 l(\boldsymbol{\theta}|\mathbf{y})}{\partial\theta_1\theta_p} \\ \vdots & \vdots\vdots\vdots & \vdots \\ \frac{\partial^2 l(\boldsymbol{\theta}|\mathbf{y})}{\partial\theta_p\theta_1} & \cdots & \frac{\partial^2 l(\boldsymbol{\theta}|\mathbf{y})}{\partial\theta_1^2} \end{bmatrix}^{-1}_{\boldsymbol{\theta}=\boldsymbol{\theta}_{n-1}} \begin{pmatrix} \frac{\partial l(\boldsymbol{\theta}|\mathbf{y})}{\partial\theta_1} \\ \vdots \\ \frac{\partial l(\boldsymbol{\theta}|\mathbf{y})}{\partial\theta_p} \end{pmatrix}_{\boldsymbol{\theta}=\boldsymbol{\theta}_{n-1}}.$$

4. If $\boldsymbol{\theta}_n = \boldsymbol{\theta}_{n-1}$ then the mode $\hat{\boldsymbol{\theta}} - \boldsymbol{\theta}_n$ and we stop, else we let $n = n + 1$ and return to step 1.

This multivariate version of the Gauss-Newton algorithm converges to the mode provided the initial starting value is close to the true value. If it is not converging modify the algorithm by only taking a small proportion of the suggested correction at each step. The curvature of the log-likelihood $l(\boldsymbol{\theta}|\mathbf{y})$ at the mode is given by second derivative matrix evaluated at the mode. When we exponentiate the second degree equation that matches the mode $\hat{\boldsymbol{\theta}}$ and the curvature at the mode we get the matched curvature *multivariate normal*$[\mathbf{m}, \mathbf{V}]$ where

$$\mathbf{m} = \hat{\boldsymbol{\theta}} \quad \text{and} \quad \mathbf{V} = \begin{bmatrix} \frac{\partial^2 l(\boldsymbol{\theta}|\mathbf{y})}{\partial\theta_1^2} & \cdots & \frac{\partial^2 l(\boldsymbol{\theta}|\mathbf{y})}{\partial\theta_1\theta_p} \\ \vdots & \vdots\vdots\vdots & \vdots \\ \frac{\partial^2 l(\boldsymbol{\theta}|\mathbf{y})}{\partial\theta_p\theta_1} & \ddots & \frac{\partial^2 l(\boldsymbol{\theta}|\mathbf{y})}{\partial\theta_1^2} \end{bmatrix}^{-1}_{\boldsymbol{\theta}=\hat{\boldsymbol{\theta}}}.$$

This matched curvature *multivariate normal* candidate density does not really have any relationship with the spread of the target. If we used it is likely that it does not dominate the target in the tails. So, instead, we use the *multivariate Student's* $t[\mathbf{m}, \mathbf{V}]$ with low degrees of freedom. First we find the lower triangular matrix L that satisfies $\mathbf{V} = \mathbf{LL}'$ by using the Cholesky decomposition. Then we draw a random sample of *Student's t* random variables with κ degrees of freedom.

$$\mathbf{t} = \begin{pmatrix} t_1 \\ \vdots \\ t_p \end{pmatrix}.$$

Then

$$\boldsymbol{\theta} = \mathbf{m} + \mathbf{Lt}$$

will be a random draw from the *multivariate Student's* $t[\mathbf{m}, \mathbf{V}]$ distribution with κ degrees of freedom.

8

Logistic Regression

Sometimes we have a binary response with only two possible outcomes, which we call "success" and "failure," respectively. The response depends on the values of a set of predictor variables. We consider these responses to come from a sequence of independent Bernoulli trials, where each trial has its own probability of success that depends on the values of the predictor variables. We define the response variable $y_i = 1$ when the ith trial is a success, and $y_i = 0$ when it is a failure. Then y_i has the *binomial*$(1, \pi_i)$ distribution, where π_i is the probability of success for the ith trial. We want to find a regression model for predicting the successes using the predictor variables, which we also observe for each observation. To do this, we need to relate the success probability to the values of p predictor variables $x_1, \ldots, x_p$. We will find that the logistic regression model does this for us. When $p = 1$ there is only a single predictor variable and it is called the simple logistic regression model. When $p > 1$ it is called the multiple logistic regression model. In Section 8.1 we will introduce the logistic regression model, determine the maximum likelihood estimators for the regression parameters using iteratively reweighted least squares, and show why confidence intervals based on these estimates may not have the desired coverage probabilities. In Section 8.2 we will develop a computational Bayesian method for drawing a sample from the true posterior distribution for this model using the Metropolis-Hastings algorithm.

In Section 8.3 we discuss the issues we face when we are modelling using the multiple logistic regression. We investigate the issue of which predictor variables to include in the model. When we include an extraneous predictor variable that does not affect the response, we will improve the fit to the given data set, but will degrade the predictive effectiveness of the model. On the other hand, when the predictors are related, they can either mask or enhance each others' effect. Simply looking at the significance of the individual predictor is not enough, as other predictors may be masking its effect. We look at how to determine which predictor variables can be removed from the model in order to get a better prediction model for new observations.

8.1 LOGISTIC REGRESSION MODEL

We are observing a sequence of independent *Bernoulli* trials where each trial has its own probability of success. The response variable $y_i = 1$ when the ith trial is a success, and $y_i = 0$ when it is a failure. Thus y_i has the *binomial*$(1, \pi_i)$ distribution. We also observe the value of a single predictor variable x for each of the trials. We want to know if we can use the value of the predictor variable to predict the occurrences of success. In other words, we want to find a regression model for predicting the response from the value of the predictor variable. We can't do this directly by equating the mean of the response (the probability of success) to the linear function of the predictor. A linear function of the predictor variable $\beta_0 + \beta_1 x$ will range from $-\infty$ to ∞ while π, the probability of success, is always between 0 and 1. Thus we can't relate the linear function of the predictor directly to the success probability. Instead we will have to find a function of the probability of success that also ranges between $-\infty$ and ∞ that we can link to the linear function of the predictor.

Suppose we decide to use the logarithm of the odds ratio as the link function.[1] The logarithm of the odds ratio is called the *logit*. We set it equal to the linear function of the predictor

$$\log_e \left(\frac{\pi}{1-\pi} \right) = \beta_0 + \beta_1 x$$

and solve for π as a function of x, where β_0 is the y-intercept, and β_1 is the slope. We are interested in the slope parameter β_1, which determines how the logarithm of the odds ratio depends on the predictor x. If $\beta_1 = 0$ then the log of the odds ratio does not depend on x so the probability of success does not either. We usually regard the intercept β_0 as a nuisance parameter.

In multiple logistic regression we are still trying to predict the successes in a sequence of independent *Bernoulli* trials each having its own probability, but now we have p predictor variables, $x_1, \ldots, x_p$, observed for each observation. We will link the logarithm of the odds to the linear function of all the predictor variables. This will give rise to the multiple logistic regression model. We can summarize the assumptions of the the logistic regression model.

[1]There are several possible link functions that are commonly used. The *logit* link that we will use leads to the logistic regression model. The inverse of the normal cumulative distribution function $\beta_0 + \beta_1 x = \Phi^{-1}(\pi)$ is called the *probit* link and leads to the probit model. The *complementary log-log* link is given by $\beta_0 + \beta_1 x = \log_e\{-\log_e(1-\pi)\}$.

Assumptions of the Logistic Regression Model

1. The ith observation has the *binomial*$(1, \pi_i)$ distribution. Each observation has its own probability of success.

2. The *logit* is linked to the linear predictor, an unknown linear function of the predictor variables.

$$\log_e \left(\frac{\pi}{1-\pi} \right) = \beta_0 + \beta_1 x_1 + \ldots \beta_p x_p \,. \tag{8.1}$$

   Exponentiating both sides we get

$$\left(\frac{\pi}{1-\pi} \right) = e^{\beta_0 + \beta_1 x_1 + \ldots \beta_p x_p} \,.$$

   When we solve this equation for π, we get the logistic equation

$$\pi = \frac{e^{\beta_0 + \beta_1 x_1 + \ldots \beta_p x_p}}{1 + e^{\beta_0 + \beta_1 x_1 + \ldots \beta_p x_p}} \tag{8.2}$$

   which relates the probability of success to the values of the predictor variables.

3. The observations are all independent of each other.

Likelihood of the Logistic Regression Model

The likelihood of a single observation y_i is the probability of a *binomial*$(1, \pi_i)$ where π_i is a function of the $p+1$ parameters $\beta_0, \ldots, \beta_p$. It is given by

$$\begin{aligned} f(y_i|\beta_0, \ldots, \beta_p) &= \pi_i^{y_i} (1-\pi_i)^{1-y_i} \\ &= \left(\frac{\pi_i}{1-\pi_i} \right)^{y_i} (1-\pi_i) \\ &= \left(e^{\beta_0 + \beta_1 x_{i1} + \ldots \beta_p x_{ip}} \right)^{y_i} \times \left(\frac{1}{1 + e^{\beta_0 + \beta_1 x_{i1} + \ldots \beta_p x_{ip}}} \right) \end{aligned}$$

where x_{ij} is the value of the jth predictor variable for the ith observation. All the observations are independent so the joint likelihood of the sample is the product of the individual likelihoods. It is given by

$$\begin{aligned} f(y_1, \ldots, y_n|\beta_0, \ldots, \beta_p) &= \prod_{i=1}^{n} f(y_i|\beta_0, \ldots, \beta_p) \qquad (8.3) \\ &= \prod_{i=1}^{n} \left(\frac{\left(e^{\beta_0 + \beta_1 x_{i1} + \ldots + \beta_p x_{ip}} \right)^{y_i}}{1 + e^{\beta_0 + \beta_1 x_{i1} + \ldots + \beta_p x_{ip}}} \right) \\ &= e^{\beta_0 \sum y_i + \sum \beta_j \sum x_{ij} y_i} \prod_{i=1}^{n} \left(\frac{1}{1 + e^{\beta_0 + \beta_1 x_{i1} + \ldots + \beta_p x_{ip}}} \right) . \end{aligned}$$

This joint likelihood cannot be factored into the product of the individual likelihoods for $\beta_0, \ldots, \beta_p$.

Logistic Regression Using the Generalized Linear Model

Nelder and Wedderburn (1972) extended the general linear model in two ways. First, they relaxed the assumption that the observations have the normal distribution to allow the observations to come from some one-dimensional exponential family, not necessarily normal. Second, instead of requiring the mean of the observations to equal a linear function of the predictor, they allowed a function of the mean to be linked to (set equal to) the linear predictor. They named this the generalized linear model and called the function set equal to the linear predictor the link function. The logistic regression model satisfies the assumptions of the generalized linear model. They are:

1. The observations y_i come from a one-dimensional exponential family of distributions. In the logistic regression case this is the binomial distribution.

2. The mean function of the observations is linked to a linear function of the predictor variable. For logistic regression the *logit* link function is used.

3. The observations are all independent of each other.

Maximum Likelihood Estimation in the Logistic Regression Model

The frequentist approach to estimation in the logistic regression model would be to find the maximum likelihood estimators. They would be the simultaneous solutions of

$$\frac{\partial \log_e f(y_1, \ldots, y_n | \beta_0, \ldots, \beta_p)}{\partial \beta_j} = 0 \quad \text{for} \quad j = 0, \ldots, p\,.$$

In general, it may be messy to find the simultaneous solution of these equations algebraically. Nelder and Wedderburn (1972) showed that in the generalized linear model, these maximum likelihood estimators could also be found by iteratively reweighted least squares. Let the observation vector and parameter vector be

$$\mathbf{y} = \begin{pmatrix} y_1 \\ \vdots \\ y_n \end{pmatrix} \quad \text{and} \quad \boldsymbol{\beta} = \begin{pmatrix} \beta_0 \\ \vdots \\ \beta_p \end{pmatrix}$$

respectively. We let the row vector of predictor values for the ith observation be

$$\mathbf{x_i} = \begin{pmatrix} x_{i0} & x_{i1} & \ldots & x_{ip} \end{pmatrix}$$

where $x_0 = 1$ is the coefficient of the intercept. We let the matrix of predictor values for all the observations be

$$\mathbf{X} = \begin{bmatrix} 1 & x_{11} & \dots & x_{1p} \\ \vdots & \vdots & \vdots & \vdots \\ 1 & x_{n1} & \dots & x_{np} \end{bmatrix}.$$

Pawitan (2001) shows that for *logistic* regression the solution of the maximum likelihood equations can be found by iterating the following steps until convergence. Let $\boldsymbol{\beta}^{(n-1)}$ be the parameter vector at step $n-1$.

1. Since the mean and variance of a *binomial*$(1, \pi_i)$ observation are π_i and $\pi_i(1-\pi_i)$ respectively, we first update the means by

$$\mu_i = \frac{e^{\mathbf{x_i}\boldsymbol{\beta}^{(n-1)}}}{1 + e^{\mathbf{x_i}\boldsymbol{\beta}^{(n-1)}}}$$

   and the variances by

$$\begin{aligned} \Sigma_{ii} &= \frac{e^{\mathbf{x_i}\boldsymbol{\beta}^{(n-1)}}}{1 + e^{\mathbf{x_i}\boldsymbol{\beta}^{(n-1)}}} \left(1 - \frac{e^{\mathbf{x_i}\boldsymbol{\beta}^{(n-1)}}}{1 + e^{\mathbf{x_i}\boldsymbol{\beta}^{(n-1)}}}\right) \\ &= \frac{e^{\mathbf{x_i}\boldsymbol{\beta}^{(n-1)}}}{(1 + e^{\mathbf{x_i}\boldsymbol{\beta}^{(n-1)}})^2}. \end{aligned}$$

2. Then we calculate the linearized observations by

$$Y_i^{(n)} = \mathbf{x_i}\boldsymbol{\beta}^{(n-1)} + \left(\tfrac{y_i - \mu_i}{\Sigma_{ii}}\right).$$

3. Then update the parameter vector to step n by

$$\boldsymbol{\beta}^{(n)} = (\mathbf{X}\boldsymbol{\Sigma}^{-1}\mathbf{X}')^{-1}\mathbf{X}\boldsymbol{\Sigma}^{-1}\mathbf{Y}^{(n)}.$$

This is the usual frequentist approach to this model. The maximum likelihood vector is

$$\hat{\boldsymbol{\beta}}_{ML} = \lim_{n\to\infty} \boldsymbol{\beta}^{(n)},$$

the limit to which the iteratively reweighted least squares procedure converged. $\mathbf{V_{ML}}$ is the "covariance matrix" of the MLE vector where its inverse is

$$\mathbf{V_{ML}^{-1}} = \mathbf{X}\boldsymbol{\Sigma}^{-1}\mathbf{X}'.$$

for $\boldsymbol{\Sigma}$ calculated at $\hat{\boldsymbol{\beta}}_{ML}$. However, this covariance matrix does not have any relationship to the spread of the likelihood function.[2] Rather, it is the covariance

[2]Actually, it is the observed Fishers information matrix of the likelihood. It relates to the *curvature* of the likelihood at the MLE, not the spread.

matrix of the multivariate normal having mean vector equal to $\hat{\beta}_{ML}$ that matches the curvature at the MLE. We shall refer to it as the matched curvature covariance matrix. It purports to be a global property of the likelihood (measuring its spread) but it was found from a local measure of the likelihood (the curvature at the mode.) If the likelihood has heavy tails, the real spread may be much larger than indicated by $\mathbf{V}_{\mathbf{ML}}$. Using the square roots from this matrix as standard deviations may lead to confidence intervals that do not have the claimed coverage probabilities.

Most statistical packages will find the maximum likelihood estimator $\hat{\beta}_{ML}$ and also the matched curvature covariance matrix $\mathbf{V}_{\mathbf{ML}}$. However they do not note that this covariance matrix does not represent the spread of the likelihood function.

8.2 COMPUTATIONAL BAYESIAN APPROACH TO THE LOGISTIC REGRESSION MODEL

In the Bayesian approach, we want to find the posterior distribution of the parameter given the data. The exact likelihood is given in Equation 8.3. We can easily find the shape of the posterior density using the proportional form of Bayes' theorem, *posterior* is proportional to the *prior* times the *likelihood*. However, there is no closed form for the integral needed to find the scale factor needed to make it an exact density. Instead of finding this scale factor by doing the integration numerically, we will use the computational Bayesian approach, where we will draw a sample from the posterior and use this sample as the basis for our Bayesian inferences. We show how to do this using the Metropolis-Hastings algorithm. First we will have to find a suitable candidate density that will have a similar shape to the posterior, but have heavier tails. We will do this by using a normal approximation to the likelihood, and using this approximate likelihood together with a *multivariate flat* prior or a *multivariate normal* prior we can find an approximate *multivariate normal* posterior using the simple updating rules given in Equations 4.11 and 4.12. Then instead of using this approximate posterior as the candidate density, we will use the *multivariate Student's t* with low degrees of freedom, which has the same mean and spread. It will have a similar shape to the true posterior, but will have heavier tails.

The Normal Approximation to the Likelihood

The vector of maximum likelihood estimators and their matched curvature covariance matrix gives us a place to start. We approximate the joint likelihood function by the *multivariate normal*$(\hat{\beta}_{ML}, \mathbf{V}_{\mathbf{ML}})$ likelihood function, which has mean vector equal to $\hat{\beta}_{ML}$, the mode of the likelihood function, and covariance matrix $\mathbf{V}_{\mathbf{ML}}$, which matches the curvature of the likelihood function at its mode.

Finding the Approximate Posterior

The joint posterior is proportional to the joint prior times the likelihood.

$$g(\beta_0, \ldots, \beta_p | y_1, \ldots, y_n) \propto g(\beta_0, \ldots, \beta_p) \times f(y_1, \ldots, y_n | \beta_0, \ldots, \beta_p) .$$

This gives the shape of the approximate posterior for any prior. However, for most priors, numerical integration would be required to obtain the scale factor needed to make it a density. However, there are two types of priors where we can find the approximate posterior without integration. These are where we use independent flat priors, and when we use multivariate normal conjugate priors. We will restrict ourselves to choosing our prior from one of these two types.

Independent flat priors for $\beta_0, \beta_1, \ldots, \beta_p$. Sometimes we have no prior knowledge about the coefficients so a flat prior is used for all coefficients. The prior gives equal weights for all values of the parameters, so it will be objective for this parameterization. For a discussion on why universal objectivity is not possible see Bolstad (2007). In this case, the posterior will be proportional to likelihood. The approximate posterior will be *multivariate normal*$(\mathbf{b_1}, \mathbf{V_1})$ where

$$\mathbf{b_1} = \hat{\beta}_{ML} \quad \text{and} \quad \mathbf{V_1} = \mathbf{V_{ML}}\,.$$

We saw in Chapter 1 that, even though the posterior has the same shape as the likelihood when flat priors are used, Bayesian inferences will not be the same as those from the likelihood approach. The Bayesian inferences will have advantages due to the probabilistic interpretation of the posterior in the Bayesian approach.

Multivariate normal conjugate priors. Suppose we use a *multivariate normal*$(\mathbf{b_0}, \mathbf{V_0})$ prior density for the parameter vector $\boldsymbol{\beta}$. For instance, we could let the prior for the ith component be *normal*(b_i, s_i^2) and let the prior for each component be independent of each other. Then

$$\mathbf{b_0} = \begin{pmatrix} b_0 \\ \vdots \\ b_1 \end{pmatrix} \quad \text{and} \quad \mathbf{V_0} = \begin{pmatrix} s_0^2 & \ldots & 0 \\ \vdots & \ddots & \vdots \\ 0 & \ldots & s_1^2 \end{pmatrix}.$$

However, our multivariate normal prior does not have to be made up from independent components. In Chapter 4 we saw that the posterior distribution for this case will be *multivariate normal*$(\mathbf{b_1}, \mathbf{V_1})$ where the mean vector and covariance matrix are found by the simple updating rules we developed in Chapter 4. In this case they give

$$\mathbf{V_1^{-1}} = \mathbf{V_0^{-1}} + \mathbf{V_{ML}^{-1}} \tag{8.4}$$

and

$$\mathbf{b_1} = \mathbf{V_1}[\mathbf{V_{ML}^{-1}}]\hat{\beta}_{ML} + \mathbf{V_1}[\mathbf{V_0^{-1}}]\mathbf{b_0}\,. \tag{8.5}$$

Choosing the Multivariate Normal Conjugate Prior

It will be advantageous in terms of finding a suitable prior if we center each predictor variable at its mean. In other words, let the centered observations for the jth predictor be $x_{ij} = x_{ij} - \bar{x}_{.j}$ where $\bar{x}_{.j}$ is the average of the jth predictor variable. Then it makes sense to have independent priors for the slopes and the intercept. The "average" observation will have all the centered predictor values equal to zero. We will find the prior mean and standard deviation for a scientist by matching two percentiles of his or her prior belief distribution.

Finding normal prior for intercept. The scientist would have some prior belief about the overall proportion of "success" to be expected for an "average" observation. We will find the *normal*(b_0, s_0^2) prior for β_0 that matches the scientist's prior 95% credible interval. Suppose that, beforehand, the scientist believes with 95% probability that the proportion of "successes' will be between l and u. The prior mean b_0 and standard deviation s_0 will be found by solving the two simultaneous equations

$$l = \frac{e^{b_0 - 1.96 s_0}}{1 + e^{b_0 - 1.96 s_0}} \quad \text{and} \quad u = \frac{e^{b_0 + 1.96 s_0}}{1 + e^{b_0 + 1.96 s_0}} .$$

The solutions are given by

$$b_0 = \frac{1}{2}\left(\log_e\left(\frac{l}{1-l}\right) + \log_e\left(\frac{u}{1-u}\right)\right) \tag{8.6}$$

and

$$s_0 = \frac{1}{3.92}\left(\log_e\left(\frac{u}{1-u}\right) - \log_e\left(\frac{l}{1-l}\right)\right) . \tag{8.7}$$

Finding normal prior for a slope. We will find the *normal*(b_j, s_j^2) prior for a slope coefficient β_j by matching the scientist prior belief about the increase in the odds ratio due to x_j. There are two methods we will use, depending on whether x_j is a 0:1 variable indicating which of two classes the observation comes from, or whether x_j is a continuous measurement variable.

Case 1: x_j is a 0:1 variable. When x_j is an indicator variable the odds of success for an observation in class $x_j = 0$ is given by

$$\begin{aligned} odds(\text{"}success\text{"}) &= \frac{P(\text{"}success\text{"})}{1 - P(\text{"}success\text{"})} \\ &= e^{\beta_0} \end{aligned}$$

while the odds of success for an observation in class $x_j = 1$ will be

$$\begin{aligned} odds(\text{"}success\text{"}) &= \frac{P(\text{"}success\text{"})}{1 - P(\text{"}success\text{"})} \\ &= e^{\beta_0 + \beta_j} . \end{aligned}$$

The odds ratio for success in an observation with $x_j = 1$ compared to an observation with $x_j = 0$ will be e^{β_j}. Suppose the scientist thinks the odds ratio for an observation in class $x_j = 1$ compared to an observation in class $x_j = 0$ is equally likely to be above and below v. This means the median of his/her belief distribution is v. This gives the equation

$$e^{b_j} = v .$$

We will choose the prior standard deviation for the slope β_j by matching the scientist's prior belief about the upper 95% point on the increase in odds level. Suppose the scientist believes with 95% prior probability, that the increase in the odds ratio will be below w. This gives the equation

$$e^{b_j + 1.645 s_j} = w\,.$$

The simultaneous solutions of these equations are

$$b_j = \log_e v \quad \text{and} \quad s_j = \frac{1}{1.645} \log_e \left(\frac{w}{v} \right)\,. \tag{8.8}$$

Case 2: x_j is a measurement variable. When x_j is a measurement variable we match the scientist's prior belief about the odds ratio for an observation where the value of x_j is at some round number approximately one standard deviation above its mean compared to an observation where the value of x_j is at its mean. Suppose she thinks the odds of success for an "average" observation where x_j is at its mean is given by

$$\begin{aligned} odds(\text{"}success\text{"}) &= \frac{P(\text{"}success\text{"})}{1 - P(\text{"}success\text{"})} \\ &= e^{\beta_0} \end{aligned}$$

while the odds for success for an observation where x is one standard deviation above its mean is given by

$$\begin{aligned} odds(\text{"}success\text{"}) &= \frac{P(\text{"}success\text{"})}{1 - P(\text{"}success\text{"})} \\ &= e^{\beta_0 + \beta_j s_x}\,. \end{aligned}$$

Suppose the scientist's prior belief is that the increase in odds ratio for an observation where x is increased by one standard deviation is equally likely to be below or above v. This gives the equation

$$e^{b_j s_x} = v\,.$$

Suppose the scientist also believes with 95% prior probability that the increase in odds ratio will be less than w, which gives the equation

$$e^{(b_j + 1.645\, s_j)\, s_x} = w\,.$$

The simultaneous solutions of these two equations are

$$b_j = \frac{\log_e v}{s_x} \quad \text{and} \quad s_j = \frac{1}{1.645\, s_x} \log_e \left(\frac{w}{v} \right)\,. \tag{8.9}$$

The Heavy-Tailed Candidate Distribution

The approximate posterior is *multivariate normal*$(\mathbf{b_1}, \mathbf{V_1})$. Let $\mathbf{L}$ be the lower triangular matrix found from the Cholesky decomposition of $\mathbf{V_1}$, and let $\mathbf{L}'$ be the transpose of $\mathbf{L}$. Then

$$\mathbf{V_1} = \mathbf{L}\mathbf{L}' .$$

Let $z_0, \ldots, z_p$ be independent *normal*$(0, 1^2)$ random variables. Then

$$\mathbf{z} = \begin{pmatrix} z_0 \\ \vdots \\ z_p \end{pmatrix}$$

is a *multivariate normal* random vector having mean vector and covariance matrix equal to

$$\begin{pmatrix} 0 \\ \vdots \\ 0 \end{pmatrix} \quad \text{and} \quad \begin{bmatrix} 1 & \ldots & 0 \\ \vdots & \vdots\vdots\vdots & \vdots \\ 0 & \ldots & 1 \end{bmatrix}$$

respectively. Then

$$\boldsymbol{\beta} = \mathbf{b_1} + \mathbf{L}\mathbf{z}$$

has the *multivariate normal*$(\mathbf{b_1}, \mathbf{V_1})$ approximate posterior distribution. However, this distribution may not have nearly as heavy tails as the true posterior. Instead we make a candidate density that has a similar shape but heavier tails. We let t_0, t_1, ..., t_p be independent *Student's t* random variables with low degrees of freedom. Then

$$\mathbf{t} = \begin{pmatrix} t_0 \\ \vdots \\ t_p \end{pmatrix}$$

has the *multivariate Student's t* distribution, and

$$\boldsymbol{\beta} = \mathbf{b_1} + \mathbf{L}\mathbf{t}$$

has the *multivariate Student's t*$(\mathbf{b_1}, \mathbf{V_1})$ distribution with similar shape to the approximate posterior, but with heavier tails. This is the independent candidate distribution we will use in the Metropolis-Hastings algorithm to find our sample from the true posterior.

Example 11 *Table 8.1 shows the the age and coronary heart disease status for 100 subjects taken from Hosmer and Lemeshow (1989). We want to determine the effect of age on coronary heart disease status. Suppose the scientist believes with 95%*

Table 8.1 Age and coronary heart disease status.

Age	CHD	Age	CHD	Age	CHD	Age	CHD	Age	CHD
20	0	34	0	41	0	48	1	57	0
23	0	34	0	42	0	48	1	57	1
24	0	34	1	42	0	49	0	57	1
25	0	34	0	42	0	49	0	57	1
25	1	34	0	42	1	49	1	57	1
26	0	35	0	43	0	50	0	58	0
26	0	35	0	43	0	50	1	58	1
28	0	36	0	43	1	51	0	58	1
28	0	36	1	44	0	52	0	59	1
29	0	36	0	44	0	52	1	59	1
30	0	37	0	44	1	53	1	60	0
30	0	37	1	44	1	53	1	60	1
30	0	37	0	45	0	54	1	61	1
30	0	38	0	45	1	55	0	62	1
30	0	38	0	46	0	55	1	62	1
30	1	39	0	46	1	55	1	63	1
32	0	39	1	47	0	56	1	64	0
32	0	40	0	47	0	56	1	64	1
33	0	40	1	47	1	56	1	65	1
33	0	41	0	48	0	57	0	69	1

prior probability that the proportion of successes is between .10 and .50. The prior mean and standard deviation for the intercept β_0 are $b_0 = -1.099$ and $s_0 = .561$, respectively. Age is a measurement variable with standard deviation 11.4 years. Suppose the scientist also believes that increasing age by ten years will increase the odds of having coronary heart disease by a factor that is equally likely to be above or below 2. She also believes with 95% prior probability that the ten year increase in the odds is less than 8. Solving the equations gives the prior mean and standard deviation for the slope β_1 as $b_1 = .069$ and $s_1 = .084$. Thus the prior mean vector and covariance matrix are

$$\mathbf{b_0} = \begin{pmatrix} -1.099 \\ .069 \end{pmatrix} \quad \text{and} \quad \mathbf{V_0} = \begin{bmatrix} .561^2 & 0 \\ 0 & .084^2 \end{bmatrix}$$

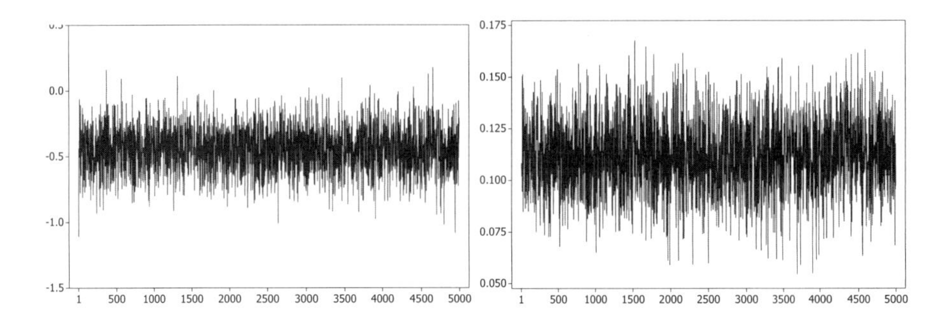

Figure 8.1 The traceplots of β_0 and β_1 for 5000 steps of the Metropolis-Hastings algorithm.

respectively. The maximum likelihood vector and its matched curvature covariance matrix are given by

$$\hat{\beta}_{ML} = \begin{pmatrix} -.386773 \\ .110921 \end{pmatrix} \quad \text{and} \quad \mathbf{V_{ML}} = \begin{bmatrix} .0574651 & -.0009865 \\ -.0009865 & .0005789 \end{bmatrix}$$

respectively. This results in the approximate multivariate normal posterior having mean vector and covariance matrix

$$\mathbf{b_1} = \begin{pmatrix} -.491952 \\ .109501 \end{pmatrix} \quad \text{and} \quad \mathbf{V_1} = \begin{bmatrix} .0485104 & -.0007712 \\ -.0007712 & .0005322 \end{bmatrix}$$

respectively.

We run the Metropolis-Hastings algorithm for 5000 steps using the multivariate Student's t density with 4 degrees of freedom to generate candidates. Figure 8.1 shows the trace plots of the intercept β_0 and the slope β_1, respectively. We see the chain is moving around the parameter space very well. We find the sample autocorrelation functions of the parameters β_0 and β_1 to help determine an adequate burn-in period. They are shown in Figure 8.2 along with the 95% bounds. We see that they are essentially indistinguishable from zero by lag 10. We ran eight chains for $2n = 40$ steps, from random starting points on the prior distribution and calculated the Gelman-Rubin statistics for the parameters β_0 and β_1. They are 1.045 and 1.0567, respectively. Then we ran six parallel chains starting from random points on the prior distribution but with identical random inputs. The traceplots for these six chains are shown in Figure 8.3. We see that the chains have clearly converged at lag 10. Looking at the traceplots, autocorrelation functions, and convergence of parallel so we decide 10 steps is a suitable burn-in time. The histograms for β_0 and β_1 together with the graphs of their matched curvature normal approximation are shown in Figure 8.4. We see that the matched curvature normal approximation to the posterior is a very good approximation to the true posterior. We use the random sample from the posterior for our inferences about the slope β_1. We estimate the probability $P(\beta_1 < 0)$ by the proportion of the posterior sample of β_1 that are less than 0. This is the estimated posterior probability of the null hypothesis when we are

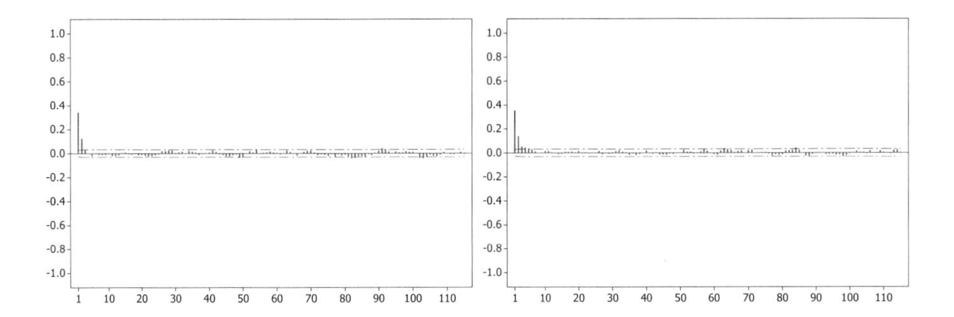

Figure 8.2 The sample autocorrelation for β_0 and β_1.

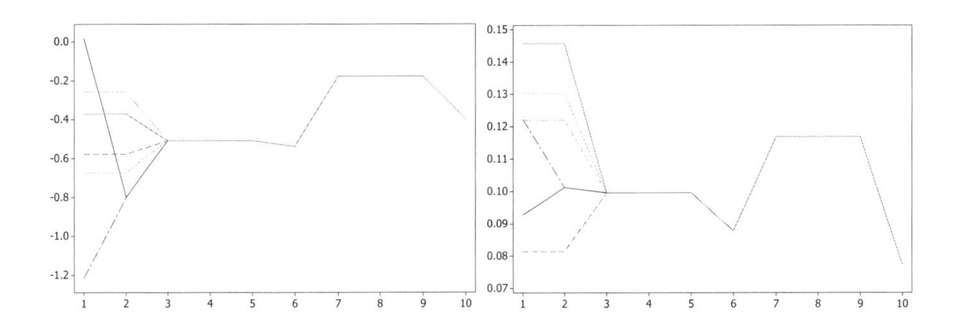

Figure 8.3 The outputs from multiple parallel Metropolis-Hastings chains for β_0 and β_1.

Table 8.2 Summary of inferences from the thinned posterior sample.

Parameter	Estimate	St. Dev.	z	Odds	95% Credible Interval	
					Lower	Upper
β_0	−.405	.2483				
β_1	.115	.0247	.000	1.122	1.070	1.178

testing

$$H_0 : \beta_1 \leq 0 \quad \text{vs.} \quad H_1 : \beta_1 > 0 \,.$$

We find a 95% credible interval for the slope β_1, *and exponentiate it to make it a 95% credible interval for the increased odds of CHD when age is increased by 1 year. The results of our inferences are presented in Table 8.2. When we examine these results, we conclude that age clearly has an effect on CHD. Each additional year of age increases the odds of CHD by a factor between 1.070 and 1.178 with 95% credibility.*

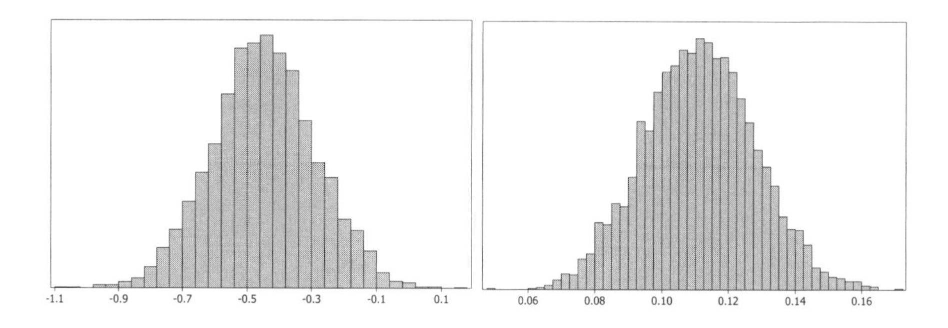

Figure 8.4 Histograms for the thinned posterior sample of β_0 and β_1 together with their matched curvature normal approximations.

8.3 MODELLING WITH THE MULTIPLE LOGISTIC REGRESSION MODEL

We use the multiple logistic regression model to help us find a model that we can use to predict successes using the predictor variables. In multiple logistic regression, as in any regression model where we are trying to determine the effect of the predictors, difficulties arise whenever the predictors are themselves related. This is known as the problem of multicollinearity. When this occurs the interpretation of the coefficients becomes more difficult since some predictors can either enhance or reduce the effect of other predictors. This makes it hard to determine which predictors have significant effects and which don't, whether a frequentist or Bayesian approach is used. Bowerman and O'Connell (1990) and Neter, Wasserman, and Kutner (1990) give a good discussion on this from the frequentist perspective.

Example 12 *Swanson et al. (2007) report on a cardiac study on the survival outcomes for 527 patients admitted with a myocardial infarction (MI) at Waikato Hospital from 1995–2005. The ten predictor variables collected on each patient are divided into three types. The demographic variables are age and sex. The medical variables include diabetes, smoking history, shock, renal failure, previous MI, and vessel number. The treatment variables are stent used and IIBIIIaInhibitor. We want to see how these predictor variables can be used to predict short-term survival of the MI episode. We define the response variable to be "survived for thirty days." We want to perform a Bayesian analysis on the data using the multiple logistic regression model to determine the effect the predictor variables have on the response variable. We will base our inferences on a sample drawn from the true posterior using the Metropolis-Hastings algorithm. Since we have no prior knowledge about the parameters, we will use independent flat priors for all the parameters. The true posterior will have the same shape as the likelihood.*

We run the Metropolis-Hastings algorithm 5000 steps using the multivariate Student's t candidate density and the trace plots of the MH sample for the intercept β_0 and the slope parameters $\beta_1, \ldots, \beta_{10}$ are shown in Figure 8.5. We see that the chain

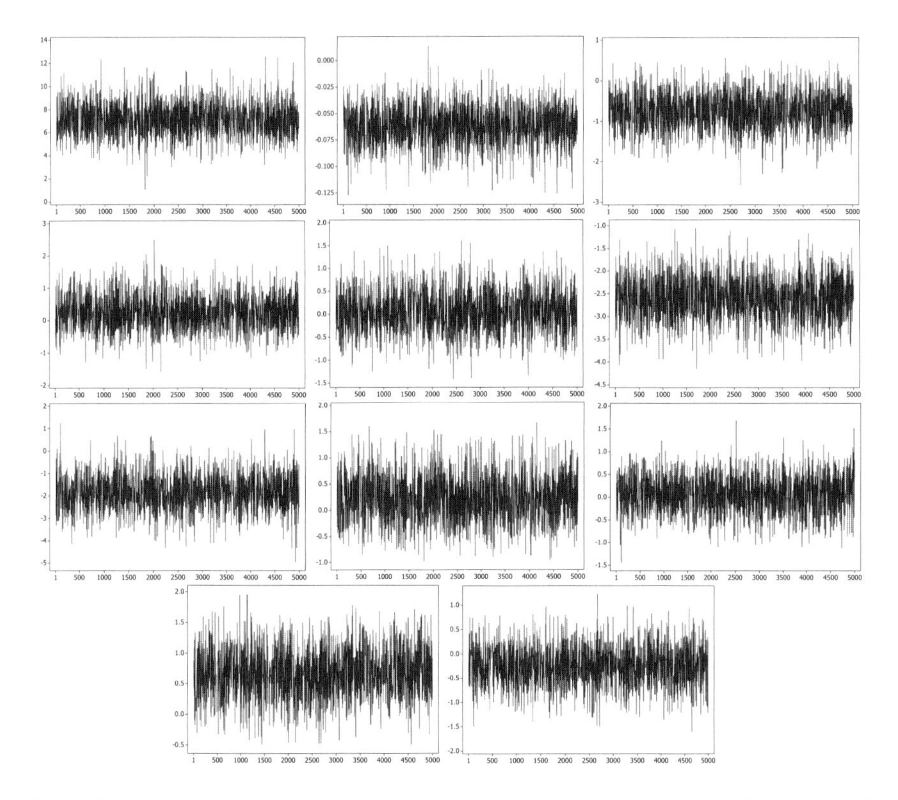

Figure 8.5 Trace plot of 5000 consecutive MH sample from posterior for the intercept β_0 and the slope coefficients $\beta_1, \ldots, \beta_{10}$.

is rapidly moving through all dimensions of the parameter space, so we won't require too long a burn-in or too much thinning. We calculate the autocorrelation function functions for all the parameters from the Metropolis-Hastings sample, and show them in Figure 8.6 along with the 95% bounds. We see that they are indistinguishable from zero after about lag 10. The occasional autocorrelation at a high lag that is outside the bounds can be attributed to random chance since we can expect 5% of them to be outside the bounds due to random chance alone. We ran six chains for $2n = 50$ *steps and calculated the Gelman-Rubin statistics for each of the parameters. The average of the Gelman-Rubin statistics over all the parameters is 1.097. We then ran six parallel chains from random starting points but with the same random inputs. We see that all the chains have converged within a few steps. From examination of the traceplots and autocorrelation functions, from the Gelman-Rubin statistics, and from the convergence of the parallel chains we decide that 25 steps is a suitable burn-in time. Then we will thin the output by only taking every* 25th *value. The resulting sample will be approximately a random sample from the posterior. We now run the chain for 50000 steps. The resulting thinned chain will give a random sample of size 2000 draws from the posterior as the basis for inference. Histograms*

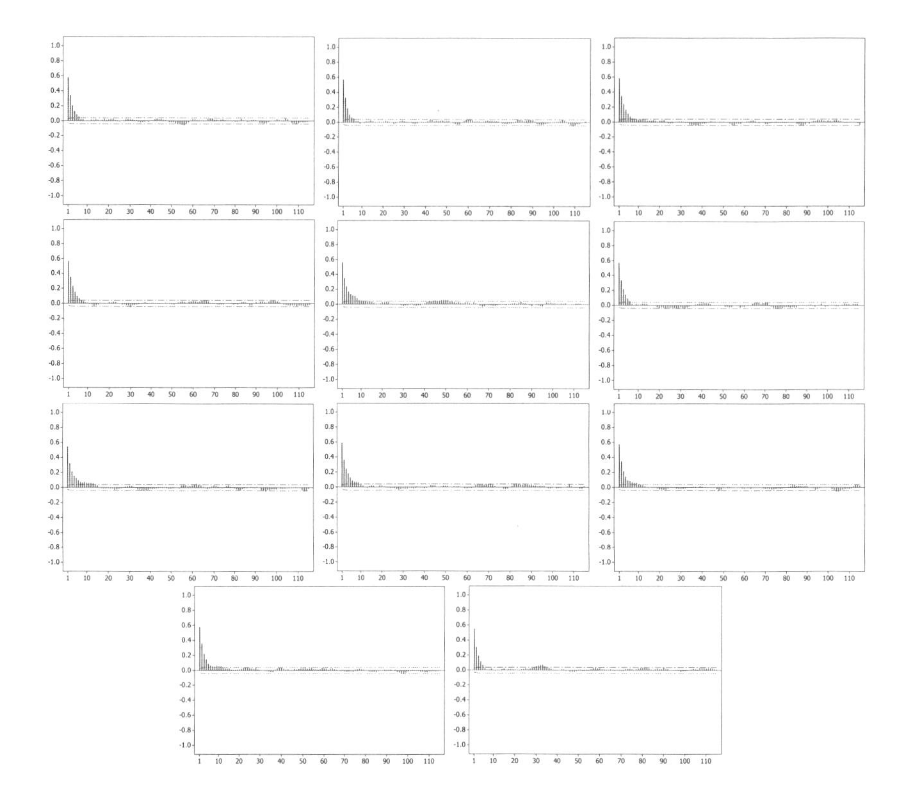

Figure 8.6 Sample autocorrelation function from posterior for the intercept β_0 and the slope coefficients $\beta_1, \ldots, \beta_{10}$.

of the draws from the posterior for the intercept β_0 and the slopes $\beta_1, \ldots, \beta_{10}$ are shown in Figure 8.8, along with the matched curvature normal approximation. The histograms summarize a random sample of draws from the true posterior. We see that the matched curvature approximation is somewhat skewed away from the true posterior. The Bayesian method will give better credible intervals than confidence intervals based on the matched curvature normal (which in this case is identical to the approximate likelihood). The summary statistics of the thinned posterior sample is given in Table 8.3.

Modelling Issues: Removing Unnecessary Variables

Often, the logistic regression model is run including all possible predictor variables that we have data for. Some of these variables may affect the response very little if at all. The true coefficient of such a variable β_j would be very close to zero. Leaving these unnecessary predictor variables in the model can complicate the determination of the effects of the remaining predictor variables. Their removal will lead to an improved model for predictions. This is often referred to as the principal of parsimony.

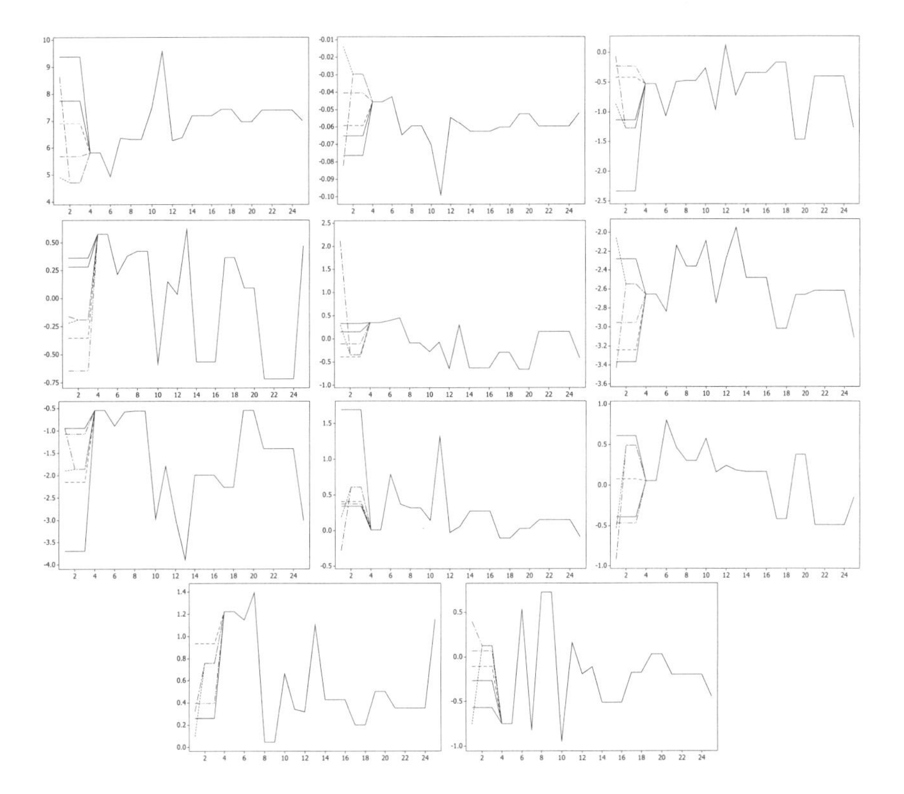

Figure 8.7 The outputs from multiple parallel Metropolis-Hastings chains for intercept β_0 and the slope coefficients $\beta_1, \ldots, \beta_{10}$.

We would like to remove all predictor variables x_j where the true coefficient $\beta_j = 0$. This is not as easy as it sounds as we do not know which coefficients are truly equal to zero. We have a random sample from the joint posterior distribution of $\beta_1, \ldots, \beta_J$. When the predictor variables $x_1, \ldots, x_J$ are correlated, some of the predictor variables can be either enhancing or masking the effect of other predictors. This means that a coefficient value estimated from the posterior sample may look very close to zero, but the effect of its predictor variable actually may be larger. Other predictor variables are masking its effect. Sometimes a whole set of predictor variables can be masking each others' effect, making each individual predictor look unnecessary (not significant), yet the set as a whole is very significant. Obviously we should not remove that set of predictors. When we have a set of predictors $x_1, \ldots, x_k$ that we consider might be removed, we should test

$$H_0 : \beta_1 = \ldots = \beta_k = 0 \quad \text{versus} \quad H_1 : \beta_j \neq 0 \text{ for some } j = 1, \ldots, k\,.$$

The null hypothesis is that all their coefficients are equal to zero simultaneously, and the alternative that not all are equal to zero. If we can't reject the null hypothesis, then we can remove all those predictors from the model safely.

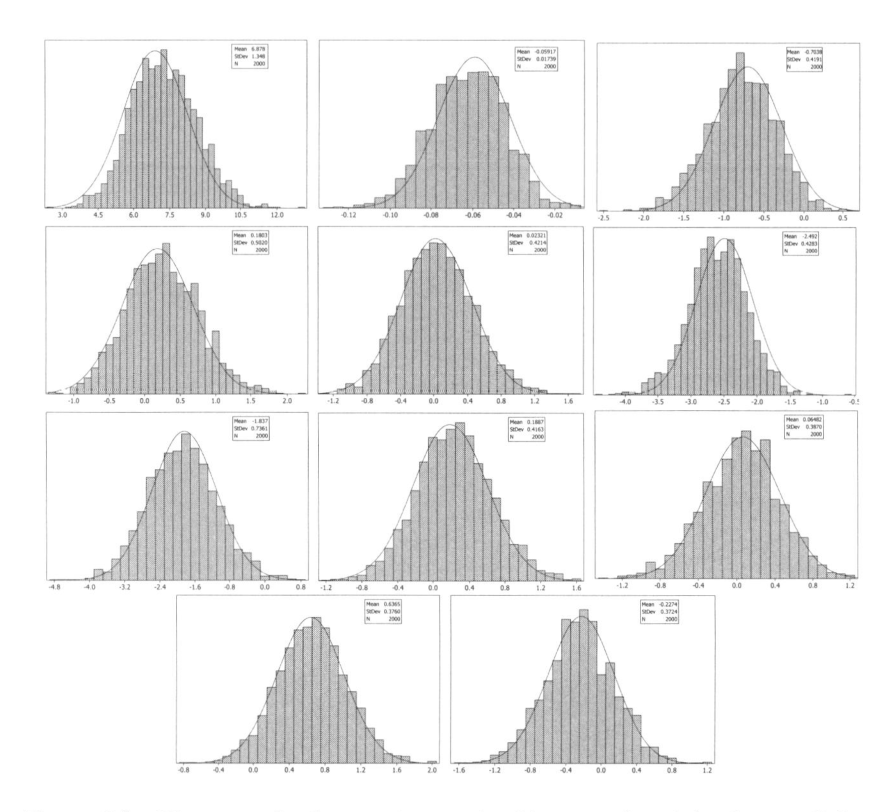

Figure 8.8 Histograms for the posterior sample of intercept β_0 and the slope coefficients $\beta_1, \ldots, \beta_{10}$ together with matched curvature normal approximation.

Table 8.3 Summary of inferences from the thinned posterior sample

Parameter	Coefficient of	Mean	St. Dev.	z
β_0	intercept	7.21871	1.40457	5.13944
β_1	age	–0.06209	0.01806	–3.43821
β_2	sex	–0.76255	0.44866	–1.69963
β_3	diabetes	0.24355	0.52439	0.46444
β_4	smoking	0.06473	0.43093	0.15021
β_5	shock	–2.61909	0.44604	–5.87180
β_6	renal failure	–1.87517	0.75939	–2.46931
β_7	previous MI	0.22615	0.42865	0.52760
β_8	vessel number	0.05692	0.40801	0.13951
β_9	stent	0.65879	0.39016	1.68854
β_{10}	IIBIIIAInhibitor	–0.23672	0.38420	–0.61615

We will give a simple approximate test[3] for this simultaneous hypothesis. Let the vector of the coefficients in question be

$$\boldsymbol{\beta} = \begin{pmatrix} \beta_1 \\ \vdots \\ \beta_k \end{pmatrix}.$$

Let the vector of means from the posterior sample and the matrix of covariances from the posterior sample for $\boldsymbol{\beta}$ be

$$\bar{\boldsymbol{\beta}} = \begin{pmatrix} \bar{\beta}_1 \\ \vdots \\ \bar{\beta}_k \end{pmatrix} \quad \text{and} \quad \hat{\mathbf{V}} = \begin{pmatrix} \hat{V}_{11} & \dots & \hat{V}_{1k} \\ \vdots & \ddots & \vdots \\ \hat{V}_{k1} & \dots & \hat{V}_{kk} \end{pmatrix}$$

respectively. The test statistic

$$U = \bar{\boldsymbol{\beta}}' \, \hat{\mathbf{V}}^{-1} \, \bar{\boldsymbol{\beta}} \tag{8.10}$$

will be approximately[4] *chi-squared* with k degrees of freedom when the null hypothesis is true. We can reject the null hypothesis at the 5% level of significance if $U > U_{.05}$ where $U_{.05}$ is the upper 5% point of the *chi-squared* distribution[5] with k degrees of freedom. If we cannot reject the null hypothesis, we can remove all the specified predictors from the model. Then we should redo the analysis without those predictors.

Example 12 (continued) *We examine the output from the thinned posterior sample. The z values, which are the ratios of the posterior means to the posterior standard deviations for the the predictors diabetes, smoking, previous MI, vessel number, and IIBIIIAinhibitor, indicate that they are candidates for removal. The posterior mean vector for the coefficients of those predictors is*

$$\bar{\boldsymbol{\beta}} = \begin{pmatrix} .243546 \\ .0647285 \\ .226153 \\ .0569198 \\ -.236724 \end{pmatrix}$$

and the posterior covariance matrix for those coefficients

$$\hat{\mathbf{V}} = \begin{bmatrix} .274996 & -.009547 & .008235 & .022694 & -.000782 \\ -.009544 & .185700 & .001890 & -.011006 & .008539 \\ .008552 & .001879 & .183747 & -.005166 & .009631 \\ .022684 & -.011006 & -.005192 & .166468 & -.008270 \\ -.000765 & .008539 & .009632 & -.008268 & .147606 \end{bmatrix}.$$

[3]Another approach is to calculate the deviance information criterion from the MCMC sample. See Spiegelhalter, et al. (2002) and Gelman, et al. (2004) for an explanation of this approach.
[4]The statistic U measures how far the zero vector is from the posterior mean vector using distance based on the covariance structure found from the posterior sample.
[5]This test will have robustness properties similar to analysis of variance tests.

Table 8.4 Summary of inferences from the final thinned posterior sample.

Parameter	Estimate	St. Dev.	z	Odds	95% Credible Interval	
					Lower	Upper
β_0	7.1095	1.2585	5.6491	1223.48	113.977	15581.2
β_1	−.0608	.0168	−3.6074	.94	.909	1.0
β_2	−.7271	.4221	−1.7224	.48	.201	1.1
β_5	−2.5884	.3868	−6.6911	.08	.034	.2
β_6	−1.7052	.7446	−2.2901	.18	.043	.8
β_9	.5479	.3599	1.5223	1.73	.863	3.4

The test statistic calculated using Equation 8.10 is $U = .93966$, which is clearly less than the chi-squared critical value. Hence we will remove those variables and redo the analysis using the remaining variables. We remove those predictor variables from the data set and run the Metropolis-Hastings algorithm another 50000 steps. We use a burn-in of 25 and thin every 25th value, which leaves us an approximately random sample from the posterior of size 2000. The summary statistics for the coefficients of the predictor variables used in the final model are given in Table 8.4.

The Predictive Distribution

Suppose we want to find the distribution for observing a new individual given some particular values of the predictor variables. The new observation given the parameters and the previous observations will not depend on the previous data since it is just another random observation from the data. Using the Bayesian approach we treat the parameters as nuisance parameters. If we knew the exact posterior distribution, we would find the joint posterior of the new observation and the parameters given the previous observations. Then we would marginalize out the parameters to find the predictive distribution of the new observation. Let y_{n+1} be the new observation. The predictive density of y_{n+1} given the observed $y_1, \ldots, y_n$ would be

$$g(y_{n+1}|y_1, \ldots, y_n) = \int \cdots \int [f(y_{n+1}|\beta_0, \ldots, \beta_p) \times g(\beta_0, \ldots, \beta_p|y_1, \ldots, y_n)] \, d\beta_0 \ldots d\beta_p \, . \tag{8.11}$$

If we have a random sample from the posterior instead of the exact posterior distribution, we draw a random sample from the predictive distribution instead of evaluating it exactly. For each draw from the posterior sample we calculate the value of the linear predictor using those particular values of the predictor variables. Then we calculate the success probability for that draw. We draw a *binomial*$(1, \pi_i)$ random variable for each of the success probabilities π_i. This gives us a random sample from the predictive distribution.

The Bayesian method for finding the predictive distribution contrasts with the plug-in predictive distribution as usually found by the frequentist approach. In the frequentist approach, the maximum likelihood estimates of $\beta_0, \ldots, \beta_p$ would be found by iteratively reweighted least squares. These would be plugged into the equation for the linear predictor using the predictor values for the new observation and this would be used to estimate the probability of success for the new observation. The predictive distribution would be the binomial with that probability of success. This underestimates the uncertainty. The parameters are treated as if they were known to be at the maximum likelihood estimates. There is additional uncertainty since the parameters are not known, and the maximum likelihood estimates are only estimates. Prediction intervals found from the frequentist predictive distribution will generally be too short and will not have the coverage probability claimed.

The Bayesian method allows for this uncertainty. If we are calculating the predictive density exactly using Equation 8.11, the posterior density of the parameters is used when they are averaged out in the marginalization. If we are calculating a random sample from the posterior, the uncertainty in the values of the parameters is allowed for by taking an observation from the predictive distribution using each draw in the posterior random sample. Prediction intervals found using the Bayesian approach will have the correct coverage probability.

Example 12 (continued) *Suppose we want to find the predictive distribution for 30-day survival of a 60-year old male patient who does not have shock or renal failure and is given a stent. Let the draws of the intercept, the slope coefficient for age, the slope coefficient for male, and the slope coefficient for stent be in columns c1, c2, c3, and c6, respectively. We let column c8 equal draws for the linear predictor for that male. Then we let column c9 draws from the exponential of c8 divided by one plus the exponential of c8 for that male. Then we let c10 be a random draw from a* binomial$(1, c10)$. *Each observation has its own success probability that is its value in c10. The summary statistics for the 2000 draws from the predictive distribution are mean .9615 and standard deviation .19245.*

Main Points

- In the logistic regression model, the observations are independent binomial observations where each observation has its own probability of success that is related to a set of predictor variables by setting the logarithm of the odds ratio equal to an unknown linear function of the predictor variables. This is known as the *logit* link function. The coefficients of the linear predictor are the unknown parameters that we are interested in.

- The logistic regression model is an example of a *generalized linear model*. The observations come from a member of one-dimensional exponential family, in this case binomial. Each observation has its own parameter value that is linked to the linear predictor by a link function, in this case the *logit* link. The observations are all independent.

- The maximum likelihood estimate of the coefficient of the linear predictors can be found by iteratively reweighted least squares. This method also finds a "covariance matrix."
- Actually it is the matched curvature covariance matrix, which is the covariance matrix of a *multivariate normal* distribution, that matches the curvature of the likelihood function at its maximum. This matched curvature covariance matrix has no relation to the spread of the likelihood function.
- We can find an normal approximation to the posterior by using the *normal* approximation to the likelihood, together with either a *normal* prior or a (improper) flat prior.
- The true posterior is given by the *binomial* likelihood times the prior. We can find a random sample from the true posterior by using the Metropolis-Hastings algorithm using a multivariate *Student's t* with low degrees of freedom that corresponds to the *normal* approximation to the posterior as the independent candidate distribution.
- Examination of the traceplots and sample autocorrelation functions is used to establish a burn-in time to also be used to thin the chain. Calculating the Gelman-Rubin statistic, and coupling with the past can also be used.
- Predictor variables may be related to each other. This can complicate the interpretation posterior.
- Predictor variables that have coefficient very close to zero should be removed from the model.
- When we have a random sample from the posterior, we can get a random sample from the predictive distribution by drawing a random draw from the observation distribution for each draw in the posterior sample.
- The sample from the predictive distribution will fully allow for the uncertainty in the parameters by drawing a value for each random draw in the posterior sample.

Exercises

8.1 The Christchurch Health and Development Study recently reported on data relating the circumcision status and the risk of sexually transmitted infection (STI) in young adult males (Fergusson, et al. 2006). This is part of a study following a birth cohort in New Zealand. In this particular study, they obtained information on the circumcision status of the young males, the number of sexual partners they had, and whether or not they had ever had a sexually transmitted infection. This study was interested in determining whether or not circumcision of males is related to the subsequent risk of catching an STI.

We don't want the effect of circumcision status to be confounded with the number of sex partners the person has had. To avoid this, the observations have been grouped into four classes having the same number of sex partners. We created membership for variables "5–9 partners," "10–15 partners" and "16 or more partners." The membership in class "0–4 partners" has 0 for all the the membership variables for the other classes. The data is in the Minitab worksheet Exercise8.1.mtw.

(a) Run a Metropolis-Hastings chain for 1000 steps using the Minitab macro BayesLogRegMH.mac or the R-function `BayesLogistic`. In this example, we do not have any prior knowledge, so we are using a joint flat prior for all the parameters. Do the traceplots show the chain has good mixing properties? Do the sample autocorrelations show the effect of the starting point quickly dying out?

(b) Run four parallel chains from different starting points, but with the same random inputs, and graph the traceplots to see how long the chains have to run before the chains have converged.

(c) Decide on a reasonable burn-in time and thinning requirement so that the thinned sample can be considered to be approximately a random sample from the posterior.

(d) Run the chain for 5000 steps and use the burn-in and thinning from (c) to get the thinned sample for inference. Summarize the thinned posterior sample. What conclusion can you draw about the effect of circumcision status on the risk of catching an STI?

8.2 The Minitab worksheet Exercise8.2.mtw contains the the data from the Waikato District Health Board cardiac study discussed in Swanson et al. (2007), which we analyzed in Example 12. We want to determine how 30-day survival for cardiac patients coming in with a myocardial infarction (MI) depends on various predictor variables collected for each patient. These include demographic variables such as sex and age; medical history and lifestyle variables such as diabetes, smoking, shock, renal failure, previous MI, and vessel number; and treatment variables such as whether a stent is used and whether IIbIIIaInhibitor is administered.

(a) Run a Metropolis-Hastings chain for 1000 steps using the Minitab macro BayesLogRegMH.mac or the R-function `BayesLogistic`. In this example, we do not have any prior knowledge, so we are using a joint flat prior for all the parameters. Do the traceplots show the chain has good mixing properties? Do the sample autocorrelations show the effect of the starting point is quickly dying out?

(b) Run four parallel chains from different starting points, but with the same random inputs, and graph the traceplots to see how long the chains have to run before the chains have converged.

(c) Decide on a reasonable burn-in time and thinning requirement so that the thinned sample can be considered to be approximately a random sample from the posterior.

(d) Run the chain for 5000 steps and use the burn-in and thinning from (c) to get the thinned sample for inference. Summarize the thinned posterior sample.

(e) Does there seem to be any variables that could be removed from the model? Test the hypothesis that the coefficients of those variables are all equal zero simultaneously. Use the 95% significance level.

(f) If you have removed the variables, then run the chain with those variables removed for 5000 steps and thin the output as determined in (c). Summarize the thinned posterior sample.

(g) State your conclusions about how the 30-day survival relates to the predictor variables.

9

Poisson Regression and Proportional Hazards Model

Sometimes, we observe a count variable y that records the number of occurrences of a rare event, and we want to relate it to a set of predictor variables $x_1, \ldots, x_p$ whose value we have recorded at each observation. We are not observing the number of successes in a fixed number of trials since the number of trials is not known. So we cannot use the logistic regression model that we looked at in the previous chapter. We saw in Chapter 4 that the *Poisson* distribution is used when we are counting the occurrence of rare events. Each observation has its own mean value. We want to develop a regression model for predicting the number of occurrences using the predictor variables. We need to relate the mean of the observed count variable to a linear function of the predictors. We cannot regress the mean directly on the predictor variables in a linear way since the mean must be nonnegative and a linear function goes from minus infinity to infinity. Instead, a function of the parameters will be linked to the linear predictor. In Section 9.1 we introduce the *Poisson regression model* to model this relationship. This is a *generalized linear model* with *Poisson* observations where each observation has its own mean value that is related to the set of predictor variables through the *log link function*. The analysis of this model will follow a similar pattern to that used for logistic regression in the Chapter 8. The only changes are due to the different observation distribution and the different link function. The frequentist approach to this model is to find the maximum likelihood estimator by iteratively reweighted least squares. Once again, the covariance matrix found by this method is not the actual covariance matrix, rather it is the covariance matrix of a multivariate normal (having mean vector equal to the MLE) that matches the curvature at the MLE. Thus it does not give a valid measure of the spread of the likelihood. We use this matched curvature normal as an approximation to the likelihood. When we use a *multivariate normal* prior or a *multivariate flat prior*, the approximation to the posterior can be found by the simple updating rules for the *multivariate normal*. This is only an approximation. We know the shape of the true posterior is given by the prior times true *Poisson* likelihood. In Section 9.2 we will use the computational Bayesian approach on this model. We will show how the Metropolis-Hastings algorithm with an independent candidate density can be used to draw a random sample from the true posterior distribution of this model. The candidate density we use is a heavy-tailed version of the matched curvature normal approximation.

In Section 9.3, we want to relate the survival time t to a set of predictor variables $x_1, \ldots, x_p$. Sometimes we don't observe the exact lifetime of an individual because that individual is still living at the time the study ends. We say that lifetime is censored. So for each individual we observe a time t, and a censoring variable w that tells us whether or not t is the exact lifetime, or whether the exact lifetime is greater than t. The *proportional hazards model* is a useful model for censored survival times. The *hazard function* is related to a linear function of the predictor variables. This also turns out to be a *generalized linear model* with *Poisson* observations. In Section 9.4 we show how a sample from the posterior is drawn using the Metropolis-Hastings algorithm in a similar way as for *Poisson regression*.

9.1 POISSON REGRESSION MODEL

We have a sequence of *Poisson* observations $y_1, \ldots, y_n$ where each observation y_i has its own mean value μ_i. We have also observed the values $(x_{i1}, \ldots, x_{ip})$ that each of the p predictor variables has for the ith observation. Let y_i come from a *Poisson*(μ_i) distribution. The likelihood is

$$f(y_i|\mu_i) \propto \mu_i^{y_i} e^{-\mu_i} .$$

The likelihood of the random sample $y_1, \ldots, y_n$ will be the product of the individual likelihoods and is given by

$$\begin{aligned} f(y_1, \ldots, y_n|\mu_1, \ldots, \mu_n) &\propto \prod_{i=1}^{n} f(y_i|\mu_i) \\ &\propto \prod_{i=1}^{n} \mu_i^{y_i} e^{-\mu_i} \\ &\propto e^{-\sum \mu_i} \prod_{i=1}^{n} \mu_i^{y_i} . \end{aligned} \tag{9.1}$$

Poisson Regression via Generalized Linear Model

The linear predictor. We want to relate the *Poisson* observations to the set of predictor variables $x_1, \ldots, x_p$. We let x_{ij} be the value of predictor j for the ith

observation y_i. Suppose that we use a linear predictor

$$\eta_i = \sum_{j=0}^{p} x_{ij}\beta_j$$

and relate it to the *Poisson* means. (Note: $x_{i0} = 1$ for all i, so we are allowing an intercept β_0 as well as the slopes β_j for $j = 1, \ldots, p$.) Let

$$\log_e(\mu_i) = \eta_i \,.$$

This is called the *log link* function and it relates the linear predictor to the parameter. Other link functions could be used, but the *log link* is the most commonly used for *Poisson* observations.[1] We use the link function to rewrite the likelihood function as a function of the parameters $\beta_0, \ldots, \beta_p$. The likelihood becomes

$$\begin{aligned} f(y_1, \ldots, y_n|\beta_1, \ldots, \beta_p) &\propto e^{(-\sum e^{\eta_i})} \prod_{i=1}^{n} (e^{\eta_i})^{y_i} \\ &\propto e^{(-\sum e^{\eta_i})} \prod_{i=1}^{n} e^{y_i \eta_i} \\ &\propto e^{(-\sum e^{\sum x_{ij}\beta_j})} \prod_{i=1}^{n} (e^{y_i \sum x_{ij}\beta_j}) \\ &\propto e^{(-\sum e^{\sum x_{ij}\beta_j})} \left[e^{\sum y_i \sum x_{ij}\beta_j}\right]. \end{aligned}$$

The maximum likelihood estimates can be found by taking derivatives with respect to the parameters, setting them equal to zero, and finding the simultaneous solutions of the resulting equations. The likelihood and its logarithm will have maximums at the same values since

$$\frac{\partial \log_e f}{\partial \theta} = \frac{1}{f} \times \frac{\partial f}{\partial \theta} \,.$$

The log-likelihood is given by

$$\log_e(f(y_1, \ldots, y_n|\beta_1, \ldots, \beta_p)) \propto \left(-\sum e^{\sum x_{ij}\beta_j}\right) + \sum y_i \sum x_{ij}\beta_j \,.$$

Since the log-likelihood is a simpler function than the likelihood function, the maximum likelihood estimates are usually found by taking derivatives of the log-likelihood with respect to the parameters, setting them equal to zero, and finding the simultaneous solutions of the resulting equations. However, even these equations can be complicated to solve.

Maximum likelihood estimation in the Poisson regression model. Since the Poisson regression model is a generalized linear model, the maximum likelihood

[1] It is the *canonical* link function for *Poisson* observations. See McCullagh and Nelder (1989).

estimates can be found using iteratively reweighted least squares instead of solving them directly. This is the same method we used for the logistic regression model. Start with initial values of the intercept and slopes $\beta_0^{(0)}, \ldots, \beta_p^{(0)}$. (Note: β_0 is the intercept, β_j is slope in direction x_j for $j = 1, \ldots, p$.) It makes it easier if we use vector and matrix notation. Let the observation vector and parameter vector be

$$\mathbf{y} = \begin{pmatrix} y_1 \\ \vdots \\ y_n \end{pmatrix} \quad \text{and} \quad \boldsymbol{\beta} = \begin{pmatrix} \beta_0 \\ \vdots \\ \beta_p \end{pmatrix}$$

respectively. We let the row vector of predictor values for the ith observation be

$$\mathbf{x_i} = (x_{i0}, x_{i1}, \ldots, x_{ip})$$

where $x_0 = 1$ is the coefficient of the intercept. We let the matrix of predictor values for all the observations be

$$\mathbf{X} = \begin{bmatrix} 1 & x_{11} & \ldots & x_{1p} \\ \vdots & \vdots & \vdots\vdots\vdots & \vdots \\ 1 & x_{n1} & \ldots & x_{np} \end{bmatrix}.$$

Following Pawitan (2001), we see that the solution of the maximum likelihood equations for *Poisson* regression can be found by iterating the following steps until convergence. At step n:

1. First update the means for each of the *Poisson* observations by

$$\mu_i^{(n)} = e^{x_i \boldsymbol{\beta}^{(n-1)}},$$

and let the adjusted observations be

$$Y_i^{(n)} = x_i \beta^{(n-1)} + (y_i - \mu_i^{(n)})/\mu_i^{(n)}.$$

The adjusted observations are each a linear function of the corresponding *Poisson* observation, so their variances are given by

$$\begin{aligned} \Sigma_{ii} &= \left(\frac{1}{\mu_i^{(n)}}\right)^2 \hat{Var}(y_i^{(n)}) \\ &= \left(\frac{1}{\mu_i^{(n)}}\right)^2 \mu_i^{(n)} \\ &= \frac{1}{\mu_i^{(n)}}. \end{aligned}$$

Let $\boldsymbol{\Sigma}$ be the diagonal matrix with i^{th} diagonal element σ_{ii} and off diagonal elements 0. $\boldsymbol{\Sigma}$ is the covariance matrix of the adjusted observations.

2. Then update the parameter vector to step n by

$$\beta^{(n)} = (\mathbf{X}\boldsymbol{\Sigma}^{-1}\mathbf{X}')^{-1}\mathbf{X}\boldsymbol{\Sigma}^{-1}\mathbf{Y}^{(n)}.$$

We recognize the equation in step 2 as the weighted least squares estimate on the adjusted observations. Iterating through these two steps until convergence finds the iteratively reweighted least squares estimates.

This is the usual frequentist approach to this model. The maximum likelihood vector is given by

$$\hat{\boldsymbol{\beta}}_{ML} = \lim_{n\to\infty} \boldsymbol{\beta}^{(n)},$$

the limit that the iteratively reweighted least squares procedure converged to, and

$$\mathbf{V}_{ML} = (\mathbf{X}\boldsymbol{\Sigma}^{-1}\mathbf{X}')^{-1}$$

is the "covariance matrix" of the MLE vector where $\boldsymbol{\Sigma}$ is calculated at $\hat{\boldsymbol{\beta}}_{\mathbf{ML}}$. However, $\mathbf{V}_{ML}$ does not have any relationship to the spread of the likelihood function.[2] It is the matched curvature covariance matrix which is the covariance matrix of the multivariate normal having mean vector equal to $\hat{\boldsymbol{\beta}}_{ML}$ that matches the curvature at the MLE. If the likelihood has heavy tails, the real spread may be much larger than indicated by $\mathbf{V}_{ML}$. Using the square roots from this matrix as standard deviations may lead to confidence intervals that have worse coverage probabilities than claimed.

9.2 COMPUTATIONAL APPROACH TO POISSON REGRESSION MODEL

In the computational Bayesian approach, we want to draw a sample from the actual posterior, not its approximation. As we noted before, we know its shape. Our approach will be to use the Metropolis-Hastings algorithm with an independent candidate density. We want a candidate density that is as close as possible to the posterior so many candidates will be accepted. We want the candidate density to have heavier tails than the posterior, so we move around the parameter space quickly. That will let us have shorter burn-in and use less thinning. We use the maximum likelihood vector $\hat{\boldsymbol{\beta}}_{ML}$ and the matched curvature covariance matrix $\mathbf{V}_{ML}$ as the starting point for our procedure.

We approximate the likelihood function by a *multivariate normal*$[\hat{\boldsymbol{\beta}}_{ML}, \mathbf{V}_{ML}]$ where $\hat{\boldsymbol{\beta}}_{ML}$ is the MLE and $\mathbf{V}_{ML}$ is the matched curvature covariance matrix that is output by the iteratively reweighted least squares. We use a *multivariate normal*$[\mathbf{b}_0, \mathbf{V}_0]$ prior for $\boldsymbol{\beta}$, or we can use a "flat" prior if we have no prior information. The approximate posterior will be

$$g(\boldsymbol{\beta}|\mathbf{y}) \quad \propto \quad g(\boldsymbol{\beta})\, f(\mathbf{y}|\boldsymbol{\beta}).$$

[2]Actually, it is the observed Fisher's information matrix of the likelihood. It relates to the *curvature* of the likelihood at the MLE, not the spread.

Since both the prior and (approximate) likelihood are *multivariate normal*, the approximate posterior will be *multivariate normal* also. The updated constants will be

$$\mathbf{V}_1^{-1} = \mathbf{V}_0^{-1} + \mathbf{V}_{\mathbf{ML}}^{-1}$$

and

$$\mathbf{b}_1 = \mathbf{V}_1 \mathbf{V}_0^{-1} \mathbf{b}_0 + \mathbf{V}_1 \mathbf{V}_{\mathbf{ML}}^{-1} \hat{\boldsymbol{\beta}}_{ML}$$

from Equations 4.12 and 4.13. (The constants of approximate posterior will be $\mathbf{b}_1 = \hat{\boldsymbol{\beta}}_{ML}$ and $\mathbf{V}_1 = \mathbf{V}_{ML}$ if we use the flat prior.) We generate candidates from a *multivariate Student's t* with low degrees of freedom that matches the approximate posterior as our candidate density using the following procedure.

1. Find the lower triangular matrix $\mathbf{L}$ such that satisfies

$$\mathbf{L}\,\mathbf{L}' = \mathbf{V}_1$$

   by using the Cholesky decomposition. If we generate $\mathbf{Z}$, a *multivariate normal*$(\mathbf{0}, \mathbf{I})$ of the right dimension by stacking independent *normal*$(0, 1^2)$ variables, then $\mathbf{w} = \mathbf{b}_1 + \mathbf{L}\,\mathbf{z}$ will be *multivariate normal*$(\mathbf{b}_1, \mathbf{V}_1)$. This is the approximate posterior.

2. Instead we stack independent *Student's t* random variables with low degrees of freedom to form $\mathbf{t}$. Applying the same transformation we get the candidate vector

$$\boldsymbol{\beta} = \mathbf{b}_1 + \mathbf{L}\,\mathbf{t}$$

   that has the *multivariate Student's t* $(\mathbf{b}_1, \mathbf{V}_1)$ distribution.

This candidate density will match the posterior near the mode, and will have heavier tails. We are generating a random sample of candidates. Since the candidates are random, you can go from anywhere to anywhere in the parameter space. If the candidate density is the true posterior, all the candidates would be accepted, and would be a random sample from the true posterior. No burn-in or thinning would be required to get the random sample to base our inferences on. In practice, the candidate density will be somewhat similar to the true posterior since its shape is matched to that of the true posterior at the mode so a good proportion of candidates will be accepted. Since it has heavier tails, it will move through the parameter space quickly. Thus, neither a long burn-in nor very much thinning will be required to get an approximately random sample from the posterior that we need for inference. We should note that this will be a sample from the true posterior, not from the approximate posterior, so the credible intervals calculated from it will have the claimed coverage probability provided the random sample is large enough.

Choosing the Multivariate Normal Conjugate Prior

The prior for the parameter vector will be *multivariate normal*$[\mathbf{b_0}, \mathbf{V_0}]$ where

$$\mathbf{b_0} = \begin{pmatrix} b_0 \\ b_1 \\ \vdots \\ b_p \end{pmatrix} \quad \text{and} \quad \mathbf{V_0} = \begin{bmatrix} s_0^2 & 0 & \dots & 0 \\ 0 & s_1^2 & \dots & 0 \\ \vdots & \vdots & \ddots & \vdots \\ 0 & 0 & \dots & s_p^2 \end{bmatrix}.$$

We will find the prior mean and standard deviation for the intercept and each of the slopes. We should always center the predictor variables at their respective means. We recode $x_{ij} = x_{ij} - \overline{x}_{.j}$ where x_{ij} is the value of the jth predictor for the ith observation and $\overline{x}_{.j}$ is the sample mean of the jth predictor. The "average observation" will have all predictor values equal to 0, so it is easier to convert our prior belief about the mean of an "average observation" to the prior for the intercept β_0 by matching percentiles.

Finding normal prior for intercept. Suppose we believe with 95% prior probability that the mean of an "average observation" will lie between l and u. We will find the *normal*(b_0, s_0^2) prior that matches this prior belief. This gives us the two simultaneous equations

$$l = e^{b_0 - 1.96\, s_0} \quad \text{and} \quad u = e^{b_0 + 1.96\, s_0}\,.$$

The solution of these two simultaneous equations is given by

$$b_0 = \frac{\log_e l + \log_e u}{2} \quad \text{and} \quad s_0 = \frac{\log_e u - \log_e l}{3.92}\,.$$

Finding normal prior for slope. We will find the *normal*(b_j, s_j^2) prior for a slope coefficient β_j by matching our prior belief about the ratio of the mean of an "average observation" when x_j is increased by 1 unit compared to an "average observation" due to increasing the predictor x_j by 1 unit. We will use one method when the predictor x_j is a 0:1 variable indicating which of two classes the observation comes from, and another method when the predictor x_j is a continuous variable.

Case 1: x_j is a 0:1 indicator variable. We match two percentiles of our prior belief distribution of the ratio of the mean of an "average observation in group labelled 1" to the mean of an "average observation in group labelled 0." Suppose we believe with 95% prior probability that the ratio will be between v and w. This will give the two equations

$$v = e^{b_j - 1.96\, s_j} \quad \text{and} \quad w = e^{b_j + 1.96\, s_j}$$

which have simultaneous solutions

$$b_j = \frac{\log_e v + \log_e w}{2} \quad \text{and} \quad s_j = \frac{\log_e w - \log_e v}{3.92}\,.$$

Table 9.1 Coal seam fractures data

Obs.	y	x_1	x_2	x_3	x_4	Obs.	y	x_1	x_2	x_3	x_4
1	2	50	70	52	1.0	23	3	65	75	68	5.0
2	1	230	65	42	6.0	24	3	470	90	90	9.0
3	0	125	70	45	1.0	25	2	300	80	165	9.0
4	4	75	65	68	0.5	26	2	275	90	40	4.0
5	1	70	65	53	0.5	27	0	420	50	44	17.
6	2	65	70	46	3.0	28	1	65	80	48	15.0
7	0	65	60	62	1.0	29	5	40	75	51	15.0
8	0	350	60	54	0.5	30	2	900	90	48	35.0
9	4	350	90	54	0.5	31	3	95	88	36	20.0
10	4	160	80	38	0.0	32	3	40	85	57	10.0
11	1	145	65	38	10.0	33	3	140	90	38	7.0
12	4	145	85	38	0.0	34	0	150	50	44	5.0
13	1	180	70	42	2.0	35	0	80	60	96	5.0
14	5	43	80	40	0.0	36	2	80	85	96	5.0
15	2	42	85	51	12.0	37	0	145	65	72	9.0
16	5	42	85	51	0.0	38	0	100	65	72	9.0
17	5	45	85	42	0.0	39	3	150	80	48	3.0
18	5	83	85	48	10.0	40	2	150	80	48	0.0
19	0	300	65	68	10.0	41	3	210	75	42	2.0
20	5	190	90	84	6.0	42	5	11	75	42	0.0
21	1	145	90	54	12.0	43	0	100	65	60	25.0
22	1	510	80	57	10.0	44	3	50	88	60	20.0

Case 2: x_j is a continuous variable. When x_j is a continuous measurement variable, we match our prior belief about the ratio of the mean of an "average observation where x_j is increased by one standard deviation s_x" to the mean of an "average observation." Suppose we believe with 95% prior probability that the ratio is between v and w. This will give the two equations

$$v = e^{(b_j - 1.96\, s_j)\, s_x} \quad \text{and} \quad w = e^{(b_j + 1.96\, s_j)\, s_x}$$

which have simultaneous solutions

$$b_j = \frac{\log_e v + \log_e w}{2\, s_x} \quad \text{and} \quad s_j = \frac{\log_e w - \log_e v}{3.92\, s_x}.$$

Example 13 *Myers et al. (2002) describes an analysis of the number of fractures that occur in the upper seam of mines in the Appalachian region of western Virginia. We want to know how y, the number of fractures that occur in the upper seam of mines, relates to the predictor variables. The predictor variables are: x_1, the inner burden thickness (in feet), x_2, the percent extraction of the lower previously mined seam, x_3, the lower seam height (in feet), and x_4, the length of time mine has been opened (in years). The data is shown in Table 9.1. Twenty iterations of the reweighted*

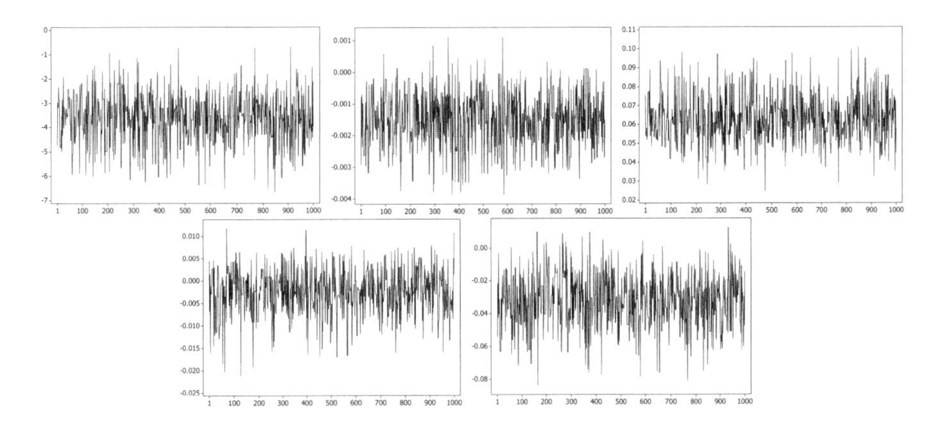

Figure 9.1 Trace plot of 1000 consecutive MH sample from posterior for the intercept β_0 and the slope coefficients $\beta_1, \ldots, \beta_4$.

least squares algorithm is enough to find the maximum likelihood vector

$$\hat{\boldsymbol{\beta}}_{ML} = \begin{pmatrix} -3.59309 \\ -.00141 \\ .06235 \\ -.00208 \\ -.03081 \end{pmatrix}$$

and its matched curvature covariance matrix

$$\mathbf{V}_{ML} = \begin{pmatrix} 1.05204 & .0000974 & -.0120656 & -.0015595 & .0020133 \\ .00010 & .0000007 & -.0000018 & -.0000008 & -.0000013 \\ -.01207 & -.0000018 & .0001510 & .0000042 & -.0000351 \\ -.00156 & -.0000008 & .0000042 & .0000257 & -.0000099 \\ .00201 & -.0000013 & -.0000351 & -.0000099 & .0002645 \end{pmatrix}.$$

Since we have no prior knowledge about the parameters, we will use "flat priors" for the intercept and the slopes. The true posterior will have the same shape as the likelihood, and the approximate posterior will have the same constants that we found for the approximate likelihood. We run the Metropolis-Hastings algorithm 1000 steps using the multivariate Student's t candidate density and the trace plots for the parameters are shown in Figure 9.1. An examination of the trace plots show the chain is moving around the parameter space very well. We calculate the sample autocorrelation to determine the burn-in and thinning that will be required. In Figure 9.2 we see the sample autocorrelation function for the parameters together with 95% confidence intervals. The autocorrelations are clearly indistinguishable from zero by lag 10. We run four chains for 20 steps with random starting points, but with identical random inputs. The traceplots of these chains coupled with the past are shown in Figure 9.3. These show that the chains have clearly converged to the long-run distribution by lag 10.

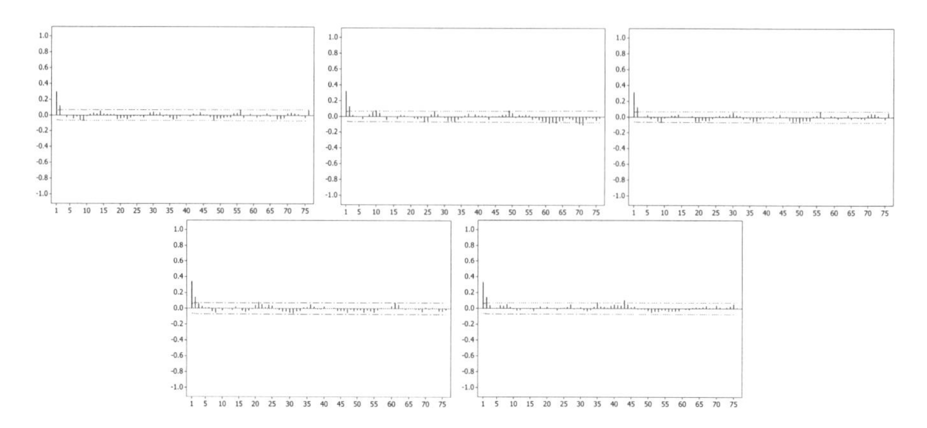

Figure 9.2 Sample autocorrelations of the intercept β_0 and the slopes $\beta_1, \ldots, \beta_4$.

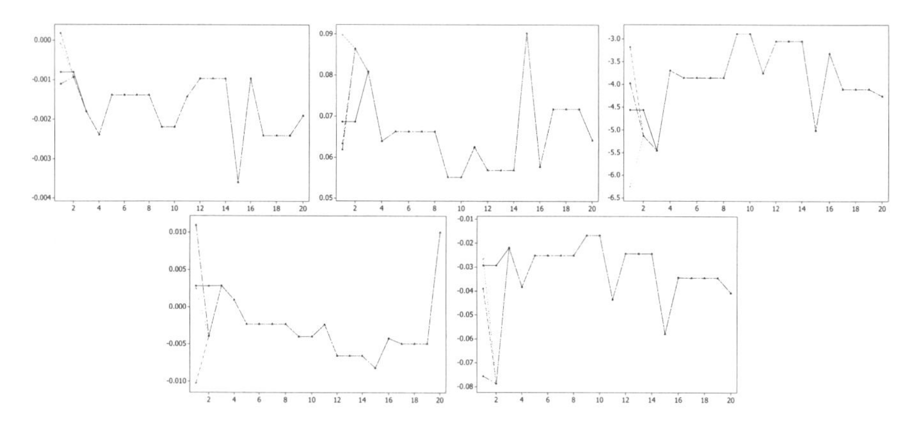

Figure 9.3 Traceplots for four parallel chains coupled with the past for the intercept β_0 and the slopes $\beta_1, \ldots, \beta_4$.

Since the chain is exhibiting very good mixing properties, we won't have to use a very long burn-in time, or thin very much. We run the chain for 40000 MH steps and discard the first 10 for the burn-in, and then use every 10th *draw. This leaves us with a sample of 4000 which we can consider to be approximately a random sample from the true posterior to use for inferences. Histograms of these samples together with the normal approximate posterior with means equal the maximum likelihood and the standard deviations from the square root of the matched curvature variances are shown in Figure 9.4. An examination of these shows that the true posteriors for* $\beta_1, \beta_3, and \beta_4$ *are skewed relative to the normal approximate posterior. This indicates that the confidence intervals found using the "covariance matrix" of the MLE will be too short on one side, and too long on the other. We calculate the mean and the median of the posterior samples, and compare them to the MLE in Table 9.2. We calculate 95% credible intervals from the posterior random samples*

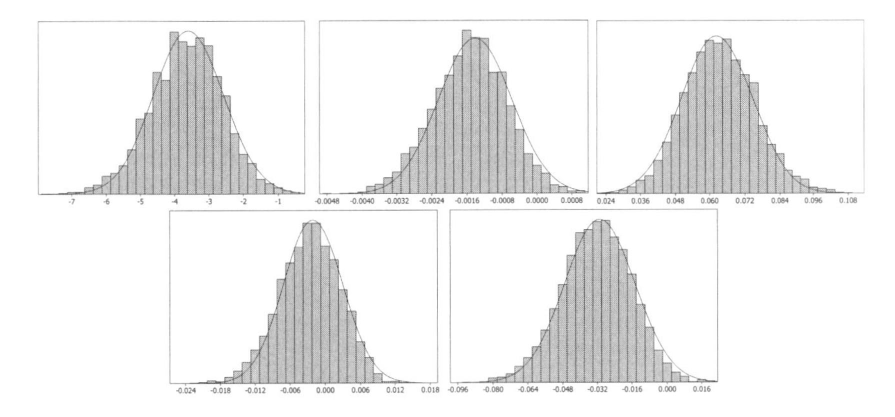

Figure 9.4 Histogram of the thinned MH sample from posterior for intercept β_0 and the slope coefficients $\beta_1, \ldots, \beta_4$.

Table 9.2 Comparing Bayesian point estimates calculated from sample with MLE

Parameter	Posterior Mean	Posterior Median	Posterior St. Dev.	MLE	St. Dev. (from MLE)
β_0	−3.644	−3.631	1.035	−3.593	1.026
β_1	−.0015	−.0015	.0085	−.0014	.0008
β_2	.0633	−.0632	.0124	.0624	.0123
β_3	−.0026	−.0235	.0051	−.0021	.0051
β_4	−.0327	−.0322	.0164	−.0308	.0163

Table 9.3 Comparing credible intervals and confidence intervals for the coefficients

Parameter	95% Credible Interval		95% Confidence Interval	
	Lower Limit	Upper Limit	Lower Limit	Upper Limit
β_0	−5.735	−1.662	−5.603	−1.583
β_1	−.00324	.000085	−.00305	.000223
β_2	.0396	.0882	.0383	.0864
β_3	−.0132	.0069	−.0120	.0079
β_4	−.0665	−.0016	−.0627	.0011

and compare them to the 95% confidence intervals from the MLE and its covariance matrix in Table 9.3. If a slope coefficient $\beta_i = 0$ then the variable x_i is not a useful predictor for mine fracture. We can use the credible intervals to test the hypothesis *that the slope*

$$H_0 : \beta_j = 0 \quad \text{vs} \quad H_1 : \beta_j \neq 0$$

for each $j = 1, \ldots, 4$. We see that β_2, the coefficient of percent extraction form lower previously mined seam, and β_4, the coefficient of the length of time the mine has been opened are both significant at the 5% level since their credible intervals do not contain 0. Those two predictors are clearly useful for the prediction of mine fractures. The other two slope coefficients, β_1 and β_3, cannot be judged significant at the 5% level since their credible intervals contain 0. It is worth noting that the more accurate Bayesian credible interval solution finds β_4 significantly different from 0 at the 95% level, while the less accurate confidence intervals based on the normal approximation to maximum likelihood estimator does not. That is because the normal approximation for the MLE only holds asymptotically, and in this example there are only 44 observations.

9.3 THE PROPORTIONAL HAZARDS MODEL

Sometimes we have observed the times until some event occurs for a sample of individuals or items. In a medical study, we could have observed the survival times of individuals in the study. In an engineering study, we could have observed the time until failure of an object operating in a controlled high stress test setting. Data of this type is called survival time data, and the event is referred to as "death" although it does not have to be an actual death. In Chapter 4 we saw that a *Poisson* process often is used to model the waiting time until an event, and when arrivals occur according to a *Poisson* process, the waiting time distribution follows the *exponential* distribution. We will use the proportional hazards model to relate the survival times to a set of predictor variables that we also observe for each individual.

The Survival Function and the Hazard Function

Let T be the random variable the time until "death" of something. Suppose its density is given by the exponential distribution

$$f(t) = \lambda e^{-\lambda t} \qquad \text{for} \qquad t > 0\,. \tag{9.2}$$

The probability of death by time t is given by the cumulative distribution function (CDF) of the random variable and is

$$\begin{aligned} F(t) &= \int_0^t f(t)dt \\ &= \int_0^t \lambda e^{-\lambda t} dt \\ &= 1 - e^{-\lambda t}\,. \end{aligned} \tag{9.3}$$

The *survival function* is the probability of surviving to time t and is given by

$$\begin{aligned} S(t) &= P(T \geq t) \\ &= 1 - F(t) \\ &= e^{-\lambda t} . \end{aligned} \tag{9.4}$$

The *hazard function* gives the instantaneous probability of death at time t given survival up until time t. It is given by

$$\begin{aligned} h(t) &= \frac{f(t)}{S(t)} \\ &= \lambda . \end{aligned} \tag{9.5}$$

Thus, when time until death follows the exponential distribution, the hazard function will be constant. We will only use this exponential model for survival times. Survival models with non-constant hazard functions are discussed in McCullagh and Nelder (1989).

Censored Survival Time Data

Suppose we have n independent observations from survival models, but each individual has their own constant hazard function. Individual i has hazard function

$$h_i(t) = \lambda e^{\eta_i} .$$

We will express the parameter η_i as a linear function of the predictor variables

$$\eta_i = \sum_{j=1}^{p} x_{i,j}\beta_j .$$

Note: we can absorb the underlying constant hazard rate λ into the linear predictor by including the intercept term $\beta_0 = \log_e \lambda$. At the end of the study, for each individual we have recorded t_i which is either time of death, or time at end of study, and an indicator variable w_i which indicates whether t_i indicates time of death, or t_i indicates the end of the study. In that case, we don't know T_i, the time of death of the ith individual, only that $T_i > t_i$. We say that observation T_i has been *censored*, and w_i is the censoring variable

$$w_i = \begin{cases} 0 & \text{observation is censored} \\ 1 & \text{observation is not censored} \end{cases}$$

Thus $w_i = 1$ when time of death observed exactly, i.e., individual i has died, and $w_i = 0$ when individual is alive at end of study.

Proportional Hazards Model: Likelihood for Censored Survival Data

The contribution to the likelihood of an individual that died is given by $f_i(t)$, and the contribution of an individual that is alive at the end of the study is $S_i(t)$. The likelihood of individual i is

$$\begin{aligned} L_i((t_i, w_i)|\eta_i) &= (f_i(t))^{w_i} (S_i(t))^{1-w_i} \\ &= (\lambda e^{\eta_i} e^{-\lambda t_i e^{\eta_i}})^{w_i} (e^{-\lambda t_i e^{\eta_i}})^{1-w_i} \\ &= (\lambda e^{\eta_i})^{w_i} \times e^{-\lambda t_i e^{\eta_i}} \\ &= e^{-\lambda t_i e^{\eta_i}} [\lambda e^{\eta_i}]^{w_i} \\ &= e^{-\lambda t_i e^{\eta_i}} [\lambda t_i e^{\eta_i}]^{w_i} \times \left(\frac{1}{t_i}\right)^{w_i} . \end{aligned}$$

Since the likelihood of the whole sample equals the product of the individual likelihoods

$$\begin{aligned} L((t_1, w_1), \ldots, (t_n, w_n)|\eta_1, \ldots, \eta_n) &= \prod_{i=1}^{n} L_i((t_i, w_i)|\eta_i) \\ &= \prod_{i=1}^{n} e^{-\lambda t_i e^{\eta_i}} [\lambda t_i e^{\eta_i}]^{w_i} \times \left(\frac{1}{t_i}\right)^{w_i} \\ &= e^{-\sum \lambda t_i e^{\eta_i}} \times \prod_{i=1}^{n} (\lambda t_i e^{\eta_i})^{w_i} \times \prod_{i=1}^{n} t_i^{-w_i} . \end{aligned}$$

Let us reparameterize to the form $\mu_i = \lambda t_i e^{\eta_i}$. The last term $\prod_{i=1}^{n} t_i^{-w_i}$ does not depend on the μ_i, so it can be absorbed into the constant. The proportional form of the joint likelihood is

$$L(w_1, \ldots, w_n|\mu_1, \ldots, \mu_n) \propto e^{-\sum \mu_i} \times \prod_{i=1}^{n} \mu_i^{w_i} . \tag{9.6}$$

We have seen this before in Equation 9.1 where it is the likelihood for a random sample of n independent *Poisson* random variables with parameters μ_i. This means that given λ, we can treat the censoring variables w_i as an independent random sample of *Poisson* random variables with respective parameters μ_i. Suppose we let $\eta_i = \sum_{j=1}^{p} x_{i,j}\beta_j$ be the linear predictor. Taking logarithm of the parameter μ_i we get

$$\log_e(\mu_i) = \log_e(\lambda t_i) + \eta_i ,$$

so the linear predictor is linked to the parameter via the log link function. The term $\log_e(\lambda t_i)$ is called an offset. The coefficient of the offset is assumed to be one and will not be estimated. We can absorb the underlying hazard rate λ by including an intercept β_0 in the linear predictor. In terms of the parameters $\beta_0, \ldots, \beta_p$ the likelihood becomes

$$L(w_1, \ldots, w_n|\beta_0, \ldots, \beta_p) = e^{-t_i \sum e^{x_{i,j}\beta_j}} \times \prod_{i=1}^{n} (t_i \sum x_{i,j}\beta_j)^{w_i} . \quad (9.7)$$

The observations of the censoring variable w_i come from the *Poisson* distribution, a member of the exponential family. The logarithm of the parameter μ, is linked to the linear predictor η. The observations are independent. Clearly the *proportional hazards model* is a generalized linear model and can be analyzed the same way as the *Poisson regression model.*

Piecewise constant hazard rates. Sometimes we know the hazard rate is not constant. We can model this by partitioning the time axis and having a constant hazard rate over each section of the axis. Since the waiting times for the first arrival for a *Poisson Process* have the *exponential* distribution

$$P(T > t + s|T > t) = P(T > s) .$$

Suppose that we have a constant hazard rate λ_0 up until time t_0 and constant hazard rate λ_1 after that. Replace each observation that has $t > t_0$ with two observations. The first one has survival time t_0 and is censored (since survival time is greater than t_0). The second one has survival time $t - t_0$, and is either censored or not depending on whether the original lifetime was observed. The contribution to the likelihood of these two new observations is equivalent to the contribution to the likelihood of the original observation. This procedure can be extended to more than two regions in the obvious manner.

We will follow the same procedure that we used for *Poisson regression*. We will use iteratively reweighted least squares to find the maximum likelihood estimate together with the matched curvature covariance matrix. We will approximate the likelihood by the *multivariate normal* having mean vector equal to the MLE, and covariance matrix matching the curvature of the likelihood at the MLE. We will use either *multivariate normal* prior, or flat priors and find the approximate posterior using the simple updating rules. We can find a heavy-tailed independent candidate distribution from this by using the *multivariate Student's t* with low degrees of freedom instead of the *multivariate normal*. This is the candidate density we will use for the Metropolis-Hastings algorithm. Its shape is similar to the shape of the posterior, so most values will be accepted. It has heavy tails, so it will quickly move through the entire parameter space.

Maximum likelihood estimation in the proportional hazards model. Finding the maximum likelihood estimators from the simultaneous solutions of the equations

$$\frac{\partial log[L(w_1, \ldots, w_n|\beta_0, \ldots, \beta_p)]}{\partial \beta_j} \quad \text{for} \quad j = 0, \ldots, p$$

may be difficult to do. Instead, since we know the proportional hazards model is a generalized linear model, we can find them by iteratively reweighted least squares. Start with initial values of the intercept and slopes $\beta_0^{(0)}, \ldots, \beta_p^{(0)}$. (Note: β_0 is the intercept, β_j is slope in direction x_j for $j = 1, \ldots, p$.) For the *proportional hazards model* the solution can be found by iterating the following steps until convergence. At step n

1. First update the following

$$\mu_i^{(n)} = t_i\, e^{x_i \boldsymbol{\beta}^{(n-1)}},$$

$$Y_i^{(n)} = x_i \beta^{(n-1)} + (y_i - \mu_i^{(n)})/\mu_i^{(n)},$$

$$\Sigma_{ii} = (1/\mu_i^{(n)})^2 \mu_i^{(n)} = 1/\mu_i^{(n)}.$$

   Note that only the first equation is slightly changed from the *Poisson* regression case due to the offset. The other two updating equations remain the same as before.

2. Then update the parameter vector to step n by

$$\boldsymbol{\beta}^{(n)} = (\mathbf{X}\boldsymbol{\Sigma}^{-1}\mathbf{X}')^{-1}\mathbf{X}\boldsymbol{\Sigma}^{-1}\mathbf{Y}^{(n)}.$$

The usual frequentist approach to this model is to use the maximum likelihood vector

$$\hat{\boldsymbol{\beta}}_{ML} = \lim_{n \to \infty} \boldsymbol{\beta}^{(n)},$$

the limit that the iteratively reweighted least squares procedure converged to, and $\mathbf{V}_{ML}$ the "covariance matrix" of the MLE vector as the basis for inferences. However, we know that this matched curvature covariance matrix does not have any relationship to the spread of the likelihood function. Instead it relates to the *curvature* of the likelihood at the MLE, not the spread. This is a local not a global property of the likelihood. It is the covariance matrix of the multivariate normal having mean vector equal to $\hat{\boldsymbol{\beta}}_{ML}$ that matches the curvature at the MLE. If the likelihood has heavy tails, the real spread may be much larger than indicated by $\mathbf{V}_{ML}$. Using the square roots from this matrix as standard deviations may lead to confidence intervals that have very different coverage probabilities than those claimed.

9.4 COMPUTATIONAL BAYESIAN APPROACH TO PROPORTIONAL HAZARDS MODEL

In the computational Bayesian approach, we want to draw a sample from the actual posterior, not its approximation. The proportional likelihood is given in Equation 9.7, and multiplying by the prior will give the true proportional posterior. Our approach will be to use the Metropolis-Hastings algorithm with an independent candidate density. We want a candidate density that has shape similar to the shape of the true posterior so many candidates will be accepted. We also want the candidate density to have heavier tails than the true posterior, so we move around the parameter space quickly. That will allow us to have a shorter burn-in time and use less thinning. Again, we will use the maximum likelihood vector $\hat{\boldsymbol{\beta}}_{ML}$ and the matched curvature covariance matrix $\mathbf{V}_{ML}$ as the starting point for our procedure.

We approximate the likelihood function by a *multivariate normal*$[\hat{\boldsymbol{\beta}}_{ML}, \mathbf{V}_{ML}]$ where $\hat{\boldsymbol{\beta}}_{ML}$ is the MLE and $\mathbf{V}_{ML}$ is the matched curvature covariance matrix that is output by the iteratively reweighted least squares. We use a *multivariate normal*$[\mathbf{b_0}, \mathbf{V_0}]$ prior for $\boldsymbol{\beta}$, or we can use a "flat" prior if we have no prior information. The approximate posterior will be

$$g(\boldsymbol{\beta}|\mathbf{data}) \quad \propto \quad g(\boldsymbol{\beta})\, f(\mathbf{data}|\boldsymbol{\beta})\,.$$

Since both the prior and (approximate) likelihood are *multivariate normal* so will the approximate posterior. The updated constants will be

$$\mathbf{V_1^{-1}} \quad = \quad \mathbf{V_1^{-1}} + \mathbf{V_{ML}^{-1}}$$

and

$$\mathbf{b_1} \quad = \quad \mathbf{V_1 V_0^{-1} b_0} + \mathbf{V_1 V_{ML}^{-1}} \hat{\boldsymbol{\beta}}_{ML}$$

from Equations 4.12 and 4.13. (The constants of approximate posterior will be the $\mathbf{b_1} = \hat{\boldsymbol{\beta}}_{ML}$ and $\mathbf{V_1} = \mathbf{V}_{ML}$ if we use the flat prior.) We generate candidates from a *multivariate Student's t* with low degrees of freedom that has a similar shape to the approximate posterior and has heavier tails using the following procedure.

1. Find the lower triangular matrix $\mathbf{L}$ such that satisfies

$$\mathbf{L}\,\mathbf{L}' = \mathbf{V_1}$$

   by using the Cholesky decomposition. If we generate a $\mathbf{Z}$ a *multivariate normal*$(\mathbf{0}, \mathbf{I})$ of the right dimension by stacking independent *normal*$(0, 1^2)$ variables, then $\mathbf{w} = \mathbf{b_1} + \mathbf{L}\,\mathbf{z}$ will be *multivariate normal*$(\mathbf{b_1}, \mathbf{V_1})$. This would a random draw from the approximate posterior.

2. Instead we stack independent *Student's t* random variables with low degrees of freedom to form $\mathbf{t}$. Applying the same transformation the candidate vector

$$\mathbf{w} = \mathbf{b_1} + \mathbf{L}\,\mathbf{t}$$

   will have the *multivariate Student's t*$(\mathbf{b_1}, \mathbf{V_1})$ distribution.

This candidate density will match the posterior near the mode, and will have heavier tails. We are generating a random sample of candidates. Since the candidates come from a heavy-tailed independent candidate distribution, you can go from anywhere to anywhere in the parameter space. If the candidate density is the true posterior, all the candidates would be accepted, and would be a random sample from the true posterior. No burn-in or thinning would be required to get the random sample to base our inferences on. In practice, the candidate density will be somewhat similar to the true posterior since its shape is matched to that of the true posterior at the mode so a good proportion of candidates will be accepted. Since it has heavier tails, it will move through the parameter space quickly. We will not have to use a long burn-in time, nor will we have to do a lot of thinning in order to get a sample from the posterior that is approximately the random sample we need for inference. We should note that this will be a sample from the true posterior, not from the approximate posterior, so the credible intervals calculated from it will have the claimed coverage probability provided the random sample is large enough.

Choosing the Multivariate Normal Conjugate Prior

The prior for the parameter vector will be *multivariate normal*$[\mathbf{b_0}, \mathbf{V_0}]$ where

$$\mathbf{b_0} = \begin{pmatrix} b_0 \\ b_1 \\ \vdots \\ b_p \end{pmatrix} \quad \text{and} \quad \mathbf{V_0} = \begin{bmatrix} s_0^2 & 0 & \dots & 0 \\ 0 & s_1^2 & \dots & 0 \\ \vdots & \vdots & \ddots & \vdots \\ 0 & 0 & \dots & s_p^2 \end{bmatrix}.$$

We will find the prior mean and standard deviation for the intercept and each of the slopes. We should always center the predictor variables at their respective means. We recode $x_{ij} = x_{ij} - \overline{x}_{.j}$ where x_{ij} is the value of the jth predictor for the ith observation. The "average individual" will have all predictor values equal to 0, so it is easier to convert our prior belief about the "average individual" to the prior for the intercept β_0 by matching percentiles.

Finding normal prior for intercept. We use our prior belief about the survival function for the "average individual" that is given by

$$S(t) = e^{-\beta_0\, t}\,.$$

We summarize our prior belief into two percentiles of the survival function. For example, suppose we believe that the 50th percentile and the 5th percentiles of the survival time distribution of the "average" individual are $ST_{.5}$ and $ST_{.05}$ respectively. (5% of "average" individuals live past $ST_{.05}$.) Then the equation for the prior mean b_0 and prior standard deviations s_0 are given by

$$.5 = e^{-b_0\, ST_{.5}} \quad \text{and} \quad .05 = e^{-(b_0 + 1.645\, s_0)\, ST_{.05}}$$

respectively. The solutions are given by

$$b_0 = \frac{-\log_e(.5)}{ST_{.5}} \quad \text{and} \quad s_0 = \frac{\log_e(.5)\frac{ST_{.05}}{ST_{.5}} - \log_e(.05)}{1.645\, ST_{.05}}\,.$$

Finding normal prior for a slope. We will find the *normal*(b_j, s_j^2) prior for the slope coefficient β_j by matching our prior belief about the increase in hazard rate due to increasing the predictor x_j by 1 unit. From Equation 9.5 the hazard rate for an individual with $x_j = 1$ will be $\beta_0 + \beta_1$ and the hazard rate for an individual with $x_j = 0$ will be β_0. Thus β_1 measures the increase in hazard rate due to an increase in the predictor x_j by 1 unit. We will use one method when the the predictor x_j is a 0:1 variable indicating which of two classes the observation comes from, and another method when the predictor x_j is a continuous variable.

Case 1: x_j is a 0:1 indicator variable. We match two percentiles of our prior belief distribution of the increase in the hazard rate due to membership in group labelled 1. We summarize our prior belief about the increase in hazard rate into two percentiles and match them to the *normal*(b_j, s_j^2) prior. Suppose we believe that membership in group 1 is equally likely to increase or decrease the hazard rate by v. That means the median of our belief distribution about the increase in hazard rate equals v. Suppose we also believe with 95% prior probability that the increase in hazard rate will be less than w. This gives us the equations

$$e^{b_j} = v \quad \text{and} \quad e^{b_j + 1.645\, s_j} = w\,.$$

The simultaneous solutions

$$b_j = \log_e v \quad \text{and} \quad s_j = \frac{\log_e w - \log_e v}{1.645}$$

give our prior mean and standard deviation for β_j.

Case 2: x_j is a continuous variable. When x_j is a continuous measurement variable, we match our prior belief about the increase in hazards ratio for an observation where the value of x_j is at one standard deviation above its mean compared to an observation where the value of $x_j = 0$. Suppose we believe that the increase in hazard ratio is equally likely to be below or above v. Suppose we also believe with 95% prior probability that the increase in hazard ratio will be less than w. These gives us the simultaneous equations

$$e^{b_j\, s_x} = v \quad \text{and} \quad e^{(b_j + 1.645\, s_j)\, s_x} = w$$

respectively, where s_x is the standard deviation of the predictor x_j. The simultaneous solutions are

$$b_j = \frac{\log_e v}{s_x} \quad \text{and} \quad s_j = \frac{\log_e w - \log_e v}{1.645\, s_x}\,.$$

Example 14 *As reported in Swanson et al. (2007), a cardiac study was done on the survival outcomes for 527 patients admitted with a myocardial infarction (MI) at Waikato Hospital from 1995 to 2005. The variables collected on each patient are divided into three groups. The demographic variables are age, sex, and ethnicity (Maori and Pacific Islander, or other). The medical variables include previous MI,*

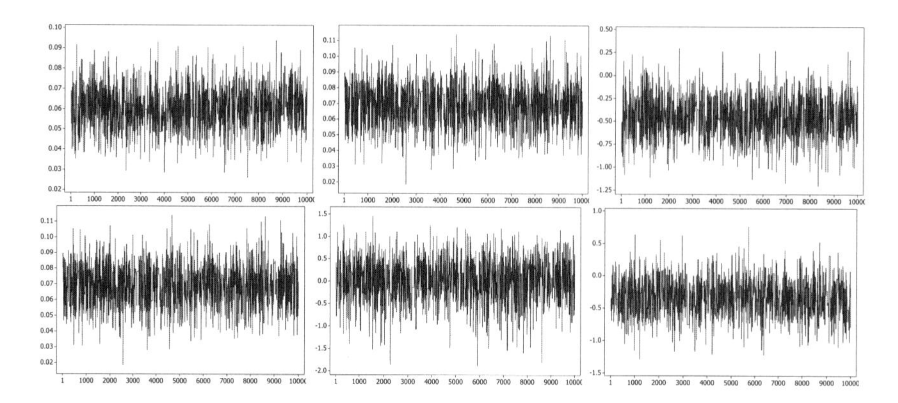

Figure 9.5 Trace plots for age, shock, and stent for first 30 days, and after 30 days, respectively.

diabetes, smoking history, shock, renal failure, and vessel number. The treatment variables are stent used and IIBIIIaInhibitor. We want to determine how these variables affect the survival time, both for the first 30 days and after surviving the first 30 days. The observation for a patient i that survived longer than 30 days is replaced by two observations: the first a censored observation for 30 days with $t_i = 30$, and the second an observation for the remainder of time $t_i = t_i - 30$ and censored or uncensored depending on whether the original observation for patient i was censored or not. This gives us 1006 observations. Also, each predictor x_j was replaced by two predictors: the first with values equal to 0 for all observations where $t_i > 30$, and the second with values equal to 0 for observations where $t_i \leq 30$. This enables the effect of the predictor to be different in the acute stage of the MI (first 30 days) and long-term stage (after 30 days). (Note: we do not include the vessel number in the second time period since it is exactly correlated with the ethnicity for that period.) We also create a new variable that has value 1 for observations in the first 30 days, and 0 otherwise to allow the two stages to have different intercepts.

Since we have no prior knowledge about the parameters, we will use "flat priors." The true posterior will have the same shape as the likelihood, and the approximate posterior will have the same constants that we found for the approximate likelihood. We run the Metropolis-Hastings algorithm 10000 steps using the multivariate Student's t candidate density and the trace plots of the MH sample for a selection of the parameters are shown in Figure 9.5, and the sample autocorrelation functions are shown in Figure 9.6. We see that the chain is moving through the parameter space rapidly, so we won't require too long a burn-in or too much thinning. Figure 9.7 gives the traceplots for four multiple parallel Metropolis-Hastings chains starting from different overdispersed starting points but with the same random inputs. We see from Figures 9.5–9.7 that a burn-in of 50 steps and thinning by taking every 50th will be sufficient to insure a random sample from the posterior.

We run the chain for 100000 MH steps, and discard the first 50 for the burn-in and then use every 50th. This leaves a thinned sample of 2000, which we can consider

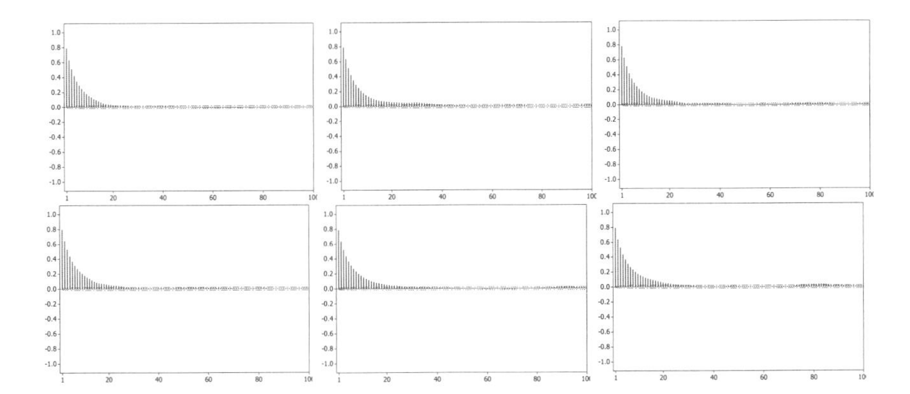

Figure 9.6 Sample autocorrelations for Age, Shock, and Stent for first 30 days, and after 30 days, respectively.

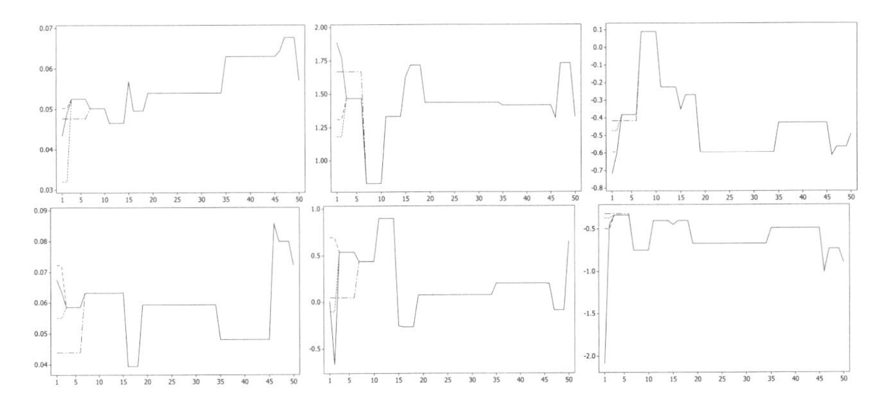

Figure 9.7 The outputs from multiple parallel Metropolis-Hastings chains for coefficients of age, shock, and stent for first 30 days, and after 30 days, respectively.

to be a random sample from the posterior. The summary statistics are shown in Table 9.4. The first two columns are the mean and standard deviation of the β coefficients. The third and fourth column are the standardized variable z and the two-tailed posterior probability for testing that the β coefficient is equal to 0. The fifth column is the odds ratio for that predictor, and the sixth and seventh column are the lower and upper limits for a 95% credible interval for the odds ratio. We see that age is significant for both surviving the initial MI (the first 30 days), and long-term (after 30 days). Ethnicity is significant for survival in the first 30 days, but not significant for long-term survival. Smoking status does not significantly affect the survival of the initial MI, but its effect on the long-term survival of the patient is almost significant at the 5% level. Entering into shock or having renal failure during the MI is very significant for surviving the initial MI, but given that a patient survives the initial MI, the shock or renal failure at the time of the MI does not significantly

Table 9.4 Summary statistics of thinned MCMC sample

Coefficient	Mean	St. Dev.	z	p		Odds	Credible Int. Lower	Upper
Intercept	–13.815	.9540	–14.481	.0000	*	.000	.000	.000
ThirtyDays	4.7670	1.2265	3.887	.0001	*	117.6	10.6	1301
Age1	.0599	.0096	6.234	.0000	*	1.062	1.04	1.08
Sex1	.3588	.2081	1.724	.0847		1.432	.952	2.15
Ethnicity1	.5947	.2914	2.041	.0412	*	1.813	1.024	3.21
Diabetes1	.0440	.2437	.180	.8569		1.045	.648	1.68
Smoking1	.2341	.2180	1.074	.2829		1.264	.824	1.94
Shock1	1.4438	.2586	5.583	.0000	*	4.237	2.552	7.03
RenalFail1	1.2703	.4049	3.137	.0017	*	3.562	1.611	7.88
PrevMI1	.1802	.2149	.839	.4018		1.197	.786	1.82
Vessel1	–.1293	.2154	–.600	.5484		.879	.576	1.34
Stent1	–.4500	.2101	–2.141	.0322	*	.638	.422	.96
Inhibitor1	.0076	.2019	.038	.9700		1.008	.678	1.50
Age2	.0662	.0124	5.329	.0000	*	1.068	1.043	1.09
Sex2	.1486	.2592	.573	.5664		1.160	.698	1.93
Ethnicity2	–.3480	.2675	–1.301	.1933		.706	.418	1.19
Diabetes2	.2927	.2893	1.012	.3116		1.340	.760	2.36
Smoking2	.5403	.2779	1.944	.0518		1.717	.996	2.96
Shock2	.0558	.4717	.118	.9058		1.057	.420	2.67
RenalFail2	.1594	.6893	.231	.8171		1.173	.304	4.53
PrevMI2	.5357	.2684	1.996	.0459	*	1.709	1.010	2.89
Stent2	–.4328	.2580	–1.678	.0934		.649	.391	1.08
Inhibitor2	.0507	.2560	.198	.8429		1.052	.637	1.74

affect his/her long-term survival. Having a stent inserted is significant for surviving the initial MI. It appears to be helping to reduce long-term risk (odds .649) but this is not seen to be significant at the 5% level.

Modelling Issues: Removing Unnecessary Variables

Including extra predictor variables that do not affect the linear predictor will lead to poor predictions. In other words, if the true coefficient $\beta_i = 0$, then we should not include variable x_i as a predictor since the parameter will not be related to it. We will get better predictions for new data if we leave out the unnecessary predictors. This is known as the principal of parsimony. We would like to remove from the predictor set all variables x_j where the coefficient $\beta_j = 0$. Of course we don't know which β_j coefficients are actually equal to zero. This is discussed from the frequentist perspective in Bowerman and O'Connell (1990) and Neter, Wasserman, and Kutner (1990).

In the computational Bayesian approach, we have a random sample from the joint posterior distribution. Since the predictor variables are often correlated, the joint posterior distribution of the coefficients cannot be factored into independent posteriors of the individual coefficients. When the predictors are correlated they can be acting as alias's to each other. This means one predictor can mask the effect of another. If we look at them one at a time it could be that the effect of each one is being masked by the others. So, if we have a set of predictors $x_1, \ldots, x_k$ that we consider could be removed, we should test the null hypothesis that all their coefficients are equal to zero simultaneously, versus the alternative that at least one of them is not equal to zero, i.e.,

$$H_0 : \beta_1 = \ldots = \beta_k \quad \text{versus} \quad H_1 : \beta_j \neq 0 \text{ for some } j = 1, \ldots, k\,.$$

We will give an approximate test for this simultaneous hypothesis. Let the vector of the coefficients in question be

$$\boldsymbol{\beta} = \begin{pmatrix} \beta_1 \\ \vdots \\ \beta_k \end{pmatrix}.$$

Let the vector of the means from the posterior sample and the matrix of covariances from the posterior sample for $\boldsymbol{\beta}$ be

$$\bar{\boldsymbol{\beta}} = \begin{pmatrix} \bar{\beta}_1 \\ \vdots \\ \bar{\beta}_k \end{pmatrix} \quad \text{and} \quad \hat{\mathbf{V}} = \begin{pmatrix} \hat{V}_{11} & \ldots & \hat{V}_{1k} \\ \vdots & \ddots & \vdots \\ \hat{V}_{k1} & \ldots & \hat{V}_{kk} \end{pmatrix}$$

respectively. The test statistic be

$$U = \bar{\boldsymbol{\beta}}' \, \hat{\mathbf{V}}^{-1} \, \bar{\boldsymbol{\beta}} \tag{9.8}$$

will be approximately[3] *chi-squared* with k degrees of freedom when the null hypothesis is true. We can reject the null hypothesis at the 5% level of significance if $U > U_{.05}$ where $U_{.05}$ is the upper 5% point of the *chi-squared* distribution[4] with k degrees of freedom. If we cannot reject the null hypothesis, then we can remove all the specified predictors from the model. Then we should redo the analysis without those predictors.

[3] U measures how far the zero vector is away from the posterior mean vector using distance measure based on covariance from the posterior sample.

[4] This test will have robustness properties similar to analysis of variance tests.

Example 14 (continued) *Looking at Table 9.3 we see that the predictor variables diabetes1,vessel1, diabetes2, shock2, and renalfail2 may have coefficient equal to 0 and are thus candidates to be removed. Other predictor variables that appear not to be significant in either time period such as sex1 and sex2 are of interest since we want to be able to model both males and females, so we would not consider removing them. We are also interested in whether a previous MI or the treatment IIBIIIAInhibitor has an effect on survival either in the short term or long term, so we do not consider removing those variables. The sample mean vector of the posterior means of the candidates for removal is*

$$\bar{\beta} = \begin{pmatrix} 0.043957 \\ -0.129312 \\ 0.292736 \\ 0.055843 \\ 0.159425 \end{pmatrix}$$

and the covariance matrix of the posterior sample of candidates for removal is

$$\hat{\mathbf{V}} = \begin{pmatrix} .0593886 & .0075251 & .0018124 & -.001303 & .002145 \\ .0075251 & .0464127 & .0003371 & -.003494 & .002140 \\ .0018124 & .0003371 & .0836990 & .006509 & .006242 \\ -.0013034 & -.0034943 & .0065093 & .222535 & .011195 \\ .0021453 & .0021398 & .0062421 & .011195 & .475136 \end{pmatrix}.$$

The test statistic

$$\begin{aligned} U &= \bar{\beta}' \hat{\mathbf{V}}^{-1} \bar{\beta} \\ &= 1.4945 \end{aligned}$$

is much less than the upper 5% tail point of the chi-squared distribution with 5 *degrees of freedom, so we cannot reject the null hypothesis. We conclude those slope coefficients all equal to zero, so we will remove the corresponding predictors and redo the analysis using the remaining predictors.*

We run a chain using only the remaining predictors 100000 steps. We use a burn-in of 50, and thin using every 50th *to get a sample of 2000 that we can consider to be a random sample from the posterior to be used for inference. Summary statistics for this sample are shown in Table 9.4. We see that the predictor variables age1, ethnicity1, shock1, renalfailure1, stent1, age2, and prevMI2 are seen to significantly affect survival at the 5% level. The magnitudes of other variables are not large enough for us to conclude from this study that they are significant at that level. However, it is best to leave them in the model for making predictions. Not finding them significant says something about the size of the effects. It does not say that the effects do not exist. In particular, the effects of smoking and of stent in the long-term period would not be judged significant at the 5% level, yet they appear almost significant at that level, so they should be left in.*

Table 9.5 Summary statistics of thinned MCMC sample with unnecessary variables removed

Coefficient	Mean	St.Dev	z	p		odds	Credible Int.	
							Lower	Upper
Intercept	−13.777	.9320	−14.782	.0000	*	.000	.000	.00
ThirtyDays	4.7118	1.1783	3.9988	.0001	*	111.2	11.0	1120
Age1	.0593	.0098	6.0254	.0000	*	1.061	1.041	1.08
Sex1	.3456	.2136	1.6176	.1058		1.413	.930	2.15
Ethnicity1	.5905	.2875	2.0540	.0400	*	1.805	1.027	3.17
Smoking1	.2334	.2195	1.0637	.2875		1.263	.821	1.94
Shock1	1.5133	.2317	6.5313	.0000	*	4.542	2.884	7.15
RenalFail1	1.2475	.4148	3.0075	.0026	*	3.482	1.544	7.85
PrevMI1	.1701	.2095	.8116	.4170		1.185	.786	1.79
Stent1	−.4518	.2170	−2.0817	.0374	*	.636	.416	.97
Inhibitor1	.0029	.2020	.0143	.9886		1.003	.675	1.49
Age2	.0675	.0123	5.5015	.0000	*	1.070	1.044	1.10
Sex2	.1543	.2556	.6039	.5459		1.167	.707	1.93
Ethnicity2	−.3920	.2536	−1.5455	.1222		.676	.411	1.11
Smoking2	.5241	.2726	1.9225	.0545		1.689	.990	2.88
PrevMI2	.5622	.2573	2.1851	.0289	*	1.754	1.060	2.91
Stent2	−.4497	.2558	−1.7581	.0787		.638	.386	1.05
Inhibitor2	.0408	.2576	.1583	.8742		1.042	.629	1.73

Survival Curves

One of the advantages of using the parametric proportional hazards model is that estimated survival curves for individuals given their predictor values can be calculated from the data. In this model, each individual has his/her own constant hazard rate. Let h_i be the hazard rate for individual i. For that individual, the time until death follows the exponential distribution given by Equation 9.3 with parameter $\lambda = h_i$. From Equation 9.4, the survival function for individual i is given by

$$\begin{aligned} S_i(t) &= e^{-\int h_i \, dt} \\ &= e^{-h_i \, t} . \end{aligned}$$

The linear predictor for i^{th} individual is given by

$$\eta_i = \sum_{j=0}^{p} \hat{\beta}_j x_{ij} .$$

We can plug in this value and compute hazard function

$$h_i = e^{\eta_i}$$

so the survival curve is given by

$$S_i(t) = e^{-e^{\eta_i}\, t}\,.$$

Example 14 (continued) *Suppose we decide to compare the estimated survival curves for a 50-year-old male who is not a Maori or Pacific Islander, who does not smoke, who does not go into shock or have renal failure, and who has not had a previous MI for the case when he was treated with a stent and not given the IIBIIIAinhibitor against the case where he was not given a stent. The intercept coefficient is*

$$\hat{\beta}_0 = \begin{cases} -13.777 + 4.7118 & \text{for the first 30 days} \\ -13.777 & \text{after 30 days} \end{cases},$$

the coefficient of age is

$$\hat{\beta}_1 = \begin{cases} .0593 & \text{for the first 30 days} \\ .0675 & \text{after 30 days} \end{cases},$$

and the coefficient for the stent is

$$\hat{\beta}_8 = \begin{cases} -.4518 & \text{for the first 30 days} \\ -.4497 & \text{after 30 days} \end{cases}.$$

All the other coefficients are equal to 0. This gives the constant hazard rate

$$h_i = \begin{cases} e^{-14.1626+5.0844+50\times.0596-.4447} & \text{for the first 30 days} \\ e^{-14.1626+50\times.0672-.2659} & \text{after 30 days} \end{cases}.$$

The two survival curves are shown in Figure 9.8. We see that being given a stent as part of the treatment for the MI is beneficial to the patients survival.

Main Points

- The *Poisson regression model* allows each observation to come from the *Poisson* distribution having its own parameter μ_i, which is linearly regressed on the known values of the predictors for that observation.
- The link function relates the linear predictor to a function of the parameter. The *log link* function is commonly used for the *Poisson regression model*.
- The *Poisson regression model* is an example of the *generalized linear model*. The maximum likelihood estimates of the coefficients of the predictors can be found by iteratively reweighted least squares. This also finds the covariance matrix of the normal distribution that matches the curvature of the likelihood

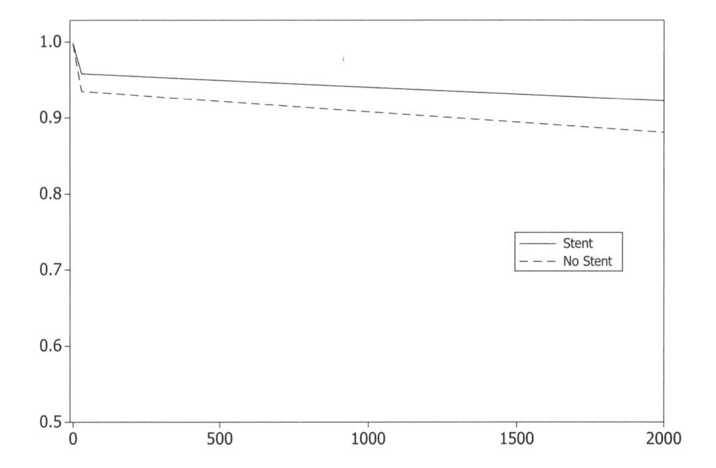

Figure 9.8 Comparing the survival curves for 50-year-old male non-Maori patient who has not had a previous MI and does not suffer shock or renal failure and was not given the Inhibitor when given a stent or not given stent. Time is in days from the MI.

at the maximum. This covariance matrix does not depend on the spread of the likelihood function, only its curvature at a single point, the maximum.

- The multivariate normal approximation to the likelihood together with either a multivariate normal prior or multivariate flat prior can be used to find a normal approximation to the posterior.

- A sample from the true posterior can be found using the Metropolis-Hastings algorithm with an independent candidate density having the same mean vector and correlation structure as the approximate posterior, but with heavier tails that come from using *Student's t* with low degrees of freedom instead of *normal*.

- This chain will move rapidly over the whole parameter space, so neither a long burn-in time nor much thinning will be required to achieve an (approximate) random sample from the true posterior in order to do inferences.

- The *proportional hazards model* is used to regress censored survival data on when the values of a set of predictor variables are known for each observation. For each observation we record t_i, its time of death, w_i the censoring variable which equals 1 if we observe the true death t_i, and 0 when the observation is censored and all we know the true lifetime is greater than t_i, and the values of the predictors $x_{i1}, \ldots, x_{ip}$ for that observation.

- A constant underlying hazard rate λ can be estimated by including an intercept β_0 is the linear predictor.

- We often use a piecewise linear hazard rate by including new variables. We can also allow the values of the coefficients of the predictor variables to change at these change points.

- The likelihood turns out to be the same as if the censoring variables w_i are a sample of *Poisson* random variables where the means are linear functions of the predictors. So this is like the *Poisson regression model.*
- The maximum likelihood and the matched curvature covariance matrix can be found by iteratively reweighted least squares.
- A multivariate normal approximation to the posterior can be found using the normal approximation to the likelihood, and either a multivariate normal prior, or a multivariate flat prior.
- A sample from the true posterior can be found using the Metropolis-Hastings algorithm with an independent candidate density having same mean vector and correlation structure, but with heavier tails from using *Student's t* with low degrees of freedom instead of the *normal*.
- This chain will have good mixing properties so an (approximately) random sample from the true posterior can be found without having to use a long a burn-in or do very much thinning.
- As in any regression models, better predictions can be found after removing any coefficients that we can't conclude are different than 0.

Exercises

9.1 The data from Table 9.1 are given in Minitab worksheet Exercise9.1.mtw. We wish to find out how y, the number of fractures that occur in an upper seam of a coal mine depends on the predictor variables x_1, the inner burden thickness, x_2, the percent extraction of the lower previously mined seam, x_3 the lower seam height, and x_4 the length of time mine has been opened. We will use the Poisson regression model. Since we don't have any particular prior knowledge we will use "flat" priors for the parameters.

   (a) Use the Minitab macro BayesPoisRegMH.mac or the equivalent R-function `BayesPois` to run 1000 steps of the Metropolis-Hastings algorithm with a heavy-tailed candidate density based on the matched curvature *normal* approximation to the posterior.

      i. Comment on the mixing properties of the chain shown in the trace-plots.

      ii. Comment on the graphs of the sample autocorrelations.

   (b) Use the Minitab macro BayesPoisRegMH.mac or the equivalent R-function `BayesPois` to run six chains from different starting points, but with the same random inputs, a process known as coupling with the past. Run each of the chains 20 steps. Plot the traceplots for each of the parameters for all six chains to see how many steps are needed until all six chains have converged.

(c) Decide on a burn-in and thinning required based on (a) and (b). Use the Minitab macro BayesPoisRegMH.mac or the equivalent R-function `BayesPois` to run the chain 10000 steps. (Do this in two blocks of 5000 each.)

(d) Use the burn-in and thinning found in (c) so the thinned sample is approximately a random sample of the posterior. Graph histograms for each parameter using the thinned sample.

(e) Calculate 95% credible intervals for $\beta_1, \ldots, \beta_4$ from the posterior sample.

(f) What conclusions can you draw about the effects of the predictors $x_1, \ldots, x_4$ on the number of fractures y.

9.2 The Minitab worksheet Exercise9.2.mtw contains the values of the response y and three predictors $x_1, \ldots, x_3$. We want to use Poisson regression to find the effects of the predictors on the response. (Note: first we want to center the predictor values at their respective means.)

(a) First we have to find our *normal* prior for each component.

i. Suppose we believe with 95% prior probability that the mean of "an average observation" will lie between 10 and 20. Calculate the prior mean and standard deviation of the intercept.

ii. Suppose we believe with 95% prior probability that the ratio of the mean of "an average observation in group $x_1 = 1$" to the mean of "an average observation in group $x_1 = 0$" lies between .5 and 2. Calculate the prior mean and standard deviation of the coefficient of x_1.

iii. Suppose we believe with 95% prior probability that the ratio of the mean of "an average observation where x_2 is increased by one standard deviation above its mean" to the mean of "an average observation where x_2 is at its mean" lies between .5 and 2. Calculate the prior mean and standard deviation of the coefficient for x_2.

iv. Suppose we believe with 95% prior probability that the ratio of the mean of "an average observation where x_3 is increased by one standard deviation above its mean" to the mean of "an average observation where x_3 is at its mean" lies between .5 and 2. Calculate the prior mean and standard deviation of the coefficient for x_3.

Put these together into a *multivariate normal* prior with independent components.

(b) Use the Minitab macro BayesPoisRegMH.mac or the equivalent R-function `BayesPois` to run 1000 steps of the Metropolis-Hastings algorithm with a heavy-tailed candidate density based on matched curvature *normal* approximation to the posterior. Use the prior we found in part(a)

i. Comment on the mixing properties of the chain shown in the trace-plots.

ii. Comment on the graphs of the sample autocorrelations.

(c) Use the Minitab macro BayesPoisRegMH.mac or the equivalent R-function `BayesPois` to run six chains from different starting points, but with the same random inputs, a process known as coupling with the past. Run each of the chains 20 steps. Plot the traceplots for each of the parameters for all six chains to see how many steps are needed until all six chains have converged.

(d) Decide on a burn-in and thinning required based on (a) and (b). Use the Minitab macro BayesPoisRegMH.mac or the equivalent R-function `BayesPois` to run the chain 10000 steps. (Do this in two blocks of 5000 each.)

(e) Use the burn-in and thinning found in (c) so the thinned sample is approximately a random sample of the posterior. Graph histograms for each parameter using the thinned sample.

(f) Calculate 95% credible intervals for $\beta_1, \ldots, \beta_3$ from the posterior sample.

(g) What conclusions can you draw about the effects of the predictors $x_1, \ldots, x_3$ on the response variable y.

9.3 Hosmer and Lemeshow (1998) describe a study on the survival times of HIV positive patients of a large HMO using the predictor variables age and intravenous drug usage. The data is on the Minitab worksheet Exercise9.3.mtw.

(a) Use the Minitab macro BayesPropHazMH.mac or the equivalent R-function `BayesCPH` to run 1000 steps of the Metropolis-Hastings algorithm with a heavy-tailed candidate density based on the matched curvature *normal* approximation to the posterior. We do not have any prior knowledge, so we will use a multivariate flat prior.

i. Comment on the mixing properties of the chain.

ii. Comment on the autocorrelations.

(b) Use the Minitab macro BayesPropHazMH.mac or the equivalent R-function `BayesCPH` to run six chains from different starting points, but with the same random inputs, a process known as coupling with the past. Run each of the chains 50 steps.

i. Plot the traceplots for each of the parameters for all six chains to see how many steps are needed until all six chains have converged.

ii. Examine the traceplots, and the autocorrelations from (a) to decide on a burn-in and the thinning required to get an approximately random sample from the posterior.

(c) Run the chain 20000 steps and thin the sample to get an approximate random sample from the posterior.

(d) Comment on which variables significantly affect the survival time.

9.4 Lawless (1982) gives a data set on the survival time for 40 patients with advanced lung cancer. We wish to determine the effect of two chemotherapy treatments on the survival times. The patients have been randomly assigned to one of the two treatment groups. The patients have been divided into four groups, squamous, small, adeno, and large, depending on their type of cancer. Other predictors included are general medical status, age, and number of months from diagnosis to entry into study. The data is in the Minitab worksheet Exercise9.4.mtw

(a) Use the Minitab macro BayesPropHazMH.mac or the equivalent R-function `BayesCPH` to run 1000 steps of the Metropolis-Hastings algorithm with a heavy-tailed candidate density based on the matched curvature *normal* approximation to the posterior. We do not have any prior knowledge, so we will use a multivariate flat prior.

i. Comment on the mixing properties of the chain.

ii. Comment on the autocorrelations.

(b) Use the Minitab macro BayesPropHazMH.mac or the equivalent R-function `BayesCPH` to run six chains from different starting points, but with the same random inputs, a process known as coupling with the past. Run each of the chains 50 steps.

i. Plot the traceplots for each of the parameters for all six chains to see how many steps are needed until all the chains have converged.

ii. Examine the traceplots, and the autocorrelations from (a) to decide on a burn-in and the thinning required to get an approximately random sample from the posterior.

(c) Run the chain 20000 steps and thin the sample to get an approximate random sample from the posterior.

(d) Does the treatment variable significantly affect the survival time?

(e) Do any of the other predictor variables significantly affect the survival time?

9.5 The data from the Waikato cardiac study reported in Swanson et. al. (2007) that we analyzed in Example 14 is given in the Minitab worksheet Exercise9.5.mtw. We want to find the effect on short-term (first 30 days) and long-term patient survival of the predictor variables age, sex, ethnicity, diabetes, smoking, shock, renal failure, previous MI, vessel number, stent, and inhibitor. (Note: vessel number is perfectly correlated with ethnicity in long-term data, so we don't use it in long-term stage.)

(a) Use the Minitab macro BayesPropHazMH.mac or the equivalent R-function `BayesCPH` to run 1000 steps of the Metropolis-Hastings algorithm with a heavy-tailed candidate density based on the matched curvature *normal* approximation to the posterior. We do not have any prior knowledge, so we will use a multivariate flat prior.

i. Comment on the mixing properties of the chain.
ii. Comment on the autocorrelations.

(b) Use the Minitab macro BayesPropHazMH.mac or the equivalent R-function `BayesCPH` to run four chains from different starting points, but with the same random inputs, a process known as coupling with the past. Run each of the chains 50 steps.

i. Plot the traceplots for each of the parameters for all four chains to see how many steps are needed until all the chains have converged.
ii. Examine the traceplots, and the autocorrelations from (a) to decide on a burn-in and the thinning required to get an approximately random sample from the posterior.

(c) Run the chain 20000 steps and thin the sample to get an approximate random sample from the posterior.

(d) Does the treatment variables stent and inhibitor significantly affect the survival time?

(e) Do any of the other predictor variables significantly affect the survival time?

10

Gibbs Sampling and Hierarchical Models

In 1876 Josiah Willard Gibbs discovered how to determine the energy states of gasses at equilibrium. He did so by cycling through the particles, drawing each one conditionally given the energy levels of all the others, over and over through time, and taking a time-average of the resulting sequence of draws. He showed this time-average approaches the true equilibrium distribution of the states of the particles. This became the basis for the field of statistical mechanics. Geman and Geman (1984) showed that an algorithm analogous to this was a good way to reconstruct images from a noisy signal, and called this technique Gibbs sampling in honor of Gibbs' contribution.

Gelfand and Smith (1990) observed that, since the long-run distribution of the Gibbs sampling algorithm is the equilibrium distribution of the states, a draw taken after letting the algorithm a long time can be considered a random draw from the equilibrium distribution. By setting up the equilibrium distribution to be the Bayesian posterior distribution, they made the Gibbs sampler into an effective tool for performing Bayesian inference. This sparked a big increase in the use of Bayesian methods in applied statistics.

In Chapter 6, we noted that Gibbs sampling is a special case of blockwise Metropolis-Hastings algorithm where the candidate distribution for each step is the correct conditional distribution. Thus, for the Gibbs sampler, every candidate will be accepted. In Section 10.1 we look at the general procedure for implementing the Gibbs sampler. In the general case, the correct conditional distribution for each block of parameters will depend on all the other parameters. This may make finding the conditional distributions the Gibbs sampler requires quite complicated. In Section 10.2 we look at using the Gibbs sampler two ways for a random sample from the *normal*(μ, σ^2) distribution with both parameters unknown. In Section 10.3 we look at Gibbs sampling in hierarchical models. The parameters and data are all related in a hierarchical structure for these models. This makes finding the conditional distributions particularly easy for those models.

10.1 GIBBS SAMPLING PROCEDURE

First, let the parameter vector be partitioned into blocks

$$\boldsymbol{\theta} = \boldsymbol{\theta}_1, \boldsymbol{\theta}_2, \ldots, \boldsymbol{\theta}_J$$

where $\boldsymbol{\theta}_j$ is the jth block of parameters. Each block contains one or more parameters. Let $\boldsymbol{\theta}_{-j}$ be the set of all the other parameters not in block j. The proportional form of Bayes theorem,

$$g(\boldsymbol{\theta}_1, \ldots, \boldsymbol{\theta}_J | y_1, \ldots, y_n) \quad \propto \quad f(y_1, \ldots, y_n | \boldsymbol{\theta}_1, \ldots, \boldsymbol{\theta}_J) \times g(\boldsymbol{\theta}_1, \ldots, \boldsymbol{\theta}_J)$$

gives the shape of the joint posterior density of all the parameters, where

$$f(y_1, \ldots, y_n | \boldsymbol{\theta}_1, \ldots, \boldsymbol{\theta}_J) \quad \text{and} \quad g(\boldsymbol{\theta}_1, \ldots, \boldsymbol{\theta}_J)$$

are the joint likelihood the joint prior density for all the parameters. This gives us the shape of the joint posterior, not its scale.

Gibbs sampling requires that we know the full conditional distribution of each block of parameters $\boldsymbol{\theta}_j$, given all the other parameters $\boldsymbol{\theta}_{-j}$ and the data $\mathbf{y} = (y_1, \ldots, y_n)$. Let the full conditional distribution of block $\boldsymbol{\theta}_j$ be denoted

$$g(\boldsymbol{\theta}_j | \boldsymbol{\theta}_{-j}, \mathbf{y}) \quad = \quad g(\boldsymbol{\theta}_j | \boldsymbol{\theta}_1, \ldots, \boldsymbol{\theta}_{j-1}, \boldsymbol{\theta}_{j+1}, \ldots, \boldsymbol{\theta}_J, \mathbf{y}) \, .$$

These full conditional distributions may be very complicated, but we must know them to run the Gibbs sampler. In Gibbs sampling, we will cycle through the parameter blocks in turn, drawing each one from its full conditional distribution given the most recent values of the other parameter blocks, and all the observed data. We noted in Chapter 6 that Gibbs sampling is a special case of the blockwise Metropolis-Hastings algorithm, where the conditional candidate density for each block of parameters is the conditional density of that block, given all the parameters in the other blocks and the data. Since the candidates are being drawn from the correct full conditional distribution, every draw will be accepted.

Steps of the Gibbs Sampler

1. At time $n = 0$ start from an arbitrary point in the parameter space $\boldsymbol{\theta}^0 = (\boldsymbol{\theta}_1^{(0)}, \ldots, \boldsymbol{\theta}_J^{(0)})$. Note: usually the starting point is chosen by taking a random draw from the joint prior distribution of the parameters.

2. For $n = 1, \ldots, N$.

- For $j = 1, \ldots, J$, draw $\boldsymbol{\theta}_j^{(n)}$ from

$$g(\boldsymbol{\theta}_j|\boldsymbol{\theta}_1^{(n)}, \ldots, \boldsymbol{\theta}_{j-1}^{(n)}, \boldsymbol{\theta}_{j+1}^{(n-1)}, \ldots, \boldsymbol{\theta}_J^{(n-1)}, \mathbf{y}).$$

3. The long-run distribution of $\boldsymbol{\theta}^{(N)} = (\boldsymbol{\theta}_1^{(N)}, \ldots, \boldsymbol{\theta}_J^{(N)})$ is the true posterior $g(\boldsymbol{\theta}_1, \ldots, \boldsymbol{\theta}_J|\mathbf{y})$. This means that for a large N the value $\boldsymbol{\theta}^{(N)} = (\boldsymbol{\theta}_1^{(N)}, \ldots, \boldsymbol{\theta}_J^{(N)})$ will be approximately a random draw from the true posterior.

10.2 THE GIBBS SAMPLER FOR THE NORMAL DISTRIBUTION

In this section we look at using the Gibbs sampler for a simple situation; where we have a random sample of size n from a *normal*(μ, σ^2) distribution with both parameters unknown. To do Bayesian inference we will need to find a joint prior density for the two parameters. Usually, we want to make inferences about the mean μ, and regard σ^2 as a nuisance parameter. There are two approaches we can take to choosing the joint prior distributions for this problem.

In the first approach, we will use independent conjugate priors for the two parameters. In Section 4.6 we noted that the joint prior made up of independent conjugate priors for μ and σ^2 is not the joint conjugate prior for *normal*(μ, σ^2) observations when both parameters are unknown. That means we cannot find the exact formula for the joint posterior. Nevertheless, we will find that we can easily set up the Gibbs sampler for this case and obtain a MCMC sample from the joint posterior.

In the second approach we will find the joint conjugate prior directly. In Section 4.6 we showed that the joint prior should have the same form as the joint likelihood in this case, which is a member of the two-dimensional exponential family. Equation 4.9 shows that the joint likelihood factors into a *normal* part containing both the mean and variance, and an *inverse chi-squared* part containing only the variance. So the joint conjugate prior will be the product of S_0 times an *inverse chi-squared* with κ_0 degrees of freedom prior for σ^2 and a *normal*$(m_0, \frac{\sigma^2}{n_0})$ prior for μ conditional on σ^2 being known.

Independent Conjugate Priors for the Two Parameters

In the first approach we use independent conjugate priors for the two parameters, and let the joint prior be the product of the two individual priors. We saw in Chapter 4 that even though each individual prior is conjugate for its parameter, given the other parameter is known, this joint prior will not be conjugate for the two parameters taken together.[1] It would be possible to find the posterior

$$g(\mu, \sigma^2|y_1, \ldots, y_n) \quad \propto \quad g_\mu(\mu) g_{\sigma^2}(\sigma^2) f(y_1, \ldots, y_n|\mu, \sigma^2)$$

[1]This joint prior is a product of parts that are conjugate for each parameter separately when the other is known, however this joint prior is not conjugate, since the likelihood does not factor into a part only containing only μ and a part only containing σ^2.

exactly. However, if we tried to find the marginal posterior for μ by marginalizing σ^2 out of the joint posterior, we couldn't get it exactly in closed form and we would have to do it numerically.[2] In the following example we show how we can draw a sample from the posterior using Gibbs sampling.

Example 15 *We have a random sample of size 10 from a normal(μ, σ^2) distribution where both μ and σ^2 are unknown parameters. The sample values are*

28.78	34.04	26.35	18.72	21.64	24.75	11.97	29.94	21.41	26.20

Suppose we let the prior distribution for μ be normal(m, s^2), where $m = 20$, and $s^2 = 5^2$. The prior distribution for σ^2 is S times an inverse chi-squared with κ degrees of freedom where $S = 11.37$ and $\kappa = 1$. The joint prior is given by

$$g_{\mu,\sigma^2}(\mu, \sigma^2) = g_\mu(\mu) \times g_{\sigma^2}(\sigma^2)\,.$$

We will draw samples from the incompletely known posterior using the Gibbs sampler. The conditional distributions for each parameter, given the other is known are:

1. *When we consider μ is known, the full conditional for σ^2 is*

$$g_{\sigma^2}(\sigma^2|\mu, y_1, \ldots, y_{10}) \propto g_{\sigma^2}(\sigma^2) f(y_1, \ldots, y_{10}|\mu, \sigma^2)\,.$$

Since we are using S times an inverse chi-squared prior with κ degrees of freedom this will be S' times an inverse chi-squared with κ' degrees of freedom where

$$S' = S + \sum_{i=1}^{10} (y_i - \mu)^2 \quad \text{and} \quad \kappa' = \kappa + n\,. \tag{10.1}$$

2. *When we consider σ^2 known, the full conditional for μ is*

$$g_\mu(\mu|\sigma^2, y_1, \ldots, y_{10}) \propto g_\mu(\mu) f(y_1, \ldots, y_{10}|\mu)\,.$$

Since we are using a normal(m, s^2) prior, We know this will be normal$(m', (s')^2)$ where

$$\frac{1}{s^2} + \frac{n}{\sigma^2} = \frac{1}{(s')^2} \quad \text{and} \quad m' = \frac{\frac{1}{s^2}}{\frac{1}{(s')^2}} \times m + \frac{\frac{n}{\sigma^2}}{\frac{1}{(s')^2}} \times \bar{y}\,. \tag{10.2}$$

To find an initial value to start the Gibbs sampler we draw σ^2 at $n = 0$ from the 11.37 times an inverse chi-squared with 1 degree of freedom by drawing the random value .0055385 from a chi-squared distribution with 1 degree of freedom. This will give $\sigma^2 = 11.37/.0055385 = 2052.91$. Then we draw μ from the normal$(20, 5^2)$ distribution by drawing the random value $z = -.786433$ from the standard normal

[2]The true posterior would be a mixture of *Student's t* with varying degrees of freedom.

Table 10.1 First five draws and updated constants for a run[a] of the Gibbs sampler using independent conjugate priors

n	S'	Random *chi-sq* 6 df	σ^2	s'	m'	Random z	μ
0			2052.9				16.068
1	1055.4	19.023	55.479	2.1308	23.584	0.1315	23.864
2	367.2	8.875	41.377	1.8842	23.757	-0.9929	21.886
3	426.7	7.992	53.393	2.0975	23.609	-1.4993	20.464
4	517.9	11.141	46.486	1.9798	23.693	0.9414	25.556
5	378.5	10.167	37.223	1.8000	23.812	-1.7247	20.707

[a] Of course, a different run would have different random values of *chi-sq* and z, so would give different Gibbs sample values for σ^2 and μ.

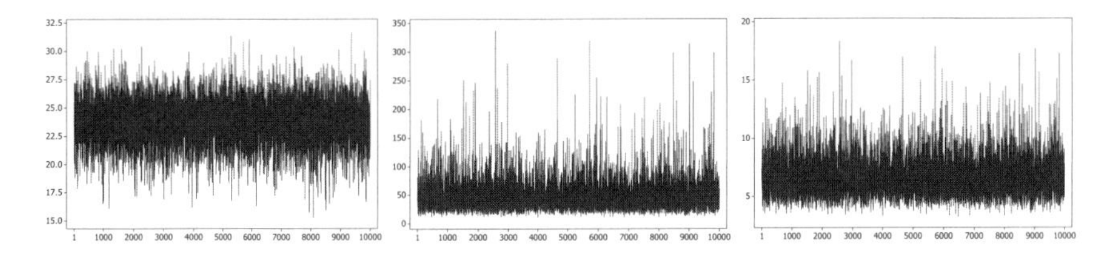

Figure 10.1 Trace plots for the mean, variance, and standard deviation when independent priors for mean and variance are used.

and letting $\mu = 20 + z \times 5 = 13.2734$*. This gives us the values to start the Gibbs sampler. We draw the Gibbs sample using the following steps:*

- *For* $n = 1, \ldots, N$.
  - *Calculate* S' *and* κ' *using Equation 10.1 where* $\mu = \mu_{n-1}$.
  - *Draw* σ_n^2 *from* S' *times an inverse chi-squared distribution with* κ' *degrees of freedom.*
  - *Calculate* $(s')^2$ *and* m' *using Equation 10.2 where* $\sigma^2 = \sigma_n^2$.
  - *Draw* μ_n *from normal*$(m', (s')^2)$.

The first five draws with updated constants are summarized in Table 10.1 We let the Gibbs sampler run for 20000 steps. The trace plots for the parameters μ, σ^2*, and* σ *are shown in Figure 10.1. We see that the Markov chain is moving through the parameter space very satisfactorily, and only a minimum burn-in and thinning will be required. The histogram of the 20000 draws from the Gibbs sampler is shown in Figure 10.2 together with the normal distribution having same mean and standard deviation as the Gibbs sample. Note the true posterior has a shape much closer to*

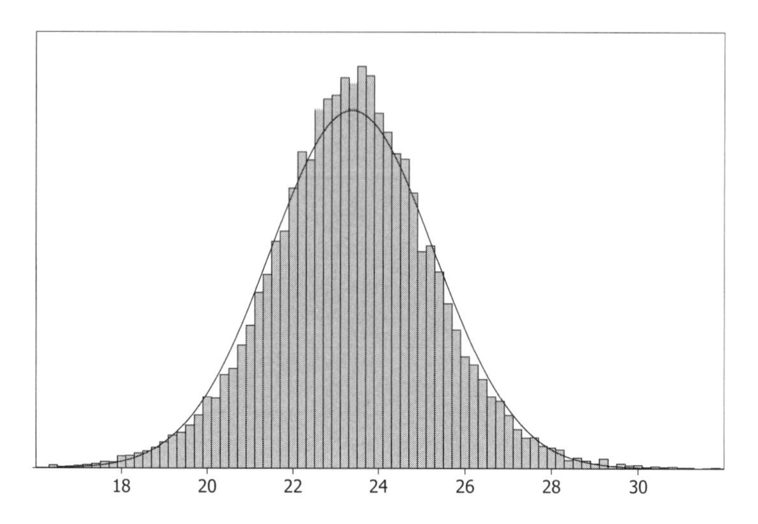

Figure 10.2 Histogram of the samples for the mean μ when independent priors for mean and variance are used.

the Student's t, as we can see from the higher peak, and longer tail of the histogram.[3] *Using the Gibbs sampler has enabled us to draw a sample from the posterior quite easily, despite it being quite difficult to find the exact posterior analytically with this prior.*

Joint Conjugate Prior for the Two Parameters

The second approach is to use joint prior for the two parameters that is conjugate to the observations from the joint distribution where both parameters are unknown. We saw in Section 4.6 that the joint conjugate prior for a sample from a normal distribution where both the mean μ and the variance σ^2 are unknown parameters is the product of a S times an *inverse chi-squared* prior for σ^2 and a *normal*$\left(m, \frac{\sigma^2}{n_0}\right)$ prior for μ conditional on the value of σ^2 where n_0 represents the equivalent sample size of the normal prior. Thus, the joint conjugate prior is given by

$$\begin{aligned} g_{\mu,\sigma^2}(\mu, \sigma^2) &= g_\mu(\mu|\sigma^2) \times g_{\sigma^2}(\sigma^2) \\ &\propto \frac{1}{(\sigma^2)^{\frac{1}{2}}} e^{-\frac{n_0}{2\sigma^2}(\mu-m)^2} \times \frac{1}{(\sigma^2)^{\frac{\kappa}{2}+1}} e^{-\frac{S}{2\sigma^2}} . \end{aligned}$$

Example 15 (continued) *Suppose we let $g_{\sigma^2}(\sigma^2)$, the conjugate prior for σ^2, be S times an inverse chi-squared with κ degrees of freedom where $S - 11.37$ and $\kappa = 1$, and we let $g_\mu(\mu|\sigma^2)$, the conditional prior for μ given the variance σ^2 be normal$\left(m, \frac{\sigma^2}{n_0}\right)$ where the prior mean $m = 20$ and the equivalent sample size*

[3]The true posterior in this case will be a mixture of *Student's t* with varying degrees of freedom.

$n_0 = 1$. *In Chapter 4 we saw how to find the joint posterior when the joint conjugate prior is used. Here we will draw a sample from the joint posterior using the Gibbs sampler. The conditional distributions of each parameter given the other are:*

1. *When we consider* μ *is known, the full conditional for* σ^2 *is*

$$g_{\sigma^2}(\sigma^2|\mu, y_1, \ldots, y_{10}) \quad \propto \quad g_{\sigma^2}(\sigma^2) f(y_1, \ldots, y_{10}|\mu, \sigma^2) .$$

Since we are using S *times an inverse chi-squared prior with* κ *degrees of freedom this will be* S' *times an inverse chi-squared with* κ' *degrees of freedom where*

$$S' = S + \sum_{i=1}^{10} (y_i - \mu)^2 \quad \text{and} \quad \kappa' = \kappa + n . \tag{10.3}$$

2. *When we consider* σ^2 *known, the full conditional for* μ *is*

$$g_\mu(\mu|\sigma^2, y_1, \ldots, y_5) \quad \propto \quad g_\mu(\mu) f(y_1, \ldots, y_5|\mu)$$

Since we are using a normal$\left(m, \frac{\sigma^2}{n_0}\right)$ *prior, we know this will be normal*$(m', (s')^2)$ *where*

$$\frac{n_0}{\sigma^2} + \frac{n}{\sigma^2} \quad = \quad \frac{1}{(s')^2} .$$

Thus

$$(s')^2 \quad = \quad \frac{\sigma^2}{n_0 + n}$$

and

$$m' \quad = \quad \frac{\frac{n_0}{\sigma^2}}{\frac{n_0+n}{\sigma^2}} \times m + \frac{\frac{n}{\sigma^2}}{\frac{n_0+n}{\sigma^2}} \times \bar{y} .$$

First we draw the initial values to start the Gibbs sampler. For step $n = 0$ *we draw* σ^2 *from* 11.37 *times an inverse chi-squared with 1 degree of freedom by drawing the value .168064 from a chi-squared distribution with 1 degree of freedom. This will give the variance* $\sigma_0^2 = 11.37/.168064 = 67.6529$ *and the standard deviation* $\sigma_0 = \sqrt{67.6529} = 8.22514$. *Then we draw* μ_0 *from the normal*$(20, \frac{8.22514^2}{1})$ *distribution by drawing the value* $z = -.371983$ *from the standard normal and letting* $\mu_0 = 20 + z \times 8.22514 = 16.9404$. *This gives us the values to start the Gibbs sampler. The first five draws with updated constants are summarized in Table 10.2. We let the Gibbs sampler run for 10000 steps. The trace plots for the parameters* μ, σ^2, *and* σ *are shown in Figure 10.3. We see that the Markov chain is moving through the parameter space very satisfactorily, and a minimum burn-in and thinning are required. The histogram of the 10000 draws of* μ *from the Gibbs sampler is together with the true marginal posterior* $g(\mu|y_1, \ldots, y_n)$ *(which is Student's t) is shown in Figure 10.4. We see that the Gibbs sampler is giving a sample from the true posterior. With a minimum burn-in time, and minimum of thinning the remaining sample from the posterior will be approximately a random sample and can be used for inferences.*

Table 10.2 First five draws and updated constants from a run[a] of the Gibbs sampler where joint conjugate prior is used

n	S'	Random *chi-sq* 6 df	σ^2	s'	m'	Random z	μ
0			67.653				16.940
1	917.9	9.920	92.532	2.9003	23.981	0.2161	24.608
2	365.1	11.896	30.691	1.6704	23.981	0.4724	24.770
3	366.1	16.132	22.695	1.4364	23.981	0.4527	24.631
4	365.2	11.551	31.620	1.6954	23.981	-1.5217	21.401
5	453.3	9.313	48.671	2.1035	23.981	0.0940	24.179

[a]Of course, a different run would have different random values of *chi-sq* and z, so would give different Gibbs sample values.

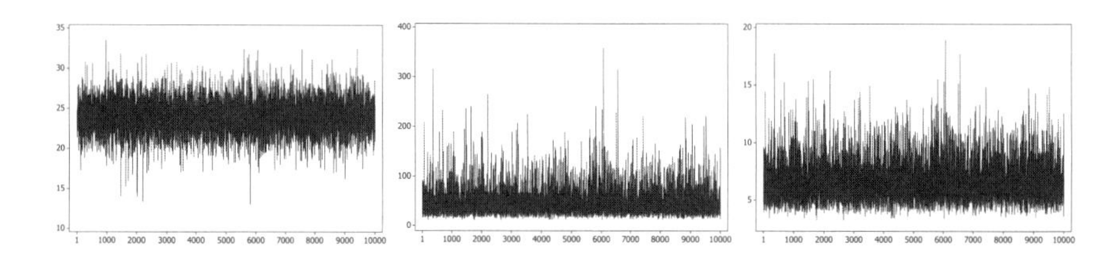

Figure 10.3 Trace plots for the mean, variance, and standard deviation when joint conjugate prior is used.

10.3 HIERARCHICAL MODELS AND GIBBS SAMPLING

In general, the conditional distributions for each block of parameters, given all the other parameters and the observed data may be very complicated. By the definition of conditional probability

$$g(\boldsymbol{\theta}_j|\boldsymbol{\theta}_{-j}, \mathbf{y}) = \frac{g(\boldsymbol{\theta}_1, \ldots, \boldsymbol{\theta}_J, \mathbf{y})}{\int g(\boldsymbol{\theta}_1, \ldots, \boldsymbol{\theta}_J, \mathbf{y})\, d\boldsymbol{\theta}_j} . \tag{10.4}$$

The integral required to evaluate this is difficult to evaluate in general. This makes the Gibbs sampling algorithm difficult to implement in general.

However, when the parameters and observations are related in a hierarchical model, the dependency relationships between all the parameters and the observations all flow in a single direction. For these hierarchical models, we can easily find the conditional distributions for each parameter, given all the other parameters and the data. The dependency relationships in a hierarchical model all flow in a single direction. Many Bayesian statistical models have the hierarchical structure that follows:

- Draw a graph with a node for each block of parameters, nodes for the data and nodes for the hyperparameters and constants. Use a circular node for a

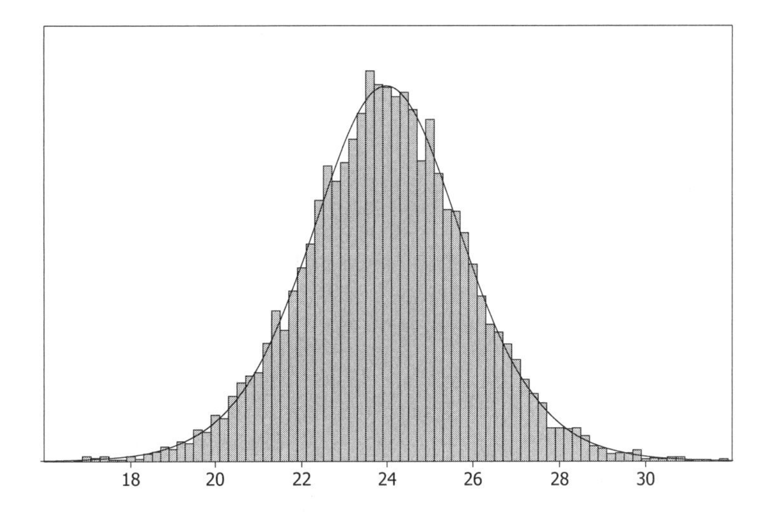

Figure 10.4 Histogram of the samples for the mean μ drawn using the Gibbs sampler along with the true *Student's t* marginal posterior $g(\mu|y_1, \ldots, y_n)$.

parameter block to indicate that it is a random variable. Use a rectangular node for the constants of the hyperparameter distributions and for the observed data, to indicated that they are fixed values in the analysis.

- Draw arrows showing how the each node depends on the other nodes.
- In a hierarchical model, because the dependencies flow one direction, the graph will have no connected loop. A graph like this is called a *directed acyclic graph* (O'Hagan, 2004).
- All nodes leading into a specific node are called the *parent* nodes of that node.
- Each node only depends on the nodes directly above it in the hierarchy. Given its parent nodes, it is independent of the parent nodes of its parent.
- All nodes that lead out of the specific node are called the *child* nodes of that node. They can be either parameter nodes, or data nodes. The only nodes that directly depend on the specific node are its child nodes.
- All other nodes that lead into a *child* node of a parent node are called *co-parent* nodes for that *child* node.

The joint distribution of all the parameters and the data can be easily found using the dependence structure shown on the graph. First start with the prior distributions of the hyperparameters at the top of the hierarchy. Then multiply by the conditional distribution of each of their child nodes given the parent and coparent nodes. Continue this working down the hierarchy of the parameters. Lastly multiply the conditional distribution of each the data value, given its parent nodes. The joint distribution of all the nodes is the product of the prior distributions of the hyperparameters times the conditional distribution of each node given its parent nodes. We substitute this into both the numerator and denominator of Equation 10.4. All the terms in the denominator that do not contain $\boldsymbol{\theta}_j$ are constants with respect to the integration, so they can be brought out in front of the integral where we see they will cancel the corresponding term in the numerator. This leaves

$$g(\boldsymbol{\theta}_j|\boldsymbol{\theta}_{-j},\mathbf{y}) \propto g(\boldsymbol{\theta}_j|\text{parents of } \boldsymbol{\theta}_j) \prod_{\boldsymbol{\theta}_k \in \text{children of } \boldsymbol{\theta}_j} f(\boldsymbol{\theta}_k|\text{parents of } \boldsymbol{\theta}_k) \tag{10.5}$$

where the parent nodes of of $\boldsymbol{\theta}_k$ includes $\boldsymbol{\theta}_j$ and the other coparent nodes of $\boldsymbol{\theta}_k$. The conditional distribution is like Bayes' theorem where the prior distribution of $\boldsymbol{\theta}_j$ is its distribution given its parent nodes and the likelihood is the distribution of all the child nodes of $\boldsymbol{\theta}_j$ given $\boldsymbol{\theta}_j$ and the other coparent nodes of those child nodes. The proportional form for the conditional distributions of one node, given all other nodes shown in Equation 10.5 comes from the hierarchical dependency structure of the model. It does not depend on the actual formulas for the densities. Thus, at least in principal, the conditional distributions for the Gibbs sampler will be relatively easily found in a hierarchical model. When the distributions are from the exponential family with conjugate priors, then we can find the exact conditional densities by the simple updating rules discussed in Chapter 4. Then we sample that node from the updated exact conditional. In other cases, we can use one of the methods discussed in Chapter 3 such as acceptance-rejection-sampling or adaptive-rejection-sampling to sample that node.

10.4 MODELLING RELATED POPULATIONS WITH HIERARCHICAL MODELS

Often we have data from several populations that we believe follow the same parametric distribution (such as the normal distribution), but may have different values of the parameter (such as the mean). The classical frequentist approach would be to analyze each population separately. The maximum likelihood estimate of the parameter for each population would be estimated from the sample from that population. Simultaneous confidence intervals such as Bonferroni, Tukey, or Scheffé intervals would be used for the difference between different population parameter values. These wider intervals would control the overall confidence level, and the overall significance level for testing the hypothesis that the differences between all the population parameters are zero. However, these intervals don't do anything about the parameter estimates themselves.

Stein (1956) and James and Stein (1961) showed that, for simultaneously estimating the means for three or more populations, the individual maximum likelihood estimates were not admissible. The estimators found by shrinking each of the individual maximum likelihood estimators back towards the overall mean would outperform the individual maximum likelihood estimators. They would have smaller total mean squared error, despite being biased. Needless to say, this was a shocking result in the statistical community because, up to that time, maximum likelihood was considered to always be the best method. Stein's result didn't seem to make sense, although it was clearly true. Because the populations are related, data from one population gives some information about the other populations as well as the one it was drawn from.

Parameter estimates found using data from related populations as well as from the population in question are called shrinkage estimators. They are found using two approaches. The first is called *empirical Bayes*. It is partly Bayesian and partly likelihood. The second approach is the fully Bayesian approach. We model the related populations by letting each one have its own parameter(s), drawn from some distribution. We call this distribution the parameters are drawn from a hyperdistribution since it is two levels above the observations. We call the parameters of this distribution hyperparameters. This sets up a hierarchical model, with the population parameters drawn from the hyper distribution, and the individual samples all drawn from the distributions given their individual population parameters. The hierarchical model gives us the framework to find and understand shrinkage estimators. We will briefly describe the empirical Bayes approach, before we look at the fully Bayesian approach. Carlin and Louis (2000) give an extensive comparison of these two approaches.

Empirical Bayes Approach to Hierarchical Model

The (parametric) empirical Bayes (EB) approach is to estimate the hyperparameters of the parameter distribution from the marginal distribution of all the data, given the hyperparameters only. Thus the parameters for the individual populations are all marginalized out, leaving only the hyperparameters, which are estimated using maximum likelihood. This is the empirical part. These hyperparameter estimates are then plugged into the prior distribution of the parameters, and the posterior distribution of each particular parameter of interest is found using Bayes theorem with that empirically found prior. This is the Bayesian part.

One problem with the empirical Bayes approach is that it overestimates the precision of the posterior parameter estimates because using the plug-in estimate for the hyperparameter does not allow for its estimation error. This results in confidence intervals that are too short. Carlin and Gelfand (1990) developed bias-corrected EB confidence intervals that do have the correct coverage probability. Efron (1996) developed an EB method for combining likelihoods for several independent experiments, each with its own related parameter, and used it for constructing a confidence interval for one of the parameters based on the combined data. In the subsequent discussion, Gelfand (1996) opined that the fully Bayesian approach will be more successful in this and similar problems, due to the more straightforward interpretation, and ease of implementation with MCMC methods. We will illustrate empirical Bayes method in the following simple example.

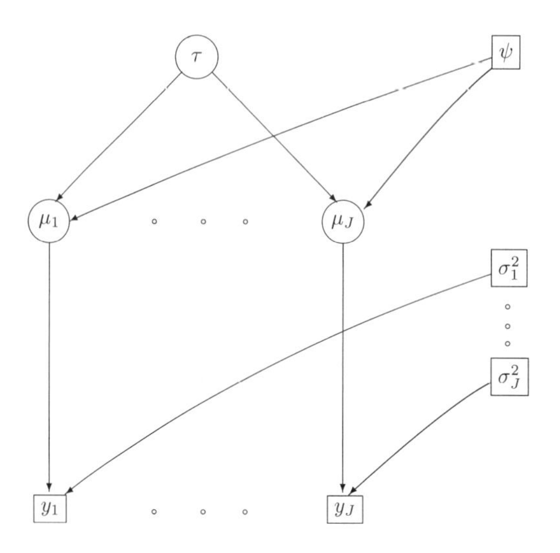

Figure 10.5 Simple hierarchical model for empirical Bayes. The parameters are shown in circles, constants in rectangles.

Example 16 *Suppose we have J independent random variables each drawn from a normally distributed population having its own mean value μ_j and known variance σ_j^2. Let y_j be drawn from a normal(μ_j, σ_j^2) for $j = 1, \ldots, J$.*

We consider the populations are related, so we model this as a hierarchical model. We consider that each μ_j is a random draw from a normal distribution having hypermean value τ and known hypervariance ψ. The directed graph for this model is shown in Figure 10.5. First we find the joint distribution of each observed y and its mean μ_j given the parameter τ by

$$\begin{aligned} f(y_j, \mu_j|\tau) &= f(y_j|\mu_j) \times f(\mu_j|\tau) \\ &\propto e^{-\frac{1}{2\sigma_j^2}(y_j-\mu_j)^2} \times e^{-\frac{1}{2\psi}(\mu_j-\tau)^2} \\ &\propto e^{-\frac{\psi+\sigma_j^2}{\sigma_j^2\psi}\left(\mu_j - \frac{\psi u_j+\sigma_j^2\tau}{\psi+\tau}\right)^2} \times e^{-\frac{1}{\sigma_j^2+\psi}(y_j-\tau)^2} . \end{aligned}$$

We then marginalize the mean μ_j out of the joint distribution. We note that only the first term contains μ_j, and it has the form of a normal density. Hence the marginal distribution of y_j given the parameter τ is given by

$$f(y_j|\tau) \propto e^{-\frac{1}{\sigma_j^2+\psi}(y_j-\tau)^2} \tag{10.6}$$

which we recognize as a normal distribution with mean τ and variance $\sigma_j^2 + \psi$. In the empirical part, we estimate the hypermean τ using maximum likelihood. We find *the MLE $\hat{\tau}$ will be the weighted sample mean of all the observations*[4]

$$\hat{\tau} = \frac{\left(\frac{1}{\sigma_1^2+\psi}\right) y_1 + \cdots + \left(\frac{1}{\sigma_J^2+\psi}\right) y_J}{\frac{1}{\sigma_1^2+\psi} + \cdots + \frac{1}{\sigma_J^2+\psi}},$$

where each observation is weighted by its precision relative to the total precision of all the observations. We plug in $\hat{\tau}$, the MLE value, as the prior mean for each of the μ_j. In the Bayesian part, we find the posterior distribution of each mean μ_j given the corresponding observation using Bayes' theorem. The prior for μ_j is normal$(\hat{\tau}, \psi)$ and the distribution of y_j given μ is normal(μ_j, σ_j^2). The posterior distribution of μ_j will be normal with mean and variance respectively equal to

$$\frac{\psi}{\sigma_j^2 + \psi} \times y_j + \frac{\sigma_j^2}{\sigma_j^2 + \psi} \times \hat{\tau} \quad \text{and} \quad \frac{\sigma_j^2 \psi}{\sigma_j^2 + \psi}. \tag{10.7}$$

The posterior mean of μ_j has been shrunk away from the jth sample value y_j towards the maximum likelihood value $\hat{\tau}$. Confidence intervals calculated from this will be too short to have the claimed coverage probability. The posterior variance is too small since it is calculated assuming we know the prior mean, but in fact we don't. Instead we have plugged in an estimate of it and acted as if that is the true value. There is additional uncertainty here that has not been allowed for.

Bayesian Approach to Hierarchical Model

The Bayesian interpretation of the empirical Bayes posterior distribution of μ_j found in Equation 10.7 is that it is actually the posterior distribution of μ, conditional on the value of the parameter τ given by $g(\mu_j|\tau, y_1, \ldots, y_J)$ evaluated at $\tau = \bar{y}$. In the fully Bayesian approach we base the inference about μ_j on the marginal posterior $g(\mu_j|y_1, \ldots, y_J)$ instead of the conditional posterior given in Equation 10.7. The fully Bayesian approach requires us to put a prior distribution on the hyperparameter(s) and evaluate the joint posterior distribution of all the parameters including the hyperparameter(s) using Bayes' theorem. The posterior distribution of any particular parameter is found by marginalizing all the other parameters out of the joint posterior distribution of all parameters. This takes the uncertainty in estimating the other nuisance parameters into account so the precision will be correct, as will the coverage probabilities for any credible interval calculated using the marginal posterior.

Example 16 (continued) *We continue with our example where we have a single random drawn from each of $j = 1, \ldots, J$ normally distributed populations where each population has its own mean and its own known variance. The observation y_j comes from a normal(μ_j, σ_j^2) distribution with known variance σ_j^2. The population distributions are related. We model this relationship between the populations by considering the population means $\mu_1, \ldots, \mu_J$ to be random draws from a normal(τ, ψ)*

[4] If the variances σ_j^2 for $j = 1, \ldots, J$ are all equal, this simplifies to $\hat{\tau} = \bar{y}$, the sample mean.

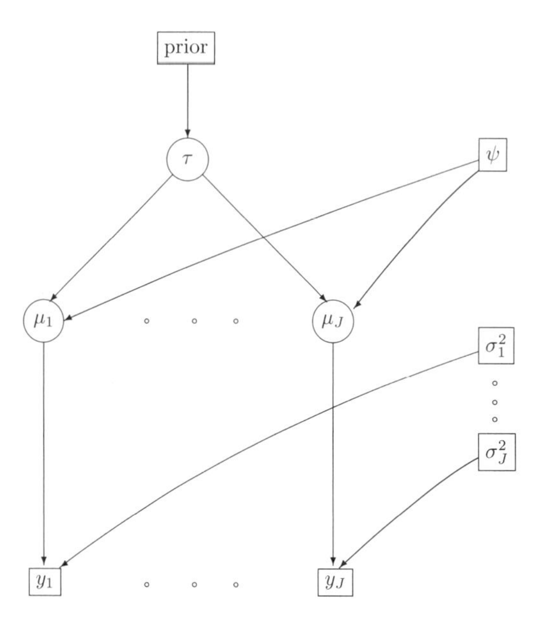

Figure 10.6 Simple hierarchical model for full Bayesian analysis. Now the parameter τ has a prior distribution.

distribution where the variance ψ is known. To do a full Bayesian analysis we need to put a prior distribution on the hypermean τ. We could either use a normal(m_0, V_0) prior or a flat prior. In this example will use an improper flat prior for τ.[5] *The flat prior does not put any information about τ into the model. That way any difference we find between the results for the fully Bayesian and the empirical Bayesian approaches must arise solely from the differing methodology and not from the prior information we put in the fully Bayesian model. The directed graph for this model is shown in Figure 10.6. The joint posterior of μ_j and τ is given by*

$$g(\mu_j, \tau, |y_1, \ldots, y_J) = g(\mu_j|\tau, y_1, \ldots, y_J) \times g(\tau|y_1, \ldots, y_J)$$

where $g(\mu_j|\tau, y_1, \ldots, y_J)$ is normally distributed with mean and variance given in Equation 10.7. The jth observed value y_j is a random draw from a $N(\tau, \sigma_j^2 + \psi)$ and the sample values are independent. Since we are using a flat prior for τ, the posterior distribution of τ will be proportional to the likelihood. This simplifies to a normal(m_1, V_1) posterior where the posterior precision and posterior mean are given by

$$\frac{1}{V_1} = \frac{1}{\sigma_1^2 + \psi} + \cdots + \frac{1}{\sigma_J^2 + \psi} \quad \text{and} \quad m_1 = \left(\frac{\frac{1}{\sigma_1^2+\psi}}{\frac{1}{V_1}}\right) y_1 + \cdots + \left(\frac{\frac{1}{\sigma_J^2+\psi}}{\frac{1}{V_1}}\right) y_J$$

[5]Using this improper flat prior will not cause any trouble in this model.

respectively. Note: since we are using a flat prior in this case, the posterior mean m_1 *equals* $\hat{\tau}$*, the MLE. Thus the joint posterior for* μ_j *and* τ *is*

$$g(\mu_j, \tau | y_1, \ldots, y_J) \quad \propto \quad e^{-\frac{\sigma_j^2+\psi}{2(\sigma_j^2\psi)}\left(\mu_j - \frac{\sigma_j^2\tau+\psi y_j}{\sigma_j^2+\psi}\right)^2} \times e^{-\frac{1}{2V_1}[\tau - m_1]^2}.$$

We collect the terms in powers of τ *and complete the square. When we marginalize* τ *out of the joint posterior we get the marginal posterior of* μ_j *to be normal with mean and variance*

$$\left(\frac{\psi}{\sigma_j^2+\psi}\right) y_j + \left(\frac{\sigma_j^2}{\sigma_j^2+\psi}\right) m_1 \quad \text{and} \quad V_1 \times \left(\frac{\sigma_j^2}{\sigma_j^2+\psi}\right)^2 + \left(\frac{\sigma_j^2\psi}{\sigma_j^2+\psi}\right)$$

and use it for our inferences on the mean μ_j*. The posterior mean of* μ_j *has been shrunk away from the value* y_j *towards the posterior mean of* τ *which in this case is equal to* $\hat{\tau}$*, the MLE. Note: in the EB case, the posterior posterior variance equals* $\frac{\sigma_j^2\psi}{\sigma_j^2+\psi}$*. In the fully Bayesian case the posterior variance equals* $V_1\left(\frac{\sigma_j^2}{\sigma_j^2+\psi}\right)^2 + \left(\frac{\sigma_j^2\psi}{\sigma_j^2+\psi}\right)$*, which is clearly larger. Credible intervals calculated using this variance will be wider and will have the correct coverage probability since we have allowed for the additional uncertainty.*

In the previous example, we found the exact marginal posterior distribution for each of the individual means analytically. They were the parameters of interest and the hyperparameter was considered a nuisance parameter. In many cases we cannot find the posterior analytically. Instead we use the Gibbs sampler to draw a sample from the joint posterior of all the parameters for the hierarchical model. The Gibbs sample for a particular parameter is a sample from its marginal posterior. We will base our inference about that parameter on the thinned sample from its marginal posterior. We don't even have to marginalize out the other parameters as looking at the sample for that particular parameter does it automatically.

Hierarchical Normal Mean Model with Regression on Covariates

Suppose we have a sample of experimental units that we randomly assign each unit into one of the J treatment groups. The random sample of n_j observations from the jth treatment group are put in the vector

$$\mathbf{y_j} = \begin{pmatrix} y_{1\,1} \\ \vdots \\ y_{n_j\,j} \end{pmatrix}.$$

The sample sizes do not have to be the same for all the treatment groups. Suppose we also measure the values of p covariates for each observation. Let

$$\mathbf{X_j} = \begin{bmatrix} X_{1\,1} & \ldots & X_{1\,p} \\ \vdots & \vdots & \vdots \\ X_{n_j\,1} & \ldots & X_{n_j\,p} \end{bmatrix}$$

be the values of the covariates for the jth treatment group. When the treatment effects are additive, the measured response for the ith observation in the jth treatment group will be *normal*$(\mu_j + \sum X_{i\,k}\beta_k, \sigma^2)$, where μ_j is the treatment effect for treatment j, β_k is the unknown regression coefficient for the kth covariate, and σ^2 is the unknown variance. Since the treatment effect is additive, the variance in the observations is due to the variation in experimental units which are randomly assigned to the treatment groups. Hence $\mathbf{y_j}$, the vector of observations in the jth treatment group, has the *multivariate normal* distribution with mean vector and covariance matrix given by

$$E(\mathbf{y_j}) = \mu_j + \mathbf{X_j}\boldsymbol{\beta} \quad \text{and} \quad Cov(\mathbf{y_j}) = \sigma^2\mathbf{I}$$

respectively, where

$$\boldsymbol{\beta} = \begin{pmatrix} \beta_1 \\ \vdots \\ \beta_p \end{pmatrix}.$$

We will consider the treatment effects $\mu_1, \ldots, \mu_J$ to be random draws from a *normal* distribution with the same (hyper) mean τ and unknown (hyper) variance ψ. We need to have a prior distribution for the hyperparameters. We don't really have too much of an idea about what values the hyperparameters could take, so we want a prior that allows for a wide range of values. We will use a *normal*(m_0, V_0) prior for the hypermean τ where V_0 is a large enough number to allow possible values over the whole range we consider possible. We will use an R_0 times an *inverse chi-squared* prior with ν_0 degrees of freedom for the hypervariance ψ. Usually we use a low value such as $\nu_0 = 1$ for the degrees of freedom, and choose R_0 to be large enough so all realistic values of ψ would be allowed. We will use an *multivariate normal*$(\mathbf{b_0}, \mathbf{B_0})$ distribution for the unknown coefficient vector $\boldsymbol{\beta}$. We use an S_0 times an *inverse chi-squared* prior with κ_0 degrees of freedom for the observation variance σ^2. We will use low degrees of freedom such as $\kappa_0 = 1$ and choose S_0 so that we match our belief about the median for the standard deviation σ. The directed acyclic graph for this model is shown in Figure 10.7. Note: all the prior distributions are chosen to be proper.

We can build the joint distribution of all the random quantities by working down using the hierarchical dependence structure. It is given by

$$g(\tau, \psi, \mu_1, \ldots, \mu_j, \sigma^2, \beta_1, \ldots \beta_p, y_{11}, \ldots, y_{n_1 1} \cdots y_{1J}, \ldots, y_{n_j J})$$
$$= g(\tau)g(\psi)g(\beta_1, \ldots, \beta_p)g(\sigma^2) \prod_{j=1}^{J} g(\mu_j|\tau, \psi) \left(\prod_{i=1}^{n_j} f(y_{ij}|\mu_j, \beta, \sigma^2) \right). \quad (10.8)$$

The full conditionals are easy to find because of the hierarchical structure. The full conditional for a parameter block given all the other parameter blocks and the data is the joint density of all the parameters and data, divided by the joint density of all the other parameters and the data. That block of parameters has been integrated out. All terms not involving that block of parameters are constants as far as the integration is concerned, so they can be brought out in front. Then they can be cancelled with the

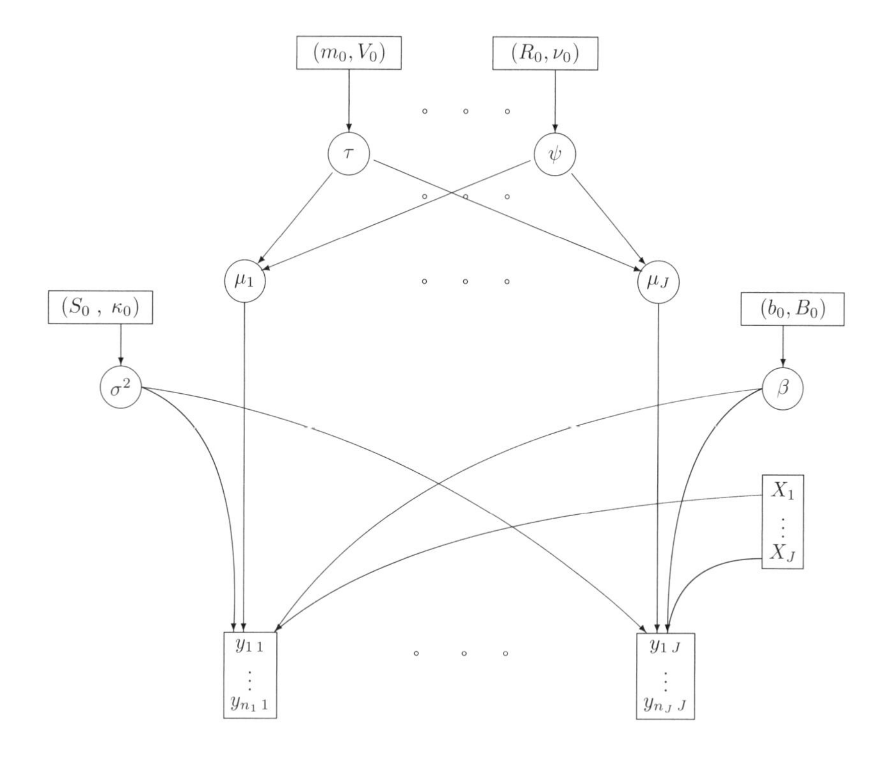

Figure 10.7 Hierarchical normal mean model with regression.

same term in the numerator. The terms that are left are the distribution of that block, given its parent nodes, times the conditional distribution of its child nodes, given the parent and the co-parent nodes.

Gibbs sampling conditional distributions for this model.

1. The prior density of the hypermean τ is *normal*(m_0, V_0), and each mean μ_j is a random draw from a *normal*(τ, ψ) distribution for $j = 1, \ldots, J$. Hence $g(\tau|\mu_1, \ldots, \mu_J, \psi, \boldsymbol{\beta}, \sigma^2, \mathbf{y_1}, \ldots, \mathbf{y_J})$, will be normally distributed with mean m_1 and variance V_1 which are found by

$$\frac{1}{V_1} = \frac{1}{V_0} + \frac{J}{\psi} \quad \text{and} \quad m' = \frac{\frac{1}{V_0}}{\frac{1}{V_1}} \times m_0 + \frac{\frac{J}{\psi}}{\frac{1}{V_1}} \times \bar{\mu}\,.$$

Thus the posterior precision is the sum of the prior precision and the sum of the precisions of the μ_j for $j = 1, \ldots, J$, and the posterior mean is the weighted sum of the prior mean m and the average of the μ_j, where the weights are the proportions of the component precisions to the posterior precision.

2. The prior density of the hypervariance is R_0 times an *inverse chi-squared* distribution with ν_0 degrees of freedom, and each mean μ_j is a random draw from a *normal*(τ, ψ) distribution for $j = 1, \ldots, J$. Hence the conditional distribution $g(\psi|\tau, \mu_1, \ldots, \mu_J, \boldsymbol{\beta}, \sigma^2, \mathbf{y_1}, \ldots, \mathbf{y_J})$ will be R_1 times an *inverse chi-squared* with ν_1 degrees of freedom where

$$R_1 = R_0 + \sum_{j=1}^{J} (\mu_j - \tau)^2 \quad \text{and} \quad \nu_1 = \nu_0 + J\,.$$

Thus the posterior degrees of freedom is the sum of the prior degrees of freedom plus the number of group means, and the posterior constant is the sum of the prior constant plus the sum of squared deviations of the group means from the hypermean in standardized units.

3. The prior density of μ_j is *normal*(τ, ψ), and given the parent node μ and the coparent node $\boldsymbol{\beta}$, the observations $y_{1j}, \ldots, y_{n_j j}$ are a random sample from a *normal*$(\mu_j + \mathbf{X_j}\beta, \sigma^2)$. Thus $\mathbf{z_j} = \mathbf{y_j} - \mathbf{X_j}\boldsymbol{\beta}$ is a *multivariate normal* random vector with mean vector $\boldsymbol{\mu_j}$ and covariance matrix $\sigma^2\mathbf{I}$. The posterior distribution of μ_j is normally distributed with variance W_j and mean u_j where

$$W_j^{-1} = (\psi)^{-1} + n_j(\sigma^2)^{-1} \quad \text{and} \quad u_j = W_j[(\psi)^{-1}\tau + n_j(\sigma^2)^{-1}\overline{z_j}]$$

and $\overline{z_j} = \overline{y_j - X_j\beta}$ is the average of the deviations away from the predicted values for the $j'th$ sample. Thus each posterior precision is the sum of the prior precision plus the precision of $\overline{z_j}$ given β, and the posterior mean is the weighted average of the prior mean and $\overline{z_j}$, where the weights are the proportions of the component precisions to the posterior precision.

4. The prior distribution of $\boldsymbol{\beta}$ is *multivariate normal*$(\mathbf{b_0}, \mathbf{B_0})$. The distribution of $(\mathbf{y_j}|\mu_j, \boldsymbol{\beta}, \sigma^2)$ is *multivariate normal*$(\mu_j + \mathbf{X_j}\boldsymbol{\beta})$. Hence the distribution of $\beta|\tau, \psi, \mu_1, \ldots, \mu_J, \sigma^2, \mathbf{y_1}, \ldots, \mathbf{y_j}$ is *multivariate normal*$(\mathbf{b_1}, \mathbf{B_1})$ where

$$\mathbf{B_1}^{-1} = \mathbf{B_0}^{-1} + \mathbf{X}'\mathbf{X}/\sigma^2 \quad \text{and} \quad \mathbf{b_1} = \mathbf{B_1}\mathbf{B_0}^{-1} \times \mathbf{b_0} + \mathbf{B_1}\frac{\mathbf{X}'\mathbf{X}}{\sigma^2} \times \mathbf{b_{LS}}$$

where $\mathbf{b_{LS}} = (\mathbf{X}'\mathbf{X})^{-1}\mathbf{X}'(\mathbf{y} - \boldsymbol{\mu})$ is the least squares regression estimate given the mean vector is known. Thus the final precision matrix is the sum of the prior precision matrix plus the precision matrix of the least squares regression estimate, and the final estimate is the weighted sum where the weights are the proportions of the final precision from each part.

5. $\sigma^2|\tau, \psi, \mu_1, \ldots, \mu_J, y_{1\,1}, \ldots, y_{n_J\,J}$ is S_1 times an inverse chi-square with κ_1 degrees of freedom where

$$\kappa_1 = \kappa_0 + (n_1 + \ldots + n_J) \quad \text{and} \quad S_1 = S_0 + SS_w$$

where $\mathbf{w} = \mathbf{y} - \boldsymbol{\mu} - \mathbf{X}\boldsymbol{\beta}$ and $SS_w = \mathbf{w}'\mathbf{w}$. Thus the posterior degrees of freedom is the sum of the prior degrees of freedom plus the degrees of freedom of the observations given μ and β, and the posterior constant is the prior constant plus the sum of squares of the observation vector minus its mean vector $\boldsymbol{\mu}$ and minus its contribution from the covariates $\mathbf{X}\beta$.

Prior Distributions for Hierarchical Model

In the hierarchical model, the parameters of the parameter distribution are known as hyperparameters. We don't really have a great deal of prior information about the hyperparameters. We might think that improper priors would be appropriate. Unexpectedly however, this approach can lead to trouble.

Hobert and Casella (1996) warn that if improper priors are used for variance components in hierarchical linear mixed models, the joint posterior will be improper. Yet the Gibbs full conditionals can easily be found by conjugate analysis. They show that the Gibbs Markov chain constructed with conditionals from an improper posterior is either null recurrent or transient, and thus does not share in the convergence properties associated with chains having proper posteriors. Although Monte Carlo output from a null Gibbs chain may appear reasonable, this is a fortunate coincidence. They describe it as "the chain getting stuck in a reasonable part of the parameter space due to the very small probability of a transition to the bad part of the space where absorption would occur." We suggest that proper priors always be used for the hyperparameters. They should be quite spread to account for our lack of prior information, but always proper. A proof that the improper Jeffrey's prior for the hypervariance can cause the joint posterior to be improper is given in an appendix at the end of this chapter.

Example 17 *We have ten observations from each of three groups, and two predictor variables recorded for each observation. The data are given in Table 10.3. We set the priors for the hyperparameters. We chose a normal*$(0, 1000^2)$ *prior for* τ*, a 500 times an inverse chi-squared prior for* ψ *and for* σ^2*, and a bivariate normal*$(\mathbf{b_0}, \mathbf{B_0})$ *prior for the vector* $\boldsymbol{\beta}$ *where*

$$\mathbf{b}_0 = \begin{pmatrix} 0 \\ 0 \end{pmatrix} \quad \text{and} \quad \mathbf{B}_0 = \begin{pmatrix} 1000 & 100 \\ 100 & 1000 \end{pmatrix}$$

and we ran the Gibbs sampling chain for 2000 steps. The traceplots are shown in Figure 10.8. We see from these traceplots that the Gibbs sampling chain is moving through the parameter space satisfactorily. However, we can see that it does not move as fast through the parameter space as the Metropolis-Hastings chains with heavy-tailed independent candidate density that we used in earlier chapters. The traceplots resemble those with a random-walk candidate density, which is not surprising since we noted in Chapter 6 that Gibbs sampler is special case of blockwise Metropolis-Hastings. We can see that the scale parameters ψ *and* σ^2 *have heavy upper tails. We also calculated the sample autocorrelations for the parameters to see what burn-in and thinning will be required to achieve near independence for our sample from the posterior. These autocorrelations are shown in Figure 10.9. We see that the*

Table 10.3 Responses y are from three groups with two predictor variables x_1 and x_2

y	Group	x_1	x_2	y	Group	x_1	x_2
351.5	1	1	1	275.5	2	6	1
233.9	1	2	2	363.6	2	7	2
350.9	1	3	3	286.7	2	8	3
281.9	1	4	4	376.6	2	9	4
280.7	1	5	5	361.4	2	10	5
262.4	1	6	1	333.6	3	1	1
283.1	1	7	2	231.5	3	2	2
419.2	1	8	3	258.3	3	3	3
354.2	1	9	4	381.0	3	4	4
322.7	1	10	5	357.7	3	5	5
263.9	2	1	1	396.4	3	6	1
291.5	2	2	2	346.4	3	7	2
314.3	2	3	3	303.1	3	8	3
333.8	2	4	4	323.1	3	9	4
313.5	2	5	5	319.2	3	10	5

autocorrelations are all tailing off slowly. They have become indistinguishable from zero by 50 steps for all the parameters. We run three chains from different starting points but with the same random inputs to see how many steps are needed until they have converged. The traceplots are shown in Figure 10.10. The Gibbs sampling chains will not become identical as the Metropolis-Hastings chains with heavy-tailed independent candidate density did. Rather, they get closer and closer to each other. After looking at the traceplots we decide that after 50 steps the chains have converged close enough to believe that there is very little effect of the starting point. We decide that a burn-in of 100 steps, and thinning by taking every 100th will give us a sample that is approximately independent to use for inference.

We run the chain for 100000 steps and thin it by including only every 100th. *We examined the autocorrelation function of the thinned sample and found none were distinguishable from zero. The traceplots of the thinned sample resembled traceplots from random samples. This gives us reason to believe that our thinned sample is a random sample of size 1000 from the posterior. Histograms of the thinned sample are shown in Figure 10.11. Summary statistics calculated from the posterior distribution are shown in Table 10.4. Both the 95% credible intervals for β_1 and for β_2 contain 0, so we cannot conclude that either x_1 or x_2 affects the response at the 95% level.*

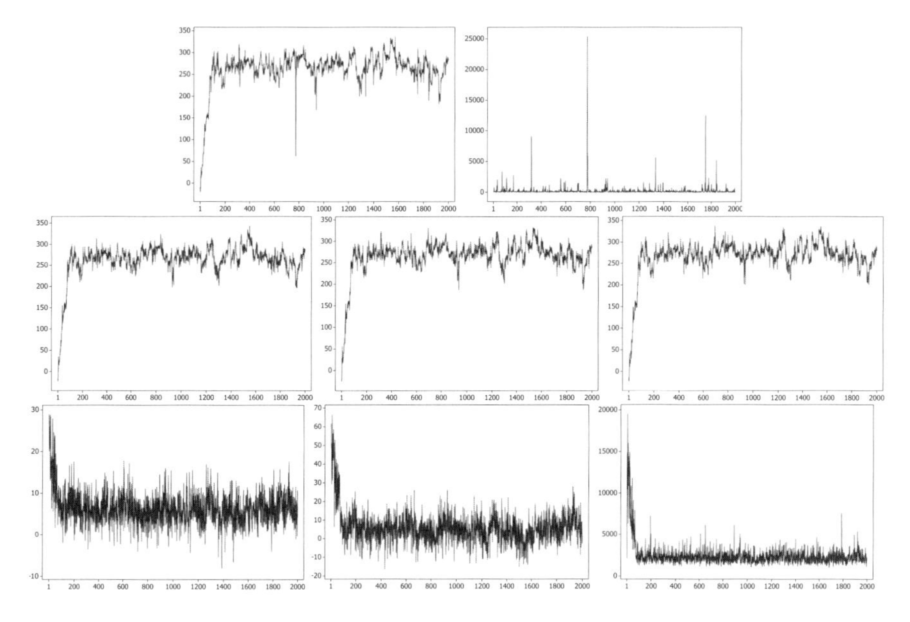

Figure 10.8 Trace plots for τ, ψ, μ_1, μ_2, μ_3, β_1, β_2, and σ^2.

Table 10.4 Summary statistics of thinned MCMC sample

Parameter	Mean	St. Dev.	Credible Int.	
			Lower	Upper
τ	285.5	26.11	231.9	338.6
ψ	708.8	1582	71.71	451.3
μ_1	282.9	23.52	238.4	329.2
μ_2	285.8	23.36	238.5	332.2
μ_3	288.5	23.79	243.2	335.9
β_1	4.922	3.372	–1.71	11.66
β_2	2.382	6.776	10.91	15.68
σ^2	2264	610.6	1324	3714

Main Points

- Gibbs sampling is the special case of the blockwise Metropolis-Hastings algorithm where the candidate distribution for each block of parameters given all other parameters not in that block is the correct full conditional density for that block. Thus all candidates will be accepted.

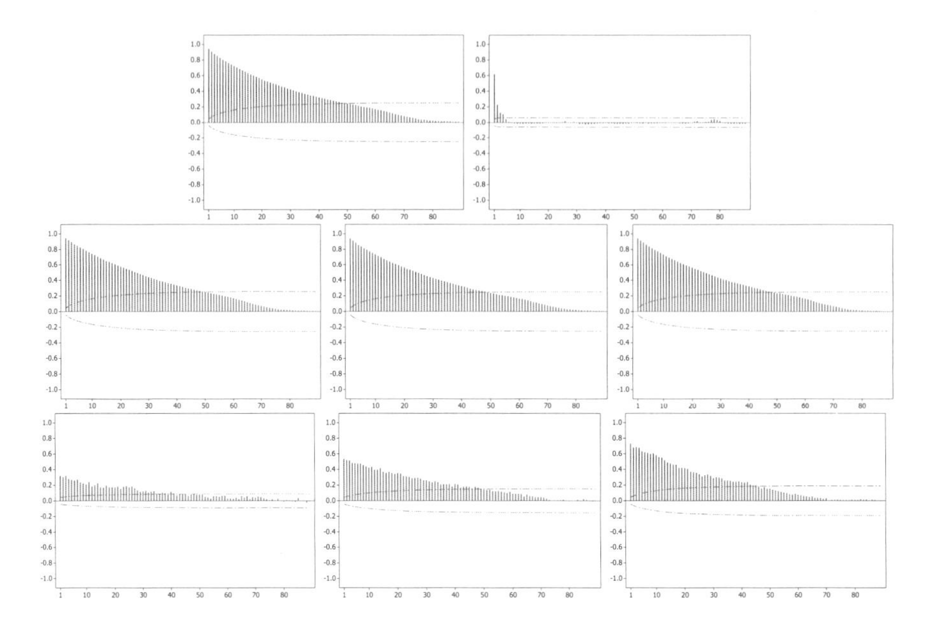

Figure 10.9 Sample autocorrelations for τ, ψ, μ_1, μ_2, μ_3, β_1, β_2, and σ^2.

- In general, the correct full conditional density for a block of parameters will depend on all other parameters and may be very complicated.
- When the parameters and observed data are related in a hierarchical structure:
  - We can draw a node for each block of parameters and block of data.
  - We can draw an arrow showing the dependency structure between the blocks.
  - The graph has no connected loops. It is a *directed acyclic graph*.
  - All nodes with an arrow going into the specified node are known as its parent nodes.
  - The destination nodes of all arrows leading out of the specified node are called its child nodes.
  - Other parents of the child nodes of a specified node are known as co-parent nodes.
- The conditional density of the parameters at a specified node, given all the other parameters and the data, will be proportional to its distribution given its parent nodes, times the joint distribution of all its child nodes given the specified node and all the co-parent nodes. For $\boldsymbol{\theta}_j$ this gives

$$g(\boldsymbol{\theta}_j|\boldsymbol{\theta}_{-j}, \mathbf{y}) \propto g(\boldsymbol{\theta}_j|\text{parents of } \boldsymbol{\theta}_j) \prod_{\boldsymbol{\theta}_k \in \text{children of } \boldsymbol{\theta}_j} f(\boldsymbol{\theta}_k|\text{parents of } \boldsymbol{\theta}_k)\,.$$

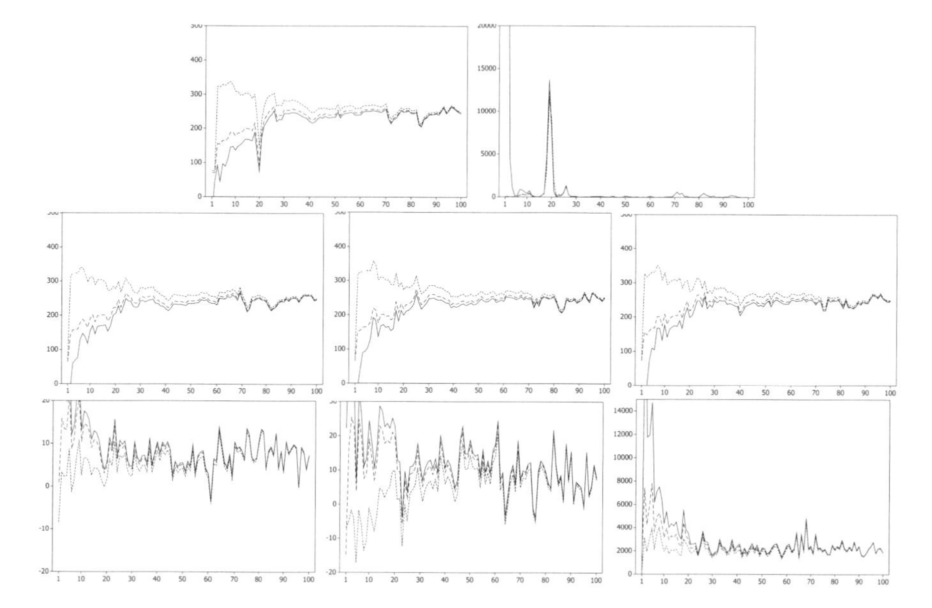

Figure 10.10 The outputs from multiple parallel Gibbs sampling chains with random starting points, and the same random inputs for τ, ψ, μ_1, μ_2, μ_3, β_1, β_2, and σ^2.

This depends only on the dependence structure of the hierarchy, not on the specific distributions.

- When the distributions of child nodes are from the exponential family, and the distribution of specified node given its parents are from the conjugate family, the exact form of these conditional densities can be found using the simple updating rules from Chapter 4.

- In other cases, we only know the proportional form of the full conditional density for that node. We would use one of the direct methods from Chapter 2 such as acceptance-rejection-sampling or adaptive-rejection-sampling to sample from that node.

- We should always use proper priors in the hierarchical model, particularly for scale parameters. When improper priors are used in the hierarchical model and the Gibbs sampler is used, each node looks like it has a proper posterior. However, overall, the posterior is improper. This means the Gibbs sampler represents a Markov chain that has all null recurrent states so it does not have a steady state distribution to converge to.

Exercises

10.1 Refer to the data given in Exercise 4.13. Use the Minitab macro *NormGibbsInd.mac* or the R-function `normGibbs` to run the Gibbs sampler 5000 steps

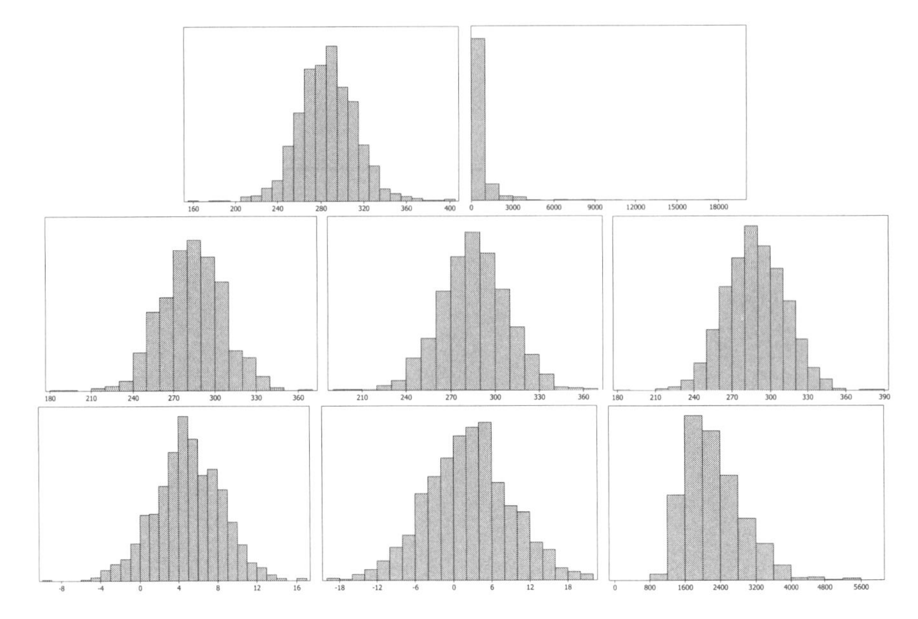

Figure 10.11 Histograms of the final thinned sample for τ, ψ, μ_1, μ_2, μ_3, β_1, β_2, and σ^2.

with the flat prior for μ and Jeffrey's prior for σ^2. Calculate the autocorrelation of the posterior sample for μ.

(a) Use a burn-in of five steps and thin the output by only taking every 5th step. Calculate the autocorrelation function of the thinned chain. Is the thinned sample approximately random?

(b) Compare the summary statistics from the posterior sample for μ with the exact solution we found in Exercise 4.13.

10.2 Refer to the data given in Exercise 4.14. Use the Minitab macro *NormGibbsInd.mac* or the R-function `normGibbs` to run the Gibbs sampler 5000 steps with the flat prior for μ and Jeffrey's prior for σ^2. Calculate the autocorrelation of the posterior sample for μ.

(a) Use a burn-in of five steps and thin the output by only taking every 5th step. Calculate the autocorrelation function of the thinned chain. Is the thinned sample approximately random?

(b) Compare the summary statistics from the posterior sample for μ with the exact solution we found in Exercise 4.14.

10.3 Refer to the data given in Exercise 4.15. Use the Minitab macro *NormGibbsJoint.mac* or the R-function `normGibbs` to run the Gibbs sampler 5000 steps with joint conjugate prior for μ and σ^2 made up from a *normal*$\left(m, \frac{\sigma^2}{n_0}\right)$ for μ given σ^2 and S_0 times an *inverse chi-squared* prior for σ^2. Let the equivalent sample size $n_0 = 1$, $S_0 = 400$, and $\kappa_0 = 1$. Calculate the autocorrelation of the posterior sample for μ.

(a) Use a burn-in of five steps and thin the output by only taking every 5th step. Calculate the autocorrelation function of the thinned chain. Is the thinned sample approximately random?

(b) Compare the summary statistics from the posterior sample for μ with the exact solution we found in Exercise 4.15.

10.4 Refer to the data given in Exercise 4.16. Use the Minitab macro *NormGibbsJoint.mac* or the R-function `normGibbs`to run the Gibbs sampler 5000 steps with joint conjugate prior for μ and σ^2 which is a product of an *normal*$(m, \frac{\sigma^2}{n_0}$ for μ given σ^2 and S_0 times an *inverse chi-squared* prior for σ^2. Let the equivalent sample size $n_0 = 1$, $S_0 = 1$, and $\kappa_0 = 3$. Calculate the autocorrelation of the posterior sample for μ.

(a) Use a burn-in of five steps and thin the output by only taking every 5th step. Calculate the autocorrelation function of the thinned chain. Is the thinned sample approximately random?

(b) Compare the summary statistics from the posterior sample for μ with the exact solution we found in Exercise 4.16.

10.5 Suppose we have independent random samples of size 8 drawn from four distributions where each distribution has its own mean value, and all three distributions have the known variance $\sigma^2 = 5^2$.

One	62.0	63.6	59.6	66.8	69.4	65.7	61.7	56.5
Two	42.1	57.7	55.8	50.6	55.0	58.3	50.1	61.3
Three	59.0	59.6	54.3	62.1	47.5	57.0	57.7	50.6
Four	59.8	70.8	59.7	65.8	59.0	65.7	60.7	58.0

(a) Find the mean and variance for the posterior distribution of each mean $\mu_1, \ldots, \mu_4$ using empirical Bayes where the known hypervariance $\psi = 5^2$.

(b) Put a *normal*$(60, 10^2)$ prior distribution for the hypermean τ. Find the mean and variance for the posterior distribution of each mean $\mu_1, \ldots, \mu_4$.

10.6 Suppose we have samples of size eight observations from four populations.

One	Two	Three	Four
626.952	588.610	582.906	640.996
589.944	598.012	604.918	618.210
591.879	605.204	595.259	586.970
590.125	564.930	583.211	580.336
657.360	600.661	604.727	644.002
621.158	652.042	643.959	613.107
570.384	615.596	606.690	609.925
661.516	590.955	633.964	610.026

We use a $normal(600, 100^2)$ prior for the hypermean τ, 500 times an *inverse chi-squared* prior with 1 degree of freedom for the hypervariance ψ, and 500 times an *inverse chi-squared* prior with 1 degree of freedom for the observation variance σ^2.

(a) Use the Minitab macro HierMeanReg.mac or the equivalent R-function `hierMeanReg` to run a Gibbs sampling chain 1000 Gibbs steps. Graph the traceplots and the autocorrelation functions of the output.

(b) Use the Minitab macro HierMeanReg.mac or the equivalent R-function `hierMeanReg` to run four Gibbs sampling chains from different starting points and with the same random inputs. Run each chain 50 steps.

(c) From the traceplots and autocorrelations in (a) and the four coupled chains in (b), decide what will be an adequate burn-in and thinning to get an approximate random sample from the posterior.

(d) Run the chain a 10000 steps and using the burn-in and thinning found in (c) so the thinned sample is approximately a random sample of the posterior. Graph histograms for each parameter using the thinned sample.

(e) Calculate 95% credible intervals for $\mu_1, \ldots, \mu_4$ from the posterior sample.

10.7 The data in Table 10.3 are given in the first four columns of the Minitab worksheet *Exercise10.7.mtw*. Use a $normal(0, 100000)$ prior for the hypermean τ, an 500 times an *inverse chi-squared* with 1 df prior for the hypervariance ψ, a *bivariate normal*$(\mathbf{b_0}, \mathbf{B_0})$ prior for the vector of slope coefficients $\boldsymbol{\beta}$, and an 500 times an *inverse chi-squared* with 1 df prior for the observation variance σ^2.

(a) Run the chain 2000 steps and examine the traceplots and sample autocorrelations.

(b) Run four parallel chains from different starting points, but with the same random inputs. Run each of the chains 100 steps. Plot the traceplots for each of the parameters for all parallel chains on the same graph.

(c) Decide on the burn-in and thinning required from the results of (a) and (b).

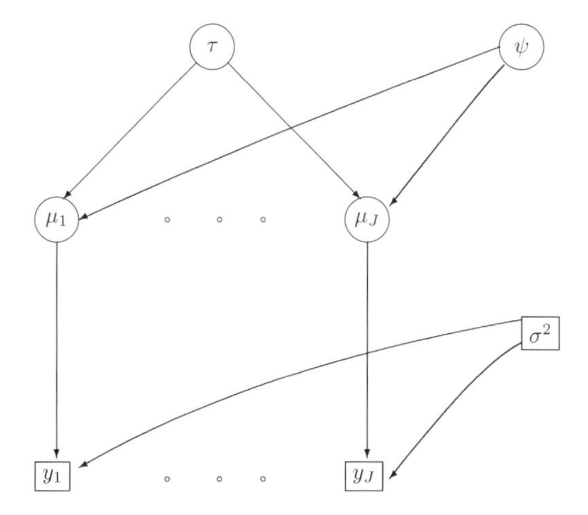

Figure A.1 Hierarchical normal mean model with regression.

(d) Run the chain 10000 steps, and use the burn-in and thinning found from (c) to get an approximately random sample from the posterior.

(e) Calculate 95% credible intervals for the means $\mu_1, \ldots, \mu_3$ and the slope β_1 and β_2 from your approximate random sample from the posterior.

Appendix: Proof That Improper Jeffrey's Prior Distribution for the Hypervariance Can Lead to an Improper Posterior

We show that an using an improper Jeffrey's prior distribution for the hypervariance can lead to an improper joint posterior for all the parameters in a hierarchical model. This is despite the Gibbs sampling conditional distribution for each node being proper We will show that using a Jeffrey's prior for a scale hyperparameter causes the joint posterior to be improper since the data is not able to drive the posterior away from the vertical asymptote at zero.

Suppose we have a hierarchical mean model. We will draw a single independent observation from normal distributions having different means but equal variance. The observation y_j comes from a *normal*(μ_j, σ^2) distribution where we may assume the variance σ^2 is known. Each of the means μ_j is an independent draw from a *normal*(τ, ψ) distribution for $j = 1, \ldots, J$. This is shown in Figure A.1. To do a full Bayesian analysis we will have to put priors on the hyperparameters. Suppose we decide to use Jeffrey's prior for the variance and an independent flat prior for the hypermean τ. If we look at each step of the Gibbs sampler, everything looks ok. The conditional density of τ given every other node is given by

$$g(\tau|everything) \quad \propto \quad g(\tau) \times \prod_{j=1,\ldots,J} f(\mu_j|\tau,\psi)\,.$$

Since all the means μ_j are normal and the prior is flat, we know the conditional density of τ given every other node will be *normal*$(\bar{\mu}, \frac{\psi}{J})$. The conditional density of ψ given every other node is given by

$$g(\psi|everything) \quad \propto \quad g(\psi) \times \prod_{j=1,\ldots,J} f(\mu_j|\tau,\psi)\,.$$

Since all the means μ_j are normal and the Jeffrey's prior for ψ is used, the conditional density of ψ given every other node will be SS_t times an *inverse chi-squared* with J degrees of freedom. The conditional distribution of mean μ_j given all other nodes is given by

$$g(\mu|everything) \quad \propto \quad g(\mu|\tau,\psi) f(y_j|\mu_j)\,.$$

Since both are normal, the conditional density of μ_j given all the other nodes will be *normal* with posterior precision and mean given by

$$\frac{1}{R_1} = \frac{1}{\psi} + \frac{1}{\sigma^2} \quad \text{and} \quad \frac{\frac{1}{\psi}}{\frac{1}{R_1}} \times \tau + \frac{\frac{1}{\sigma^2}}{\frac{1}{R_1}} \times y_j$$

respectively. Thus each of the conditional densities looked at individually is proper. One might think this would mean that the joint posterior of all the parameters would be proper. We will see that this is not the case.

The joint posterior can be written

$$\begin{aligned} g(\tau,\psi,\mu_1,\ldots,\mu_J|y_1,\ldots,y_J) \quad \propto \quad & g(\psi|y_1,\ldots,y_j) \times g(\tau|\psi,y_1,\ldots,y_J) \\ & \times g(\mu_1,\ldots,\mu_J|\tau,\psi,y_1,\ldots,y_J)\,, \end{aligned}$$

so if we can show that $g(\psi|y_1,\ldots y_J)$ is improper when we use an independent flat prior

$$g(\tau) = 1$$

and Jeffrey's prior

$$g(\psi) \quad \propto \quad \frac{1}{\psi}$$

then the joint posterior must be improper. The distribution of the observation y_j given the hypermean τ and hypervariance ψ is *normal*$(\tau, \psi + \sigma^2)$. The observations $y_1, \ldots, y_j$ form a random sample from that distribution. The joint posterior distribution of the hyperparameters is given by

$$g(\tau,\psi|y_1,\ldots,y_J) = g(\tau|\psi,y_1,\ldots,y_J) \times g(\psi|y_1,\ldots,y_J)\,.$$

The equation defining the joint posterior of τ and ψ can be solved to find the posterior density of ψ given by

$$g(\psi|y_1,\ldots,y_J) = \frac{g(\tau,\psi|y_1,\ldots,y_J)}{g(\tau|\psi,y_1,\ldots,y_J)}. \tag{A.1}$$

Since we are using the improper flat prior for τ the conditional posterior of τ given ψ is *normal* with mean and variance given by

$$\tau^* = \left(\frac{\frac{y_1}{\sigma^2+\psi}+\ldots+\frac{y_j}{\sigma^2+\psi}}{\frac{1}{\sigma^2+\psi}+\ldots+\frac{1}{\sigma^2+\psi}}\right) \quad \text{and} \quad V^* = \frac{1}{\sigma^2+\psi}+\ldots+\frac{1}{\sigma^2+\psi}$$

respectively. Equation A.1 can be simplified to

$$g(\psi|y_1,\ldots,y_J) \quad \propto \quad \frac{g(\psi)\left(\prod_{j=1}^{J}\frac{1}{\sigma^2+\psi}\right)e^{-\frac{1}{2(\sigma^2+\psi)}\sum(\tau-y_j)^2}}{\frac{1}{\sqrt{V^*}}e^{-\frac{1}{2V^*}(\tau-\tau^*)^2}}. \tag{A.2}$$

From the left-hand side of Equation A.2, we know the right-hand side cannot be a function of τ. That means the right-hand side must have the same ratio for all values of τ. So we substitute in $\tau = \tau^*$ which gives

$$g(\psi|y_1,\ldots,y_J) \quad \propto \quad \frac{g(\psi)\left(\prod_{j=1}^{J}\frac{1}{\sigma^2+\psi}\right)\frac{1}{\sigma^2+\psi}e^{-\frac{1}{2(\sigma^2+\psi)}\sum(\tau^*-y_j)^2}}{\frac{1}{\sqrt{V^*}}}. \tag{A.3}$$

This is a complicated function of ψ. Note

$$\lim_{\psi\to 0}\frac{g(\psi|y_1,\ldots,y_J)}{g(\psi)} = k$$

Thus if we use the improper non-informative Jeffrey's prior $g(\psi) = \frac{1}{\psi}$ recommended for a scale parameter the posterior $g(\psi|y_1,\ldots,y_J)$ will also have a vertical asymptote at zero. Since

$$\lim_{a\to 0}\int_a^b \frac{1}{\psi}d\,\psi = \infty$$

for all b the same will hold true for the posterior $g(\psi|y_1,\ldots,y_J)$. Hence it is improper, and the full joint posterior must also be improper. The Gibbs sampling Markov chain must be null recurrent rather than positive recurrent. Making inferences from its output is problematic. That is why most users of the Gibbs sampler decide to use proper priors for the hyperparameters which insures the Gibbs sampler has a proper long-run distribution.

11

Going Forward with Markov Chain Monte Carlo

In the previous chapters we have seen how Markov chain Monte Carlo methods can be used to draw samples from the posterior distribution, even when we only know it's unscaled form. After we have allowed for sufficient burn-in and thinning, the thinned sample will approximate a random sample from the posterior so we can base our inferences on it. The Metropolis-Hastings algorithm is the most basic algorithm for MCMC and is based on balancing the flow between every pair of states. In Bayesian statistics, the states are the parameters. We have a great deal of choice in how we apply this algorithm. We can update all the states at once, or we or we can cycle through blocks of states updating each block in turn. We can choose a candidate density that is centered at the current state or we can choose a candidate density that is independent of the state. The Gibbs sampling algorithm is a special case of blockwise Metropolis-Hastings where the candidate density for each block is its correct conditional density given the most recent values of the other blocks. Nevertheless, the Gibbs sampler was developed independently of the Metropolis-Hastings algorithm, initially for reconstructing a noisy image.

We saw in Chapter 10 that Gibbs sampling is particularly well adapted for models with a hierarchical parameter structure. The conditional distribution for a particular node will be proportional to the distribution of that node, given its parent nodes, times the product of the distribution of all the child nodes of that particular node, given it and the coparent nodes. When the child distributions are from the exponential family and the distribution given the parents is conjugate then the conditional distribution is another conjugate family member and is easily found by applying simple formulas. The Gibbs sampler is particularly easy to use in these cases.

Gibbs Sampling at a Non-Conjugate Node

When we have a hierarchical model, but for some particular node, the distribution of its child nodes, given it and its coparents, and the distribution of that node, given its parents are not from the exponential family and its conjugate family, respectively, the Gibbs sampler can still be used. The conditional distribution of that particular node is still proportional to the distribution of that node, given its parent nodes, times the product of the distribution of all the child nodes of that particular node, given it and the coparent nodes. However, we will have to find some other way to sample from that node because of the non-conjugacy.

If the node is a single parameter and its conditional distribution is log-concave, we can draw an observation from the conditional distribution using the adaptive rejection sampling algorithm described in Chapter 2. Generally it takes only a few steps before we get an accepted draw from the conditional distribution, since we are tightening the candidate density with every unaccepted draw.

Another alternative we can use when the conditional distribution of a node is not log-concave is to use a Metropolis-Hastings step for that node. We draw a candidate for that node from a candidate distribution, then either accept it and move to the candidate, or we reject it and stay at the current value. This is sometimes called "Metropolis within Gibbs." However, this is a misnomer since we saw in Chapter 6 that the Gibbs sampler is a special case of the Metropolis-Hastings algorithm. Of course, we must be sure that the candidate density of that node dominates its conditional density in the tails. If it fails to do so, there will be a region that will not be visited in the proper proportion. Tierney (1991) suggested that adaptive rejection sampling can be used even when the conditional distribution (target) is not log-concave by appending a Metropolis-Hastings step after the final accepted value. We run the adaptive rejection algorithm until a value is accepted, even though the envelope distributions do not dominate the target. When we finally accept a value from the adaptive rejection step, this is used as the candidate value for a Metropolis-Hastings step. We either accept the candidate and move to that value, or we reject the candidate and remain at the current value.

Speeding Up Convergence of the Gibbs Sampler

Our main concern in MCMC modelling is designing a Markov chain that converges to the long-run distribution as quickly as possible. That way we won't have to discard too many observations in the burn-in and thinning stages. In the Gibbs sampler, the model could be reparameterized so the reparameterized model has uncorrelated parameters. Alternatively, we should put parameters that are highly correlated together into a single node and draw them from their multivariate distribution given the most recent values of the parameters in other nodes. Either of these will usually speed up the convergence a great deal. When we have regression on covariates we should recode the covariates so they measure the distance away from their respective means. This also reparameterizes the model, changing the meaning of the mean parameters in such

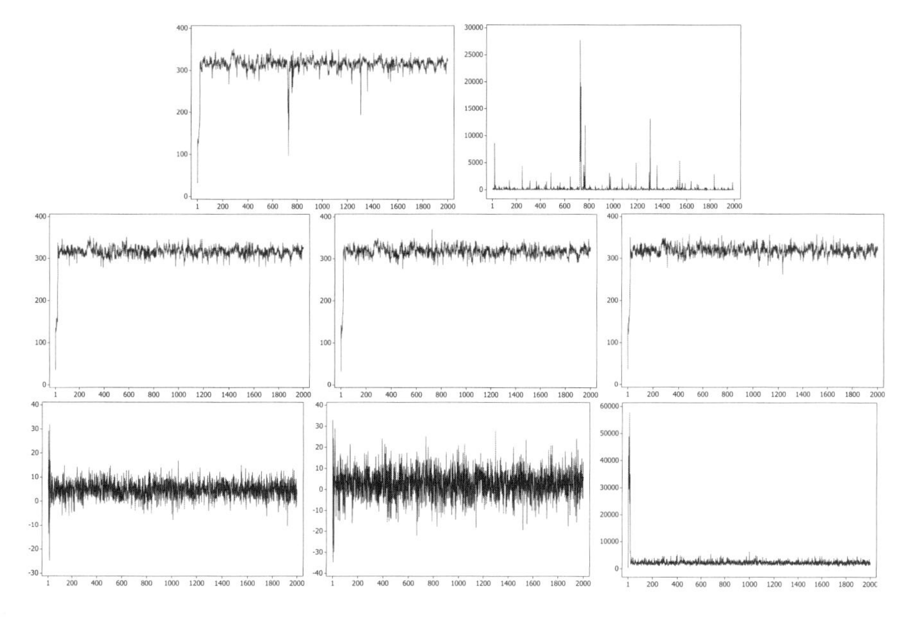

Figure 11.1 The traceplots for τ, ψ, μ_1, μ_2, μ_3, β_1, β_2, and σ^2.

a way that it reduces the correlation between the slopes and the mean parameters. This speeds up the convergence.

Example 17 (continued) *We note in Table 10.3 that the covariates x_1 and x_2 are not centered around their means. Let us recode them to be centered around their respective means. We note that this reparameterizes the mean parameters $\tau, \mu_1, \ldots, \mu_3$. We will run the chain for 2000 steps using the same priors as before. The traceplots are shown in Figure 11.1. When we compare them with the corresponding traceplots in Figure 10.8 where the covariates were not centered around their means we see that the centered chain is moving through the parameter space faster. The autocorrelations are shown in Figure 11.2. When we compare them with the corresponding autocorrelations for the chain that is not centered shown in Figure 10.9, we see that a shorter burn-in time and less thinning will be required for the centered chain. We run three parallel chains from different starting positions but with the same random inputs to see how many steps are needed until they have converged. The traceplots are shown in Figure 11.3. We compare them to the corresponding traceplots in Figure 10.10 where the covariates were not centered about their means. We see the effect of the starting point wears off much more quickly for the centered chain. For this chain, we could get an approximately random thinned sample by using a burn-in of 25 steps and thinning by using every 25th step. That means, out of the same length run of the chain, if we center the covariates, we could get the final sample four times as large which would lead to standard errors being halved.*

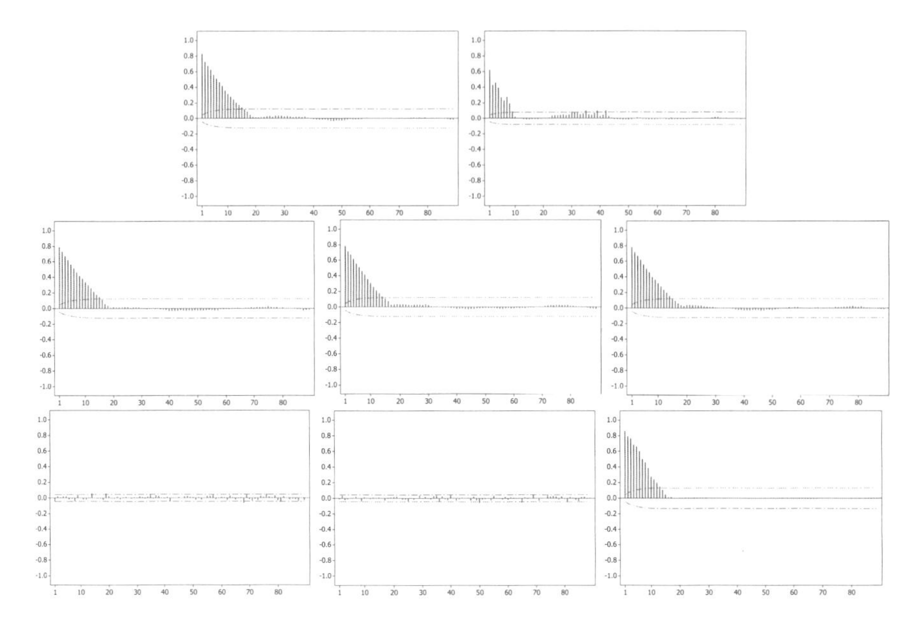

Figure 11.2 The autocorrelation functions for τ, ψ, μ_1, μ_2, μ_3, β_1, β_2, and σ^2.

Gibbs Sampling Packages

The availability of computer software is often the factor that determines how often a method gets used. The Medical Research Council Biostatistics Unit Cambridge has undertaken the BUGS (Bayesian Using Gibbs Sampling) project. Originally they developed the BUGS progam to implement the Gibbs sampler. Following on this, they developed WinBUGS which allows the user to set up his/her model by drawing a hierarchical graph for the parameters and observations, see Lunn et al. (2000). The OpenBUGS project at the University of Helsinki also provides an open source code for the core of BUGS. Gilks et al. (1996) and Gill (2008) use the BUGS program for Bayesian inference.

In Chapter 6, we observed that the Gibbs sampling algorithm is a special case of the Metropolis-Hastings algorithm. We noted that the traceplots for the output from a Gibbs sampling chain look more like the traceplots of an MH chain with a random-walk candidate density than those from an MH chain with an independent candidate density. It should not surprise us then that the Gibbs sampling chain may not be as efficient at moving through the parameter space as an MH chain with an independent candidate density that has a similar shape to the target, but with heavier tails.

Metropolis-Hastings Algorithm with Multiple Modes

When the posterior has multiple widely separated modes the rate of convergence to the long-run distribution will slow way down when the Gibbs sampler is used. This

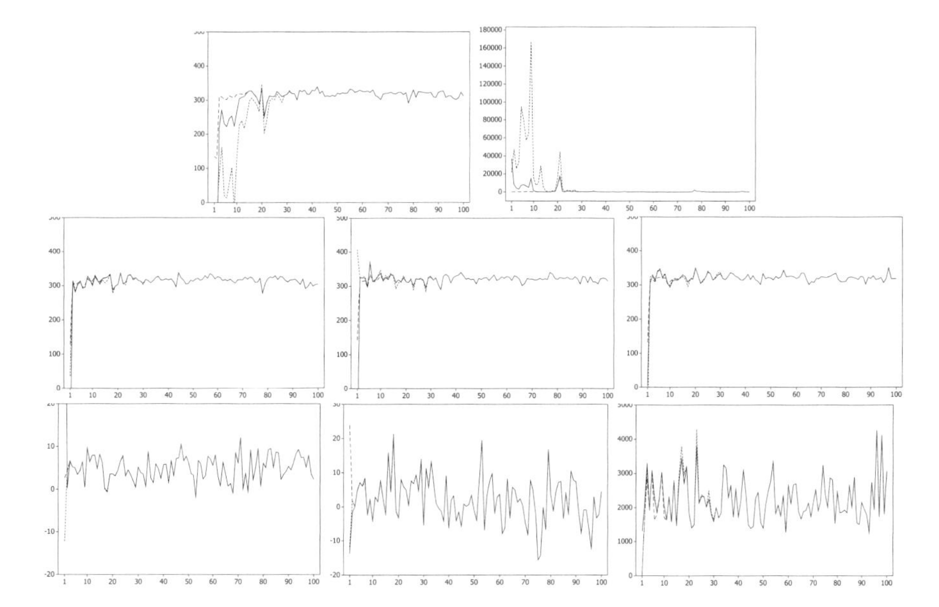

Figure 11.3 The traceplots from multiple parallel Gibbs sampling chains with random starting points, and the same random inputs for τ, ψ, μ_1, μ_2, μ_3, β_1, β_2 and σ^2.

will mean that a very long burn-in time and a great deal of thinning will be required. Otherwise, the region around the subsidiary node won't be adequately explored, and certainly the Gibbs sample won't be a random one from the posterior. We saw in Chapters 7–9 that the Metropolis-Hastings algorithm with an independent candidate distribution, which has similar shape to the target but with heavy tails, gave us very fast convergence to the long-run distribution. When the posterior has a single mode we can find a suitable candidate by finding the *multivariate normal* distribution that matches the curvature of the posterior at the mode. Then we use the corresponding *multivariate Student's t* distribution with low degrees of freedom as the candidate density.

We suggest that when the posterior has multiple nodes a similar strategy be used. At each mode, we find the *multivariate normal* distribution that matches the curvature of the target at that mode. The candidate density should be a mixture of the *multivariate Student's t* distributions with low degrees of freedom where each component of the mixture corresponds to the *multivariate normal* found for that node. Using this mixture density as the independent candidate density will give fast convergence to the posterior.

More Advanced Uses for MCMC: Jumping Between Models

In this book, we have only considered using MCMC methods to find the posterior probability distribution for a given parametric model. Carlin and Chib (1995) noted that MCMC approach is so flexible and easy to use that the class of candidate models now appears to be limited only by the users imagination. They developed a method for using these methods for choosing between M competing models which may not be nested. Carlin and Chib (1995) set up a supermodel that contains all M competing models. An indicator parameter is included that indicates which model is used. The prior $p(\boldsymbol{\theta_j}|M = j)$ and likelihood $f(\mathbf{y}|\boldsymbol{\theta_j})$ for each model to be conditional on the indicator parameter. A linking density or pseudoprior $p(\boldsymbol{\theta_j})|M \neq j)$ is created since it is needed to make the model completely specified. The Gibbs sampler is implemented where the model choice parameter is updated as well as the other parameters in its turn. Thus we will draw a sample from the posterior model probabilities as well as the posterior probability of the parameters for each model. Even though the form of the pseudoprior is irrelevant, a bad choice can lead to poor convergence. This is discussed in Gamerman (1997).

Green (1995) took another approach to this problem. He set up reversible jump Markov chains that make the probabilities for a jump from each point in one model space to another point in another model space satisfy the detailed balance equation. This allows the long-run distribution of the reversible jump Markov chain include the posterior probability of each model as well as the posterior distribution of the parameters within each model.

Nonparametric Bayesian Statistics

In this text we have only considered parametric observation models. In other words, the density is indexed by a finite dimensional parameter. Bayesian inference is based on the posterior distribution of those parameters. Nonparametric Bayesian models use distributions with infinitely many parameters, as the probability model is defined on a function space, not a finite dimensional parameter space. The random probability model on the function space is often generated by a Dirichlet process. Interested readers are referred to Dey et al. (1998).

Exercises

11.1 Redo Exercise 10.7 but this time centering the covariates around their mean values.

A

Using the Included Minitab Macros

Minitab macros for performing the Bayesian analysis and Markov chain Monte Carlo simulations shown in this text are included on a website maintained for this text. A link to the web address can be found on the web page for this text on the site <www.wiley.com>. The Minitab Macros are zipped up in a package called *Bolstad.Minitab.Macros.zip*. Some Minitab worksheets are also included at that site. First, define a directory named *BAYESMAC* on your hard disk to store the macros. The best place is inside the Minitab directory, which is often called *MINITAB 15* on PCs running Microsoft Windows. For example, on my PC, *BAYESMAC* is inside *MINITAB 15*, which is within *Program Files*, which is on drive *C*. The correct path I need to invoke to use these macros is *C:/progra~1/MINITA~1/BAYESM~1/*. (Note that the the filenames are truncated at six characters). You should also define a directory BAYESMTW for the Minitab worksheets containing the data sets. The best place is also inside the Minitab directory, so you can find it easily.

Chapter 3: Bayesian Inference

Calculating the Numerical Posterior and Its CDF

The proportional posterior density is easily found numerically by multiplying the prior density times the likelihood. Multiplying either by a constant does not matter,

Table A.1 Minitab commands for calculating numerical posterior CDF. In this example, the likelihood is proportional to $\theta^{(5/2-1)} \times e^{-\theta/(2*2)}$ and a $normal(6, 3^2)$ prior is used.

Minitab Commands	Meaning
Set c1	θ range
.001:40/.001	
end	
Let c2= exp(-(c1-6)**2/(2*3**2))	$normal(6, 3^2)$ proportional prior
Let c3=c1**(5/2-1)*exp(-c1/(2*2))	proportional likelihood
Let c4=c2*c3	proportional posterior
%<*path*>tintegral c1 c4;	find scale factor k5
output k5 c5.	
let c4=c4/k5	numerical posterior
%<*path*>tintegral c1 c4;	find posterior CDF
Output k5 c5.	posterior CDF in c5

because the constant will cancel out when we find the exact numerical posterior density by

$$g(\theta|y) = \frac{g(\theta)\, f(y|\theta)}{\int_{-\infty}^{\infty} g(\theta)\, f(y|\theta)\, d\theta}\,.$$

The Minitab macro *tintegral.mac* is used to evaluate the denominator numerically. We want to do an inference such as finding a credible interval for the parameter, or calculating the posterior probability of a one-sided null hypothesis about the parameter. The cumulative distribution function (CDF) of the posterior is the definite integral of the numerical posterior density. It is given by

$$G(\theta|y) \quad = \quad \int_{-\infty}^{y} g(\theta|y)\, d\theta\,.$$

The CDF will be evaluated using the Minitab macro *tintegral.mac*. Table A.1 shows the Minitab commands for calculating the posterior CDF numerically for an example where the likelihood is proportional to $\theta^{(5/2-1)} \times e^{-\theta/(2*2)}$ and a $normal(6, 3^2)$ prior is used.

Credible Intervals from the Numerical Posterior CDF

The macro *CredIntNum.mac* calculates a lower, upper, or two-sided credible interval from the numerical posterior CDF. The Minitab commands for invoking this macro are shown in Table A.2. We can test a two-sided hypothesis

$$H_0 : \theta = \theta_0 \quad \text{vs} \quad H_1 : \theta \neq \theta_0$$

Table A.2 Minitab commands for calculating credible interval from a numerical posterior CDF

Minitab Commands	Meaning
%<*path*>CredIntNum c1 c5 .95;	θ in c1, CDF in c5,
	95% credible level
type ktype.	ktype=-1, Lower credible bound
	ktype=0, Two-sided credible interval
	ktype=+1, Upper credible bound

Table A.3 Minitab commands for calculating probability of one-sided null hypothesis from a numerical posterior CDF

Minitab Commands	Meaning
%<*path*>PNullHypNum c1 c5 k1;	θ in c1, CDF in c5,
	null value θ_0 in k1
type ktype.	ktype=-1, $H_0 : \theta \leq \theta_0$
	ktype=1, $H_0 : \theta \geq \theta_0$

at the α level of significance by observing whether or not the null value θ_0 lies inside the two-sided $(1 - \alpha) \times 100\%$ credible interval.

Testing a One-Sided Hypothesis from the Numerical Posterior CDF

We can test a one sided hypothesis about the parameter

$$H_0 : \theta \leq \theta_0 \quad \text{vs} \quad H_1 : \theta > \theta_0$$

by calculating the posterior probability of the null hypothesis. If this probability is less than the level of significance α, then we can reject the null hypothesis at the α level. The macro *PNullHypNum.mac* calculates the probability of a one-sided null hypothesis from a numerically calculated posterior CDF. The Minitab commands for this macro are shown in Table A.3.

Bayesian Inference from a Posterior Random Sample

Sometimes we only know the shape of the posterior, not the exact posterior density. Nevertheless, we have a random sample from the posterior, despite not knowing it exactly. We might have obtained this sample using a direct sampling method such as acceptance-rejection-sampling or sampling-importance-resampling. Or, we might have obtained this sample by sampling from a Markov chain that has the posterior as its long-run distribution. This is known as a *Markov chain Monte Carlo* method. *Gibbs sampling* and the *Metropolis-Hastings* algorithm are special cases of this.

Table A.4 Minitab commands for calculating credible interval from a posterior sample

Minitab Commands	Meaning
%<*path*>CredIntSamp c1 .95;	sample in c1, 95% credible level
type ktype.	ktype=-1, lower credible bound
	ktype=0, two-sided credible interval
	ktype=+1, upper credible bound

Note, we would have to burn-in and thin the MCMC sample from the posterior to make it approximately random before we use it for inference.

Finding a credible interval. We want to use our random sample from the posterior to do an inference such as finding a credible interval for the parameter, or calculating the posterior probability of a one-sided null hypothesis about the parameter. The macro *CredIntSamp.mac* calculates a lower, upper, or two-sided credible interval from a random sample from the posterior. The Minitab commands for running this macro are given in Table A.4. We can test a two-sided hypothesis

$$H_0 : \theta = \theta_0 \quad \text{vs} \quad H_1 : \theta \neq \theta_0$$

at the α level of significance by observing whether or not the null value θ_0 lies inside the two-sided $(1 - \alpha) \times 100\%$ credible interval.

Testing a one-sided hypothesis using the posterior random sample. We can use the random sample from the posterior to test a one sided hypothesis about the parameter

$$H_0 : \theta \leq \theta_0 \quad \text{vs} \quad H_1 : \theta > \theta_0$$

by calculating the proportion in the posterior sample that satisfies the null hypothesis. If this proportion is less than the level of significance α, then we can reject the null hypothesis at the α level. The macro *PNullHypSamp.mac* calculates the proportion of a one-sided null hypothesis from a numerically calculated posterior. The Minitab commands for running this macro are given in Table A.5

Chapter 6: Markov chain Monte Carlo Sampling from the Posterior

A mixture of two normal distributions. The macros *NormMixMHRW.mac* and *NormMixMHInd.mac* use the Metropolis-Hastings algorithm to draw a sample from a univariate target distribution that is a mixture of two normal distributions using an independent *Normal* candidate density and a random-walk *normal* candidate density, respectively. Table A.6 shows the Minitab commands to set up and run these two macros.

Table A.5 Minitab commands for calculating probability of one-sided null hypothesis from a posterior sample

Minitab Commands	Meaning
%<*path*>PNullHypSamp c1 k1;	c1 contains sample from posterior,
	k1 is null hypothesis value.
type ktype.	ktype=-1, $H_0 : \theta \leq \theta_0$
	ktype=1, $H_0 : \theta \geq \theta_0$

Table A.6 Minitab commands for drawing Markov chain Monte Carlo sample from a mixture of *normal* distributions using either a *normal* random-walk candidate density or an independent candidate density

Minitab Commands	Meaning
let k1=0	mean of first component
let k2=1	standard deviation of first component
let k3=3	mean of second component
let k4=2	standard deviation of second component
let k5=.8	mixture proportion of first component
%<*path*>NormMixMHRW ;	random-walk chain
Target k1 k2 k3 k4 k5;	
MH 1000;	number of MH steps
Candidate 0 .5;	*normal*$(0, .5^2)$ random-walk density
MCMCsample m1.	Markov chain Monte Carlo sample
%<*path*>NormMixMHInd ;	independent chain
Target k1 k2 k3 k4 k5;	
MH 1000;	number of MH steps
Candidate 0 3;	*normal*$(0, 3^2)$ independent density
MCMCsample m2.	Markov chain Monte Carlo sample

A correlated bivariate normal distribution. The macros *BivNormMHRW.mac* and *BivNormMHInd.mac* use the Metropolis-Hastings algorithm to draw a sample from a correlated bivariate normal target density using a random-walk candidate and an independent candidate density, respectively, where we are drawing both parameters in a single draw. The macros *BivNormMHblock.mac* and *BivNormMHGibbs.mac* use the blockwise Metropolis-Hastings algorithm and Gibbs sampling, respectively, to draw a sample from the correlated bivariate normal target. Table A.7 gives the Minitab commands to set up and run these macros.

Table A.7 Minitab commands for drawing Markov chain Monte Carlo sample from a correlated *bivariate normal* distribution using either a *normal* random-walk candidate density or an independent candidate density

Minitab Commands	Meaning
%<*path*>BivNormMHRW .9;	random-walk chain
MH 1000;	number of MH steps
MCMCsample m1.	Markov chain Monte Carlo sample
%<*path*>BivNormMHIND .9;	independent chain
MH 1000;	number of MH steps
MCMCsample m2.	Markov chain Monte Carlo sample
%<*path*>BivNormMHblock .9;	blockwise MH chain
MH 1000;	number of MH steps
MCMCsample m3.	Markov chain Monte Carlo sample
%<*path*>BivNormMHGibbs .9;	Gibbs sampling chain
MH 1000;	number of MH steps
MCMCsample m4.	Markov chain Monte Carlo sample

Table A.8 Minitab commands for thinning the output of a MCMC process

Minitab Commands	Meaning
%<*path*>ThinMCMC m2 k1 k2 ;	m2 contains output of MCMC chain, k1 is number of output variables, and k2 is thinning constant
output m12.	m12 contains thinned sample

Chapter 7: Statistical Inference from a Markov chain Monte Carlo Sample

Sequential draws from a Markov chain are serially dependent. A Markov chain Monte Carlo sample will not be suitable for inference until we have discarded the draws from the burn-in period and thinned the sample so the thinned sample will approximate a random sample from the posterior. In Table A.8 we give the Minitab commands for thinning the output of a MCMC process using the macro *ThinMCMC.mac*.

The Gelman-Rubin statistic is used to indicate the amount of improvement possible if n, the burn-in, was increased. What it actually measures is how well the chain is moving through the parameter space. It is an analysis of variance approach comparing the between chain variation to the within chain variation for the next n steps after burn-in of n steps for multiple chains started from an overdispersed starting distribution.

Table A.9 Minitab commands for calculating the Gelman-Rubin statistic for four parameters from four runs of a Markov chain, each run being $2n$ steps

Minitab Commands	Meaning
copy m1 c11-c14	copy output of first run in columns
copy m2 c21-c24	output of second run in columns
copy m3 c31-c34	output of third run in columns
copy m4 c41-c44	output of fourth run in columns
copy c11 c21 c31 c41 m6	first parameter copied into matrix
%<*path*>GelmanRubin 4 m6 .	
copy c12 c22 c32 c42 m7	second parameter copied into matrix
%<*path*>GelmanRubin 4 m7 .	
copy c13 c23 c33 c43 m8	third parameter copied into matrix
%<*path*>GelmanRubin 4 m8 .	
copy c14 c24 c34 c44 m9	fourth parameter copied into matrix
%<*path*>GelmanRubin 4 m9 .	

Table A.9 shows the commands for calculating the Gelman-Rubin statistic from the outputs of four runs of the chain given in matrices $m_2, \ldots, m_5$. Each output matrix has four columns, one for each parameter. The corresponding columns from each output matrix have to be copied into a new matrix for the samples of that parameter which is input into the macro.

Coupling with the past is where we run chains with different starting points, but with the same random inputs and see how long it takes the chains converge. At that time, the effect of the starting point has died out. This is sometimes called perfect sampling. When a Metropolis-Hastings chain with a random-walk candidate density is used, the coupled runs eventually get very close, but are not exactly identical. However, when a suitable independent candidate density is used, the coupled runs converge exactly.

Table A.10 gives the Minitab commands for three Metropolis-Hastings chains with a random-walk candidate density coupled with the past using the macro *NormMixMHRW.mac* for a mixture of normal distributions four Markov chain Monte Carlo sample from a mixture of *normal* distributions using either a *normal* random-walk candidate density or an independent candidate density. If we want to use an independent candidate density, the macro *NormMixMHInd.mac* is used instead and the constants for the candidate density changed.

Table A.10 Minitab commands for three Markov chain Monte Carlo samples from a mixture of *normal* distributions using a *normal* random-walk candidate density or an independent candidate density coupled with the past. The chains have different starting points, but all have the same random inputs.

Minitab Commands	Meaning
random 1 c99;	this draws random seed
integer 1 100000.	
let k1=c99(1)	
%<*path*>NormMixMHRW ;	first random-walk chain
Target 0 1 3 2 .8;	mixture of normals target
MH 100;	number of MH steps
candidate 0 1;	*normal*$(0, 1^2)$ candidate density
start -2 ;	start first chain at -2
rstart k1 ;	seed
MCMCsample m1.	MCMC sample from first chain
%<*path*>NormMixMHRW ;	second random-walk chain
Target 0 1 3 2 .8;	mixture of normals target
MH 100;	number of MH steps
candidate 0 1;	*normal*$(0, 1^2)$ candidate density
start 1 ;	start second chain at 1
rstart k1 ;	seed
MCMCsample m2.	MCMC sample from second chain
%<*path*>NormMixMHRW ;	third random-walk chain
Target 0 1 3 2 .8;	mixture of normals target
MH 100;	number of MH steps
candidate 0 1;	*normal*$(0, 1^2)$ candidate density
start 4 ;	start third chain at 4
rstart k1 ;	seed
MCMCsample m3.	MCMC sample from third chain
copy m1 c1	
copy m2 c2	
copy m3 c3	
tsplot c1 c2 c3 ;	
connect:	
overlay.	

Table A.11 Minitab commands for logistic regression

Minitab Commands	Meaning
%<*path*>BayesLogRegMH k1 k2 c1 m1;	k1 is number of obs , k2 is number of
	parameters including intercept,
	response y are in c1, covariates in m1
#MLEin c20 m20;	we don't know $\hat{\beta}_{ML}$ and $\mathbf{V}_{\mathrm{ML}}$ yet
MH 1000;	run 1000 steps of MH algorithm
prior c10 m10;	*MVN*$(c10, m10)$ prior
	(default is flat prior)
graphs;	output traceplots and autocorrelations
MCMCsample m3;	Monte Carlo sample
MLEout c20 m20.	save $\hat{\beta}_{ML}$ and $\mathbf{V}_{\mathrm{ML}}$

Chapter 8: Logistic Regression

The macro *BayesLogRegMH.mac* draws a random sample from the posterior distribution for the logistic regression model. The logistic regression model is an example of a generalized linear model, and the maximum likelihood can be found by iteratively reweighted least squares. First, *BayesLogRegMH.mac* finds an approximate normal likelihood function for the logistic regression model where the mean matches the maximum likelihood estimator found using iteratively reweighted least squares. The covariance matrix is found that matches the curvature of the likelihood function at its maximum. Details of this are found in Myers et al. (2002) and Jennrich (1995). The approximate normal posterior by applying the usual normal updating formulas with a normal conjugate prior. If we used this as the candidate distribution, it may be that the tails of true posterior are heavier than the candidate distribution. This would mean that the accepted values would not be a sample from the true posterior because the tails would not be adequately represented. We can make the candidate distribution have heavier tails by using a *Student's t* with low degrees of freedom in place of the normal. This will make the accepted sample be from the posterior. The Minitab commands for logistic regression are given in Table A.11. The MCMC sample of $\beta_0, \ldots, \beta_p$ values is output in matrix $M3$. Discard the values in the burn-in time, and the remainder can constitute a sample from the posterior. Thin the accepted sample enough so they can be considered to be a independent random sample from the posterior, and use the thinned sample for your inferences.

Table A.12 gives the commands we use to run three parallel Metropolis-Hastings chains with different starting points, but with the same random inputs for the logistic regression model with two parameters (one predictor). Here we want to see how long we have to run the chains before they converge to the same values. At that point, the effect of the starting point has finished, so the draw can be considered a random draw from the posterior.

Table A.12 Minitab commands for three parallel runs of Metropolis-Hastings algorithm for the logistic regression model coupled with the past. The runs have different starting points, but all have the same random inputs. In this example there are two parameters (one predictor) and we are using a flat prior.

Minitab Commands	Meaning
random 2 c5-c7	random starting values
random 1 c99;	random seed
integer 1 100000.	
let k1=c99(1)	
%<*path*>BayesLogRegMH 100 2 c1 m1;	first run of chain
MH 20;	
start c5 ;	starting values for first run
rbase k1;	seed
MLEout c20 m20;	save $\hat{\boldsymbol{\beta}}_{ML}$ and $\hat{\mathbf{V}}_{\mathbf{ML}}$
MCMCsample m2.	
%<*path*>BayesLogRegMH 100 2 c1 m1;	second run of chain
MLEin c20 m20;	input $\hat{\boldsymbol{\beta}}_{ML}$ and $\hat{\mathbf{V}}_{\mathbf{ML}}$
MH 20;	
start c6 ;	starting values for second run
rbase k1 ;	
MCMCsample m3.	
%<*path*>BayesLogRegMH 100 2 c1 m1;	third run of chain
MLEin c20 m20;	input $\hat{\boldsymbol{\beta}}_{ML}$ and $\hat{\mathbf{V}}_{\mathbf{ML}}$
MH 20;	
start c7;	starting values for third run
rbase k1 ;	
MCMCsample m4.	
copy m2 c6 c7	
copy m3 c8 c9	
copy m4 c10 c11	
tsplot c6 c8 c10;	
connect;	
overlay.	
tsplot c7 c9 c11;	
connect;	
overlay.	

Table A.13 Minitab commands for *Poisson* regression

Minitab Commands	Meaning
%<*path*>BayesPoisRegMH k1 k2 c1 m1;	k1 is number of obs, k2 is number of
	parameters including intercept,
	response y are in c1, covariates in m1
#MLEin c20 m20;	we don't know $\hat{\beta}_{ML}$ and $\mathbf{V}_{\mathrm{ML}}$ yet
MH 1000;	1000 steps of MH algorithm
prior c10 m10;	$MVN(c10, M10)$ prior
	(default is flat prior)
graphs;	output traceplots and autocorrelations
MCMCsample m2;	Markov chain Monte Carlo sample
MLEout c20 m20.	save $\hat{\beta}_{ML}$ and $\hat{\mathbf{V}}_{\mathrm{ML}}$

Chapter 9: Poisson Regression and Proportional Hazards Model

The macro *BayesPoisRegMH.mac* draws a random sample from the posterior distribution for the Poisson regression model. The macro finds the maximum likelihood estimator by iteratively reweighted least squares, and then finds a normal approximation to the likelihood function by matching the curvature at the maximum likelihood estimator. The approximate posterior is found using the normal updating rules using either a multivariate normal prior or a multivariate flat prior. This approximate posterior may have lighter tails than the true posterior, so we use a matching *multivariate Student's t* with low degrees of freedom as the candidate density. It has similar shape to the true posterior, and it will have heavy tails, so it will be a very effective independent candidate density. The Minitab commands for using this macro are given in Table A.13. Table A.14 gives the Minitab commands for running three parallel chains for the Poisson regression model coupled with the past. We want to determine how many steps are needed before the effect of the starting point has vanished and the parallel chains with the same random inputs have converged to the same point.

Table A.14 Minitab commands for three parallel runs of the Metropolis-Hastings algorithm for the Poisson regression model with five parameters (four predictors) coupled with the past. The chains have different starting points, but all have the same random inputs. Note: we are using flat prior in this example.

Minitab Commands	Meaning
random 5 c7-C9	random starting values
random 1 c99;	random seed
integer 1 100000.	
let k1=c99(1)	
%<*path*>BayesPoisRegMH 44 5 c1 m1;	first run
MH 20;	
start c7 ;	starting values for first run
rbase k1;	seed
MLEout c20 m20;	save $\hat{\boldsymbol{\beta}}_{ML}$ and $\hat{\mathbf{V}}_{\mathbf{ML}}$
MCMCsample m2.	
%<*path*>BayesPoisRegMH 44 5 c1 m1;	second run
MLEin c20 m20;	input $\hat{\boldsymbol{\beta}}_{ML}$ and $\hat{\mathbf{V}}_{\mathbf{ML}}$
MH 20;	
start c8 ;	starting values for second run
rbase k1 ;	
MCMCsample m3.	
%<*path*>BayesPoisRegMH 44 5 c1 m1;	third run
MLEin c20 m20;	input $\hat{\boldsymbol{\beta}}_{ML}$ and $\hat{\mathbf{V}}_{\mathbf{ML}}$
MH 20;	
start c9;	starting values for third run
rbase k1 ;	
MCMCsample m4.	
copy m2 c14-c18	
copy m3 c24-c28	
copy m4 c34-c38	
tsplot c14 c24 c34 ;	
connect;	
overlay.	
tsplot c15 c25 c35 ;	
connect;	
overlay.	
tsplot c16 c26 c36 ;	
connect;	
overlay.	

Table A.15 Minitab commands for *Proportional hazards model*

Minitab Commands	Meaning
%<*path*>BayesPropHazMH k1 k2 c1 c2 m1;	k1 observations, k2 parameters, response in c1, time in c2, predictors in m1
#MLEin c20 m20;	we don't know $\hat{\beta}_{ML}$ and $\mathbf{V}_{\mathrm{ML}}$ yet
MH 1000;	1000 steps of MH algorithm
prior c10 m10;	*MVN*$(c10, M10)$ prior (default is flat prior)
graphs;	output traceplots and autocorrelations
MCMCsample m2;	Markov chain Monte Carlo sample
MLEout c20 m20.	store $\hat{\beta}_{ML}$ and $\mathbf{V}_{\mathrm{ML}}$

When we have censored survival times data, and we relate the linear predictor to the hazard function we have the proportional hazards model. The macro *BayesPropHazMH.mac* draws a random sample from the posterior distribution for the proportional hazards model. First, the macro finds an approximate normal likelihood function for the proportional hazards model. The (multivariate) normal likelihood matches the mean to the maximum likelihood estimator found using iteratively reweighted least squares. Details of this are found in Myers et al. (2002) and Jennrich (1995). The covariance matrix is found that matches the curvature of the likelihood function at its maximum. The approximate normal posterior by applying the usual normal updating formulas with a normal conjugate prior. If we used this as the candidate distribution, it may be that the tails of true posterior are heavier than the candidate distribution. This would mean that the accepted values would not be a sample from the true posterior because the tails would not be adequately represented. Table A.15 gives the Minitab commands needed to find a random sample from the posterior using the Metropolis-Hastings algorithm for the proportional hazards model. Table A.16 gives the Minitab commands for running three parallel chains for the Proportional hazards model coupled with the past. We want to determine how many steps are needed before the effect of the starting point has vanished and the parallel chains with the same random inputs have converged to the same point.

Chapter 10: Hierarchical Models and Gibbs Sampling

Normal(μ, σ^2) with both parameters unknown. When we have a random sample from a *normal*(μ, σ^2) where both parameters are unknown, we have a choice in the priors we can use with the Gibbs sampler. We can use independent conjugate priors for the two parameters. The prior for μ would be a *normal*(m, s^2) prior and the prior for σ^2 would be S times an *inverse chi-squared*. (The defaults are flat prior

Table A.16 Minitab commands for three Markov chain Monte Carlo sample from *Proportional hazards model* with three parameters (two predictors) coupled with the past. The chains have different starting points, but all have the same random inputs. Note: in this example we are using flat prior.

Minitab Commands	Meaning
random 3 c31-c34;	
t 4.	
random 1 c99;	
integer 1 100000.	
%<*path*>BayesPropHazMH 100 3 c1 c2 m1;	
MH 20;	
start c31;	# column of initial values
rbase k1;	# seed
MCMCsample m2;	
MLEout c20 m20;	
%<*path*>BayesPropHazMH 100 3 c1 c2 m1;	
MLEin c20 m20;	
MH 20;	
start c32;	# column of initial values
rbase k1;	# seed
MCMCsample m3.	
%<*path*>BayesPropHazMH 100 3 c1 c2 m1;	
MLEin c20 m20;	
MH 20;	
start c33;	# column of initial values
rbase k1;	# seed
MCMCsample m4.	
copy m2 c41-c43	
copy m3 c51-c53	
tsplot c41 c51 c61;	
connect;	
overlay.	
tsplot c42 c52 c62;	
connect;	
overlay.	
tsplot c43 c53 c63;	
connect;	
overlay.	

Table A.17 Minitab commands for Gibbs sampler on *Normal*(μ, σ^2) model with independent conjugate priors for the two parameters

Minitab Commands	Meaning
%<*path*>NormGibbsInd c1;	data in column c1
GS 5000;	5000 steps of Gibbs sampler
Primu 10 3;	*Normal*$(10, 3^2)$ prior
PriVar 25 1;	25 times *inverse chi-squared* prior
	with 1 degree of freedom
MCMCsample m2.	MCMC sample in m2

Table A.18 Minitab commands for Gibbs sampler on *Normal*(μ, σ^2) model with joint conjugate prior for the two parameters

Minitab Commands	Meaning
%<*path*>NormGibbsJoint c1;	data in column c1
GS 5000;	5000 steps of Gibbs sampler
Primu 10 3;	*normal*$(10, \sigma^2/3)$ prior
	(equivalent sample size is 3)
PriVar 25 1;	25 times *inverse chi-squared* prior
	with 1 degree of freedom
MCMCsample m2.	MCMC sample in m2

for μ and Jeffrey's prior for σ^2.) The joint prior would be the product of these two priors since we are using independent priors. We noted in Chapter 4 that this joint prior will not be conjugate to the joint likelihood. Nevertheless we can draw a sample from the posterior using the Gibbs sample quite easily. The steps to draw the sample using macro *NormGibbsInd.mac* are shown in Table A.17. We can also use a joint conjugate prior for μ and σ^2. This will be a *normal*$(m, \sigma^2/n_0)$ prior for μ given the variance σ^2, and an S times an *inverse chi-squared* prior for σ^2. The steps to draw the Gibbs sample using the Minitab macro *NormGibbsJoint.mac* are given in Table A.18.

Hierarchical mean model with regression on covariates. We have independent random samples from J related populations where each population has its own mean value. We model this by having the population means $\mu_1, \ldots, \mu_J$ being an independent draw from a *normal*(τ, ψ) hyperdistribution. We also have the values for some covariates for each of the observation. We give independent conjugate priors for the hypermean, hypervariance, regression coefficients, and observation variance. The steps to draw the Gibbs sample using the Minitab macro *HierMeanRegGS.mac*

Table A.19 Minitab commands for Gibbs sampler on hierarchical normal mean model with regression on covariates

Minitab Commands	Meaning
%<*path*>HierMeanRegGS 3 2 m1;	3 populations, 2 predictors, data in matrix $m1$
Gibbs 5000;	5000 steps of Gibbs sampler
pritau 100 250000;	$normal(100, 500^2)$ prior for τ
pripsi 100 1;	$100\times$ *inverse chi-squared*1 for ψ
privar 100 1;	$100\times$ *inverse chi-squared*1 for σ^2
pribeta c10 m10;	$multivariate\ normal(c10, m10)$ prior for $\boldsymbol{\beta}$ (regression coefficients)
graphs;	
MCMCsample m2.	MCMC sample in m2

Table A.20 Minitab commands for three Gibbs sampler chains coupled with the past on hierarchical normal mean model with regression on covariates

Minitab Commands	Meaning
random 1 c99;	random seed
integer 1 100000.	
let k99=c99(1)	
let k1=3	number of populations
let k2=2	number of predictors
random 3 c30;	random starting values for τ
norm 0 1.	
random 3 c31;	random starting values for ψ
chis 1.	

are given in Table A.19. Table A.20 gives the Minitab commands for running three parallel Gibbs sampling chains for the hierarchical normal mean model with regression on covariates coupled with the past. We want to determine how many steps are needed before the effect of the starting point has vanished and the parallel chains with the same random inputs have converged to the same point.

Table A.20 (Continued)

random 3 c32-c36;	random starting values for
norm 0 1.	$\mu_1, \ldots, \mu_{k1}$ and $\beta_1, \ldots, \beta_{k2}$
random 3 c37;	random starting values for σ^2
chis 1.	
copy c30-c37 m30	
trans m30 m31	
copy m31 c21-c23	starting values in columns
%<*path*>HierMeanRegGS k1 k2 m1;	data in m1
Gibbs 100;	number of Gibbs steps
pritau 0 10000;	prior for hypermean τ
pripsi 50 1;	prior for hypervariance ψ
privar 50 1;	prior for observation variance σ^2
pribeta c16 m16;	prior for $\beta_1, \ldots, \beta_{k2}$
start c21;	initial values for first chain
rbase k99;	base for random number sequence
MCMCsample m2.	MCMC sample for first chain
copy m2 c51-c58	copy output to columns
%<*path*>HierMeanRegGS k1 k2 m1;	data in m1
Gibbs 100;	number of Gibbs steps
pritau 0 10000;	prior for hypermean τ
pripsi 50 1;	prior for hypervariance ψ
privar 50 1;	prior for observation variance σ^2
pribeta c16 m16;	prior for $\beta_1, \ldots, \beta_{k2}$
start c22;	initial values for second chain
rbase k99;	base for random number sequence
MCMCsample m3.	MCMC sample for second chain
copy m2 c51-c58	copy output to columns

Table A.20 (Continued)

%<*path*>HierMeanRegGS k1 k2 m1;	data in m1
Gibbs 100;	number of Gibbs steps
pritau 0 10000;	prior for hypermean τ
pripsi 50 1;	prior for hypervariance ψ
privar 50 1;	prior for observation variance σ^2
pribeta c16 m16;	prior for $\beta_1, \ldots, \beta_{k2}$
start c23;	initial values for third chain
rbase k99;	base for random number sequence
MCMCsample m4.	MCMC sample for third chain
copy m2 c51-c58	copy output to columns
tsplot c51 c61 c71;	traceplots for τ
overlay.	
tsplot c52 c62 c72;	traceplots for ψ
overlay.	
tsplot c53 c63 c73;	traceplots for μ_1
overlay.	
tsplot c54 c64 c74;	traceplots for μ_2
overlay.	
tsplot c55 c65 c75;	traceplots for μ_3
overlay.	
tsplot c56 c66 c76;	traceplots for β_1
overlay.	
tsplot c57 c67 c77;	traceplots for β_2
overlay.	
tsplot c58 c68 c78;	traceplots for σ^2
overlay.	

B

Using the Included R Functions

James Curran
University of Auckland

Installing the R package *Bolstad2* from the Comprehensive R Archive Network (CRAN)

R functions for performing the Bayesian analysis and Markov chain Monte Carlo simulations shown in this text are included on a website maintained for this text. The functions written for this book have been put into an R package (or library if you're unfamiliar with R) called `Bolstad2`. The latest version of the package can always be found on CRAN, and for all platforms (Windows, Mac, Linux). This package should not be confused with the `Bolstad` package, which contains a wholly different set of functions described in Bolstad (2007).

Installation using Windows. To install the library under Windows select **Install Package(s)...** from the **Package** Menu. R will display a dialog box asking you to choose a mirror. A mirror is simply a location where the package may be downloaded. Choose a mirror that is either the country you currently are residing in or geographically close. This will make the download faster. For example, we, of course, choose *New Zealand*. A second dialog box will appear with a list of all the packages available from the mirror. Simply scroll down the list until you find `Bolstad2`. Select it by clicking on it with the mouse, and then click on the **OK** button. R should then download and install the package for you. If you are unable to locate the package, it is possible that the mirror has not updated recently. In this circumstance repeat the process and select a different mirror.

Installing using Linux or Mac OSX. To install the library under Linux or Mac OSX, start R and type `install.packages()` at the command prompt. R will display a dialog box asking you to choose a mirror. A mirror is simply a location where the package may be downloaded. Choose a mirror that is either the country you currently are residing in or geographically close. This will make the download faster. For example, we, of course, choose *New Zealand*. A second dialog box will appear with a list of all the packages available from the mirror. Simply scroll down the list until you find `Bolstad2`. Select it by clicking on it with the mouse, and then click on the **OK** button. R should then download and install the package for you. If you are unable to locate the package, it is possible that the mirror has not updated recently. In this circumstance repeat the process and select a different mirror.

Installing the R package *Bolstad2* from a `.zip` file

The `Bolstad2` package be downloaded from CRAN and saved rather than installed directly. This file will also be available from the Web page for this text on the site. The address of the text web site can be found on <www.wiley.com>. The R Functions are zipped up in a package called *Bolstad2-x.y-z.zip* where `x`,`y` and `z` are the major, minor and revision numbers respectively. To install the package directly from the `zip` file, select **Install package(s) from local zip file...** from the **Packages** menu. This will bring up a file location dialog. Navigate to the location of the `.zip` file, select it, and click on **Open**. R should then install the package.

Mac OSX users and Linux users should download the `.tar.gz` version rather than the `.zip` version and refer to section 6.3 of the *R Installation and Administration Manual* for instructions on how to install downloaded packages.

Using the Package

To use the `Bolstad2` package, you must instruct R to load it each time you start a new R session. Simply type `library(Bolstad2)` to load the functions at the command prompt or select the item **Load package...** from the **Packages** menu. To get a list of all the functions and data files in the package type `library(help=Bolstad2)`. This should bring up the following list:

AidsSurvival.df	HIV Survival data
BayesCPH	Bayesian Cox Proportional Hazards Modelling
BayesLogistic	Bayesian Logistic Regression
BayesPois	Bayesian Poisson Regression
bivnormMH	Metropolis Hastings sampling from a Bivariate Normal distribution
c10ex16.df	Chapter 10 Example 16 data
credInt	Calculate a credible interval from a numerically specified posterior CDF or from a sample from the posterior
describe	Give simple descriptive statistics for a matrix or a data frame
GelmanRubin	Calculate the Gelman-Rubin statistic
hierMeanReg	Hierarchical Normal Means Regression Model
hiermeanRegTest.df	Test data for hierMeanReg
logisticTest.df	Test data for BayesLogistic
normGibbs	Draw a sample from a posterior distribution of data with an unknown mean and variance using Gibbs sampling
normMixMH	Sample from a normal mixture model using Metropolis-Hastings
pNull	Test a one-sided hypothesis from a numerically specified posterior CDF or from a sample from the posterior
poissonTest.df	A test data set for BayesPois
sintegral	Numerical integration using Simpson's Rule
thin	Thin a MCMC chain

This command will also give you the version number of the library, which is useful when communicating with the authors should you find any bugs. You may get a help file on any function by typing either `?function` or `help(function)`, where `function` is the name of the package function you want help on. For example, `?BayesCPH` will give you help on the Bayesian Cox Proportional Hazards model command. HTML-based help is also available. To use HTML help, select **Html help** from the **Help** menu. Click on the `Packages` link, and then the link for `Bolstad2`. This will bring up an index page where you may select the help file for the function you're interested in.

If the help file contains an examples section, and most of them do, then you can have R execute the examples by typing `example(function)` where `function` is the name of the package function you would like to see the examples for. For example `example(describe)` will show you an example of how the `describe` function works.

Each help file has a standard layout, which is as follows:

Title: a brief title that gives some idea of what the function is supposed to do or show

Description: a fuller description of the what the function is supposed to do or show

Usage: the formal calling syntax of the function

Arguments: a description of each of the arguments of the function

Values: a description of the values (if any) returned by the function

See also: a reference to related functions

Examples: some examples of how the function may be used. These examples may be run either by using the `example` command (see above) or copied and pasted into the R console window

Tips for the Unwary: Optional Arguments, Argument Ordering, and Invisible Lists

The R language has two special features that may make it confusing to users of other programming and statistical languages: default or optional arguments, and variable ordering of arguments. An R function may have arguments for which the author has specified a default value. Let us take the function `credInt` as an example. The syntax of `credInt` is `credInt(theta, cdf = NULL, conf = 0.95, type = "twosided")`. The function takes four arguments `theta`, `cdf`, `conf`, and `type`. However, the author has specified default values for `cdf`, `conf`, and `type`, namely `cdf = NULL`, `conf= 0.95` and `ret = "twosided"`. This means that the user only has to supply the argument `theta`. Therefore, the arguments `cdf`, `conf`, and `type` are said to be optional or default. In this example, `theta` contains a random sample of 1,000 observations from a $Beta(6, 8)$ posterior which could be selected by typing `theta <- rbeta(1000,6,8)`. The simplest example for `credInt` is given as `credInt(theta)`. If the user wanted to change the confidence level of the credible interval, say to 0.9, then they would type `credInt(theta, conf = 0.9)`. There is a slight catch here, which leads into the next feature. Assume that the user wanted to use a one-sided lower bound credible interval. One might be tempted to type `credInt(theta,"lower")`. This is incorrect. R will think that the value `"lower"` is the value being assigned to the parameter `cdf`, and convert it from a string, `"lower"`, to the numerical equivalent of a missing value, `NA`, which will of course give an error because if `cdf` is not NULL, then it must be the same length as `theta` and not contain missing values. The correct way to make such a call is to use named arguments, such as `credInt(theta, type = "lower")`. This specifically tells R which argument is to be assigned the value `"lower"`. This feature also makes the calling syntax more flexible because it means that the order of the arguments does not need to be adhered to. For example, `credInt(type = "lower", theta = theta)` would be a perfectly legitimate function call.

Any function in the `Bolstad2` package that has a *Values* section in its help file returns a result. The result, however, is invisible. This means that it will not be printed out to screen by default, nor will it be stored. For example, the function `credInt` returns a list with the elements `lower.bound` and `upper.bound`, depending on whether the user has asked for a two-sided interval or a one-sided lower/upper bound interval. To store the result of a function you can assign a function call to a variable. Following the example above, if `theta` contains a sample of 1,000 observations from a $Beta(6, 8)$ distribution, then the credible interval obtained from `credInt` can be stored by typing something like `ci<-credInt(theta)`. The new variable `ci` now contains a two-sided credible interval for this sample. The lower bound of the interval can be obtained by typing `ci$lower.bound` and the upper bound by typing `ci$upper.bound`.

Chapter 3: Bayesian Inference

Calculating the numerical posterior and its CDF. The proportional posterior density is easily found numerically by multiplying the prior density times the likelihood. Multiplying either by a constant does not matter, because the constant will cancel out when we find the exact numerical posterior density by

$$g(\theta|y) = \frac{g(\theta)\, f(y|\theta)}{\int_{-\infty}^{\infty} g(\theta)\, f(y|\theta)\, d\theta}\,.$$

The R function `sintegral` is used to evaluate the denominator numerically. We want to do an inference such as finding a credible interval for the parameter, or calculating the posterior probability of a one-sided null hypothesis about the parameter. The cumulative distribution function (CDF) of the posterior is the definite integral of the numerical posterior density. It is given by

$$G(\theta|y) \quad = \quad \int_{-\infty}^{y} g(\theta|y)\, d\theta\,.$$

The CDF will be evaluated using the R function `sintegral`. In this example we show the R commands for calculating the posterior CDF numerically where the likelihood is proportional to $\theta^{(5/2-1)} \times e^{-\theta/(2*2)}$ and a $normal(6, 3^2)$ prior is used.

Firstly we define a reasonable set of values for θ. We let θ go from 0 to 40 in 0.001 increments

```
theta <- seq(0, 40, by = 0.001)
```

Next we calculate the prior density $N(6, 3^2)$ across the range of θ values

```
prior <- dnorm(theta, 6, 3)
```

and the proportional likelihood and proportional posterior

```
ppn.like <- theta^(5/2 - 1) * exp(-theta/(2 *
    2))
ppn.post <- ppn.like * prior
```

To find the constant of proportionality for the posterior we need to integrate the proportional posterior over θ

```
k <- sintegral(theta, ppn.post)$int
```

The numerical posterior density is then

```
post <- ppn.post/k
```

and the numerical CDF can be obtained by integrating the numerical density over θ

```
integral <- sintegral(theta, post)
post.cdf <- with(integral, list(x = x, y = y))
plot(post.cdf, type = "l")
```

Credible intervals from the numerical posterior CDF. The function *credInt* calculates a lower, upper, or two-sided credible interval from the numerical posterior CDF or from a sample of random values from the posterior.

We can test a two-sided hypothesis

$$H_0 : \theta = \theta_0 \quad \text{vs} \quad H_1 : \theta \neq \theta_0$$

at the α level of significance by observing whether or not the null value θ_0 lies inside the two-sided $(1-\alpha) \times 100\%$ credible interval. For example, we may be interested in testing the hypothesis $H_0 : \theta = 15$ where the CDF is defined in the previous section. Assuming the CDF is stored in a variable called `post.cdf` with values x and y then the R commands for calculating credible interval from a numerical posterior CDF are

```
with(post.cdf, credInt(x, y))
```

This function will return an interval from approximately 2 to 11, showing that $\theta = 15$ is not in the credible range.

Testing a one-sided hypothesis from the numerical posterior CDF. We can test a one sided hypothesis about the parameter

$$H_0 : \theta \leq \theta_0 \quad \text{vs} \quad H_1 : \theta > \theta_0$$

by calculating the posterior probability of the null hypothesis. If this probability is less than the level of significance α, then we can reject the null hypothesis at the α level. The function `pNull` calculates the probability of a one-sided null hypothesis from a numerically calculated posterior CDF. Assuming the same example as the previous section the R commands for this function are shown

```
theta <- post.cdf$x
cdf <- post.cdf$y
pNull(15, theta, cdf)
```

These commands should return a probability of approximately 6×10^{-4}.

Bayesian inference from a posterior random sample. Sometimes we only know the shape of the posterior, not the exact posterior density. Nevertheless, we have a random sample from the posterior, despite not knowing it exactly. We might have obtained this sample using a direct sampling method such as acceptance-rejection-sampling or sampling-importance-resampling. Or, we might have obtained this sample by sampling from a Markov chain that has the posterior as its long-run distribution. This is known as a *Markov Chain Monte Carlo* method. *Gibbs sampling* and the *Metropolis-Hastings* algorithm are special cases of this. Note, we would have to burn-in and thin the MCMC sample from the posterior to make it approximately random before we use it for inference.

Finding a credible interval. We want to use our random sample from the posterior to do an inference such as finding a credible interval for the parameter, or calculating the posterior probability of a one-sided null hypothesis about the parameter. The function `credInt` calculates a lower, upper, or two-sided credible interval from a random sample from the posterior. Assuming that the vector `theta` contains a random sample from the posterior density we found in the previous example, then

```
credInt(theta)
```

will return a 95% two-sided interval,

```
credInt(theta, type = "lower")
```

will return a 95% lower credible bound and

```
credInt(theta, type = "upper")
```

will return a 95% upper credible bound. Note that `credInt` will accept `'t'`, `'l'` and `'u'` as valid input for the `type` parameter.

We can test a two-sided hypothesis

$$H_0 : \theta = \theta_0 \quad \text{vs} \quad H_1 : \theta \neq \theta_0$$

at the α level of significance by observing whether or not the null value θ_0 lies inside the two-sided $(1 - \alpha) \times 100\%$ credible interval.

Testing a one-sided hypothesis using the posterior random sample. We can use the random sample from the posterior to test a one sided hypothesis about the parameter

$$H_0 : \theta \leq \theta_0 \quad \text{vs} \quad H_1 : \theta > \theta_0$$

by calculating the proportion in the posterior sample that satisfies the null hypothesis. If this proportion is less than the level of significance α, then we can reject the null hypothesis at the α level. The function `pNull` calculates the proportion of a one-sided null hypothesis from a random sample posterior. Assuming that `theta` contains a random sample from the posterior and `theta0` is the hypothesized value θ_0, then

```
pNull(theta0, theta)
```

will return $\Pr(\theta \geq \theta_0)$ and

```
pNull(theta0, theta, type = "lower")
```

will return the lower tail probability $\Pr(\theta \leq \theta_0)$

Chapter 6: Markov Chain Monte Carlo Sampling from the Posterior

A mixture of two normal distributions. The function `normMixMH` can use the Metropolis-Hastings algorithm to draw a sample from a univariate target distribution that is a mixture of two normal distributions using an independent *normal* candidate density or a random-walk *normal* candidate density respectively. To use this function we need to specify the mean and standard deviation of each component of the mixture, as well as the mixture proportion. In this example we use `theta0` to hold the mean and standard deviation of the first component of the mixture, and `theta1` to hold the mean and standard deviation of the second. The mixture proportion is stored in a variable called `p`

```
theta0 <- c(0, 1)
theta1 <- c(3, 2)
p <- 0.8
```

By default, `normMixMH` samples from an independent candidate density and uses 1,000 steps in the Metropolis-Hastings algorithm. Therefore to sample using an independent $N(0, 3^2)$ candidate density we type

```
candidate <- c(0, 3)
MCMCsampleInd <- normMixMH(theta0, theta1, p,
    candidate)
```

If we wish to use the alternative random-walk $N(0, 0.5^2)$ candidate density we type

```
candidate <- c(0, 0.5)
MCMCsampleRW <- normMixMH(theta0, theta1, p, candidate,
    type = "rw")
```

The results of each of these function calls is stored in a variable called `MCMCsampleInd` and `MCMCsampleRW` for the "independent" and "random-walk" cases, respectively.

A correlated bivariate normal distribution. The function `bivnormMH` can use the Metropolis-Hastings algorithm to draw a sample from a correlated bivariate normal target density using either an independent candidate density or a random-walk candidate density when we are drawing both parameters in a single draw. Also, bivnormMH can either use the blockwise Metropolis-Hastings algorithm or the Gibbs sampling algorithm to draw a sample from the correlated bivariate normal target when we are drawing the parameters blockwise (one parameter at a time).

For an independent chain an independent chain we type

```
MCMCSampleInd <- bivnormMH(0.9)
```

for a random-walk chain we type

```
MCMCsampleRW <- bivnormMH(0.9, type = "rw")
```

for a blockwise Metropolis-Hastings chain we type

```
MCMCSampleBW <- bivnormMH(0.9, type = "block")
```

and for a Gibbs sampling chain we type

```
MCMCSampleGibbs <- bivnormMH(0.9, type = "gibbs")
```

Chapter 7: Statistical Inference from a Markov Chain Monte Carlo Sample

Sequential draws from a Markov chain are serially dependent. A Markov chain Monte Carlo sample will not be suitable for inference until we have discarded the draws from the burn-in period and thinned the sample so the thinned sample will approximate a random sample from the posterior. The function thin thins the output of a MCMC process. If x is a vector, matrix or data.frame from an MCMC process then thin(x,k) will return every k^{th} element or row of x.

The Gelman-Rubin statistic is used to indicate the amount of improvement possible if n, the burn-in, was increased. What it actually measures is how well the chain is moving through the parameter space. It is an analysis of variance approach comparing the between chain variation to the within chain variation for the next n steps after burn-in of n steps for multiple chains started from an overdispersed starting distribution. The example below shows the commands for calculating the Gelman-Rubin statistic from the outputs of four runs of the chain given in vectors $v_1, \ldots, v_4$. These vectors are combined into a single matrix that is used to call the GelmanRubin function. Note that the Gelman-Rubin statistic can only be calculated for a single parameter at a time. If you have multiple parameters in a chain, then the column vectors corresponding to the parameter of interest must be copied from each chain into a matrix.

```
theta0 <- c(0, 1)
theta1 <- c(3, 2)
p <- 0.6
candidate <- c(0, 3)
v1 <- normMixMH(theta0, theta1, p, candidate,
     steps = 200)
```

```
v2 <- normMixMH(theta0, theta1, p, candidate,
    steps = 200)
v3 <- normMixMH(theta0, theta1, p, candidate,
    steps = 200)
v4 <- normMixMH(theta0, theta1, p, candidate,
    steps = 200)
theta <- cbind(v1, v2, v3, v4)
GelmanRubin(theta)
```

Coupling with the past is where we run chains with different starting points, but with the same random inputs and see how long it takes the chains converge. At that time, the effect of the starting point has died out. This is sometimes called perfect sampling. When Metropolis-Hastings chains with random-walk candidate density are used, the chains eventually get very close, but not identical. However, when a suitable independent candidate density is used, the chains converge exactly. In the example that follows, the R commands for three Metropolis-Hastings chains with a random-walk candidate density coupled with the past using the function `normMixMH` for a mixture of normal distributions four Markov chain Monte Carlo sample from a mixture of *normal* distributions using either a *normal* random-walk candidate density or an independent candidate density. If we want to use an independent candidate density, the `type = 'ind'` option must be used instead and the constants for the candidate density should be changed.

The chains have different starting points, but all have the same random inputs. First we draw a random seed from $U[1, 10^6]$

```
rseed <- floor(1e+06 * runif(1)) + 1
```

The first chain is a random-walk chain with

- the first mixture component a $N(0, 1)$,
- the second a $N(3, 2)$,
- a mixture proportion of 0.8,
- a candidate density of $N(0, 1)$
- 100 Metropolis-Hastings steps
- an initial starting value of -2 and
- the random seed set to `rseed`

The function arguments are specifically given in the function calls below for clarity rather than necessity

```
chain1 <- normMixMH(theta0 = c(0, 1), theta1 = c(3,
    2), p = 0.8, candidate = c(0, 1), steps = 100,
    type = "rw", randomSeed = rseed, startValue = -2)
```

The second random-walk chain has all the same input values, except that the chain starts at the value 1:

```
chain2 <- normMixMH(theta0 = c(0, 1), theta1 = c(3,
    2), p = 0.8, candidate = c(0, 1), steps = 100,
    type = "rw", randomSeed = rseed, startValue = 1)
```

The third random-walk chain starts at 4:

```
chain3 <- normMixMH(theta0 = c(0, 1), theta1 = c(3,
    2), p = 0.8, candidate = c(0, 1), steps = 100,
    type = "rw", randomSeed = rseed, startValue = 4)
```

Now we can plot each of the chains on the same plot:

```
yRange <- range(c(chain1, chain2, chain3))
plot(chain1, type = "l", col = "red", ylim = yRange)
lines(chain2, col = "blue")
lines(chain3, col = "green")
```

Chapter 8: Logistic Regression

The function `BayesLogistic` draws a random sample from the posterior distribution for the logistic regression model. The logistic regression model is an example of a generalized linear model, and as such, the maximum likelihood can be found by iteratively reweighted least squares. First, `BayesLogistic` finds an approximate normal likelihood function for the logistic regression model where the mean matches the maximum likelihood estimator found using iteratively reweighted least squares. The covariance matrix is found that matches the curvature of the likelihood function at its maximum. Details of this are found in Myers et al. (2002) and Jennrich (1995). The approximate normal posterior by applying the usual normal updating formulas with a normal conjugate prior. If we used this as the candidate distribution, it may be that the tails of true posterior are heavier than the candidate distribution. This would mean that the accepted values would not be a sample from the true posterior because the tails would not be adequately represented. Instead we make the candidate distribution have heavier tails by using a *Student's t* with low degrees of freedom in place of the normal. This will make the accepted sample be from the posterior. The user must supply a vector of responses `y`, and a matrix (or vector) of covariate(s), `x`. The default call to the function is then

```
BayesLogistic(y, x)
```

Note that formula notation of functions like `lm` and `glm` have not yet been implemented, but may be added in a later version. The function returns a list containing:

1. the MCMC sample of $\beta_0, \ldots, \beta_p$ values is a data frame `beta`
2. the matched curvature mean `mleMean`

3. and the matched curvature covariance matrix `mleVar`

The latter two items can be used as starting values for future calls. The user is advised to discard the values in the burn-in time, and the remainder can constitute a sample from the posterior. Thin the accepted sample enough so they can be considered to be a independent random sample from the posterior, and use the thinned sample for your inferences.

In the example below we give the R commands we use to run three parallel Metropolis-Hastings chains with different starting points, but with the same random inputs for the logistic regression model. Here we want to see how long we have to run the chains before they converge to the same values. At that point, the effect of the starting point has finished, so the draw can be considered a random draw from the posterior. We assume that the vectors `y` and `x` hold the response and covariate data respectively for 100 observations. The chains have different starting points, but all have the same random inputs. Note that we are using flat prior in this example. Firstly we select some random starting values

```
rseed <- floor(1e+06 * runif(1)) + 1
rstart <- matrix(rnorm(6), nc = 3)
```

The first chain is

```
chain1 <- BayesLogistic(y, x, steps = 100, startValue
      = rstart[, 1], randomSeed = rseed)
```

The second chain is

```
chain2 <- BayesLogistic(y, x, steps = 100, startValue
      = rstart[, 2], mleMean = chain1$mleMean, mleVar
      = chain1$mleVar, randomSeed = rseed)
```

The third chain is

```
chain3 <- BayesLogistic(y, x, steps = 100, startValue
      = rstart[, 3], mleMean = chain1$mleMean, mleVar
      = chain1$mleVar, randomSeed = rseed)
```

Note that the second and third chains used $\hat{\beta}_{ML}$ and $\hat{V}_{ML}$ from the first chain. We can now plot the chains for each of the regression coefficients

```
yRange <- range(c(chain1$beta[, 1], chain2$beta[,
    1], chain3$beta[, 1]))
plot(chain1$beta[, 1], ylab = expression(beta[0]),
    type = "l", ylim = yRange)
lines(chain2$beta[, 1], col = "red")
lines(chain3$beta[, 1], col = "blue")
yRange <- range(c(chain1$beta[, 2], chain2$beta[,
    2], chain3$beta[, 2]))
plot(chain1$beta[, 2], ylab = expression(beta[1]),
```

```
    type = "l", ylim = yRange)
lines(chain2$beta[, 2], col = "red")
lines(chain3$beta[, 2], col = "blue")
```

Chapter 9: Poisson Regression and Proportional Hazards Model

The function `BayesPois` draws a random sample from the posterior distribution for the Poisson regression model. The function finds the maximum likelihood estimator by iteratively reweighted least squares, and then finds a normal approximation to the likelihood function by matching the curvature at the maximum likelihood estimator. The approximate posterior is found using the normal updating rules using either a multivariate normal prior or a multivariate flat prior. This approximate posterior may have lighter tails than the true posterior, so we use a matching *multivariate Student's t* with low degrees of freedom as the candidate density. It has similar shape to the true posterior, and it will have heavy tails, so it will be a very effective independent candidate density. We assume that the vectors `y` and `x` hold the response and covariate data respectively for 100 observations. The R command for using this function are to carry out a simple Poisson regression is

```
BayesPois(y, x)
```

The example below shows the R commands for running three parallel chains for the Poisson regression model coupled with the past. We want to determine how many steps are needed before the effect of the starting point has vanished and the parallel chains with the same random inputs have converged to the same point. The chains have different starting points, but all have the same random inputs. Note, once again, we are using flat prior in this example. Firstly we select some random starting values

```
rseed <- floor(1e+06 * runif(1)) + 1
rstart <- matrix(rnorm(6), nc = 3)
```

The first chain is

```
chain1 <- BayesPois(y, x, steps = 100, startValue
     = rstart[, 1], randomSeed = rseed)
```

The second chain is

```
chain2 <- BayesPois(y, x, steps = 100, startValue
     = rstart[, 2], mleMean = chain1$mleMean, mleVar
     = chain1$mleVar, randomSeed = rseed)
```

The third chain is

```
chain3 <- BayesPois(y, x, steps = 100, startValue
     = rstart[, 3], mleMean = chain1$mleMean, mleVar
     = chain1$mleVar, randomSeed = rseed)
```

Note that the second and third chains used $\hat{\beta}_{ML}$ and $\hat{V}_{ML}$ from the first chain. We can now plot the chains for each of the regression coefficients

```
yRange <- range(c(chain1$beta[, 1], chain2$beta[,
    1], chain3$beta[, 1]))
plot(chain1$beta[, 1], ylab = expression(beta[0]),
    type = "l", ylim = yRange)
lines(chain2$beta[, 1], col = "red")
lines(chain3$beta[, 1], col = "blue")
yRange <- range(c(chain1$beta[, 2], chain2$beta[,
    2], chain3$beta[, 2]))
plot(chain1$beta[, 2], ylab = expression(beta[1]),
    type = "l", ylim = yRange)
lines(chain2$beta[, 2], col = "red")
lines(chain3$beta[, 2], col = "blue")
```

When we have censored survival times data, and we relate the linear predictor to the hazard function we have the proportional hazards model. The function `BayesCPH` draws a random sample from the posterior distribution for the proportional hazards model. First, the function finds an approximate normal likelihood function for the proportional hazards model. The (multivariate) normal likelihood matches the mean to the maximum likelihood estimator found using iteratively reweighted least squares. Details of this are found in Myers et al. (2002) and Jennrich (1995). The covariance matrix is found that matches the curvature of the likelihood function at its maximum. The approximate normal posterior by applying the usual normal updating formulas with a normal conjugate prior. If we used this as the candidate distribution, it may be that the tails of true posterior are heavier than the candidate distribution. This would mean that the accepted values would not be a sample from the true posterior because the tails would not be adequately represented. Assuming that `y` is the Poisson censored response vector, `time` is time, and `x` is a vector of covariates then

```
BayesCPH(y, time, x)
```

will find a random sample from the posterior using the Metropolis-Hastings algorithm for the proportional hazards model. The example below gives the R commands for running three parallel chains for the proportional hazards model coupled with the past. We want to determine how many steps are needed before the effect of the starting point has vanished and the parallel chains with the same random inputs have converged to the same point. Firstly we select some random starting values

```
rseed <- floor(1e+06 * runif(1)) + 1
rstart <- matrix(rt(6, 4), nc = 3)
```

The first chain is

```
chain1 <- BayesCPH(y, time, x, steps = 100, startValue
      = rstart[, 1], randomSeed = rseed)
```

The second chain is

```
chain2 <- BayesCPH(y, time, x, steps = 100, startValue
     = rstart[, 2], mleMean = chain1$mleMean, mleVar
     = chain1$mleVar, randomSeed = rseed)
```

The third chain is

```
chain3 <- BayesCPH(y, time, x, steps = 100, startValue
     = rstart[, 3], mleMean = chain1$mleMean, mleVar
     = chain1$mleVar, randomSeed = rseed)
```

Note that the second and third chains used $\hat{\beta}_{ML}$ and $\hat{V}_{ML}$ from the first chain. We can now plot the chains for each of the regression coefficients

```
yRange <- range(c(chain1$beta[, 1], chain2$beta[,
    1], chain3$beta[, 1]))
plot(chain1$beta[, 1], ylab = expression(beta[0]),
    type = "l", ylim = yRange)
yRange <- range(c(chain1$beta[, 2], chain2$beta[,
    2], chain3$beta[, 2]))
plot(chain1$beta[, 2], ylab = expression(beta[1]),
    type = "l", ylim = yRange)
```

Chapter 10: Hierarchical Models and Gibbs Sampling

Normal*(μ, σ^2) *with both parameters unknown. When we have a random sample from a *normal*(μ, σ^2) where both parameters are unknown, we have a choice in the priors we can use with the Gibbs sampler. We can use independent conjugate priors for the two parameters. The prior for μ would be a *normal*(m, s^2) prior and the prior for σ^2 would be S times an *inverse chi-squared*. (The defaults are flat prior for μ and Jeffrey's prior for σ^2.) The joint prior would be the product of these two priors since we are using independent priors. We noted in Chapter 4 that this joint prior will not be conjugate to the joint likelihood. Nevertheless we can draw a sample from the posterior using the Gibbs sample quite easily. Assuming that our data is stored in a vector `y`, the steps to draw the sample using function `normGibbs` with a $N(10, 3^2)$ prior for μ and a 25 times *inverse chi-squared* with one degree of freedom prior σ^2 are

```
MCMCSample <- normGibbs(y, steps = 5000, priorMu = c(10,
    3), priorVar = c(25, 1))
```

We can also use a joint conjugate prior for μ and σ^2. This will be the product of a *normal*$(m, \sigma^2/n_0)$ prior for μ given the variance σ^2 times an S times an *inverse chi-squared* prior with one degree of freedom for σ^2. Suppose $m = 10$, $n_0 = 1$, and $S = 25$. The steps to draw the Gibbs sample with the joint conjugate prior using the R function `normGibbs` are given below.

```
MCMCSample <- normGibbs(y, steps = 5000, type = "conj",
    priorMu = c(10, 1), priorVar = c(25, 1))
```

Hierarchical mean model with regression on covariates. We have independent random samples from J related populations where each population has its own mean value. We model this by having the population means $\mu_1, \ldots, \mu_J$ being an independent draw from a *normal*(τ, ψ) hyperdistribution. We also have the values for some covariates for each of the observation. We give independent conjugate priors for the hypermean, hypervariance, regression coefficients, and observation variance. The steps to draw the Gibbs sample using the R function `hierMeanReg` are given in below. The data is a list with elements `y`, the response vector, `group`, the grouping vector and `x`, a matrix of covariates or NULL if there are no covariates. In this example we will use the data set `hierMeanRegTest.df`. We need to reorganise this data into a list to use it.

```
data(hiermeanRegTest.df)
design <- with(hiermeanRegTest.df, list(y = y,
    group = group, X = cbind(x1, x2)))
```

In this example we will use

- 5000 steps in the Gibbs sampler
- a $N(100, 1000^2)$ prior for τ
- a $500 \times inverse\ chi - squared$ prior with 1 df for ψ
- a $500 \times inverse\ chi - squared$ prior with 1 df for σ^2
- a multivariate normal prior for β (the regression coefficients) with

$$\mu_0 = \begin{pmatrix} 0 \\ 0 \end{pmatrix}$$

  and

$$\Sigma_0 = \begin{pmatrix} 1000 & 100 \\ 100 & 1000 \end{pmatrix}$$

```
priorTau <- list(tau0 = 0, v0 = 1000)
priorPsi <- list(psi0 = 500, eta0 = 1)
priorVar <- list(s0 = 500, kappa0 = 1)
priorBeta <- list(b0 = c(0, 0), bMat = matrix(c(1000,
    100, 100, 1000), nc = 2))
MCMCSample <- hierMeanReg(design, priorTau, priorPsi,
    priorVar, priorBeta, steps = 5000)
```

The example below gives the R commands for running three parallel Gibbs sampling chains for the hierarchical normal mean model with regression on covariates coupled with the past. We want to determine how many steps are needed before the effect of the starting point has vanished and the parallel chains with the same random inputs have converged to the same point. Firstly we choose some random starting values for each of the parameters.

```
rSeed <- floor(1e+06 * runif(1)) + 1
rTau <- rnorm(3)
rPsi <- rchisq(3, 1)
rMu <- matrix(rnorm(9), nc = 3)
rSigma <- rchisq(3, 1)
rBeta <- matrix(rnorm(6), nc = 3)
```

Next we set up the priors that will be used in every chain.

```
priorTau <- list(tau0 = 0, v0 = 10000)
priorPsi <- list(psi0 = 50, eta0 = 1)
priorVar <- list(s0 = 50, kappa0 = 1)
priorBeta <- list(b0 = c(0, 0), bMat = matrix(c(1000,
    100, 100, 1000), nc = 2))
```

And finally we can start each of the chains.

```
startValue <- list(tau = rTau[1], psi = rPsi[1],
    mu = rMu[, 1], sigmaSq = rSigma[1], beta = rBeta[,
        1])
chain1 <- hierMeanReg(design, priorTau, priorPsi,
    priorVar, priorBeta, steps = 100, startValue
    = startValue, randomSeed = rSeed)
startValue <- list(tau = rTau[2], psi = rPsi[2],
    mu = rMu[, 2], sigmaSq = rSigma[2], beta = rBeta[,
        2])
chain2 <- hierMeanReg(design, priorTau, priorPsi,
    priorVar, priorBeta, steps = 100, startValue
    = startValue, randomSeed = rSeed)
startValue <- list(tau = rTau[3], psi = rPsi[3],
    mu = rMu[, 3], sigmaSq = rSigma[3], beta = rBeta[,
        3])
chain3 <- hierMeanReg(design, priorTau, priorPsi,
    priorVar, priorBeta, steps = 100, startValue
    = startValue, randomSeed = rSeed)
```

And then we can plot the results.

```
par(mfrow = c(3, 3))
yRange <- range(chain1$tau, chain2$tau, chain3$tau)
plot(ts(chain1$tau), xlab = "", ylab = expression(tau),
      type = "l", ylim = yRange)
lines(chain2$tau, col = "red")
lines(chain3$tau, col = "blue")
yRange <- range(chain1$psi, chain2$psi, chain3$psi)
plot(ts(chain1$psi), xlab = "", ylab = expression(psi),
    type = "l", ylim = yRange)
```

```
lines(chain2$psi, col = "red")
lines(chain3$psi, col = "blue")
yRange <- range(chain1$mu.1, chain2$mu.1, chain3$mu.1)
plot(ts(chain1$mu.1), xlab = "", ylab = expression(mu.1),
    type = "l", ylim = yRange)
lines(chain2$mu.1, col = "red")
lines(chain3$mu.1, col = "blue")
yRange <- range(chain1$mu.2, chain2$mu.2, chain3$mu.2)
plot(ts(chain1$mu.2), xlab = "", ylab = expression(mu.2),
    type = "l", ylim = yRange)
lines(chain2$mu.2, col = "red")
lines(chain3$mu.2, col = "blue")
yRange <- range(chain1$mu.3, chain2$mu.3, chain3$mu.3)
plot(ts(chain1$mu.3), xlab = "", ylab = expression(mu.3),
    type = "l", ylim = yRange)
lines(chain2$mu.3, col = "red")
lines(chain3$mu.3, col = "blue")
yRange <- range(chain1$beta.1, chain2$beta.1,
    chain3$beta.1)
plot(ts(chain1$beta.1), xlab = "", ylab
     = expression(beta.1), type = "l", ylim = yRange)
lines(chain2$beta.1, col = "red")
lines(chain3$beta.1, col = "blue")
yRange <- range(chain1$beta.2, chain2$beta.2,
    chain3$beta.2)
plot(ts(chain1$beta.2), xlab = "", ylab
    = expression(beta.2), type = "l", ylim = yRange)
lines(chain2$beta.2, col = "red")
lines(chain3$beta.2, col = "blue")
yRange <- range(chain1$sigmaSq, chain2$sigmaSq,
    chain3$sigmaSq)
plot(ts(chain1$sigmaSq), xlab = "", ylab
    = expression(sigmaSq), type = "l", ylim = yRange)
lines(chain2$sigmaSq, col = "red")
lines(chain3$sigmaSq, col = "blue")
```

References

1. Abraham, B., and Ledolter, J., (1983), *Statistical Methods for Forecasting*, John Wiley & Sons, New York.

2. Bartlett, M. S. (1946) On the theoretical specification of sampling properties of autocorrelated time series, *Journal of the Royal Statistical Society, Series B*, Vol. 8, pp. 27-41.

3. Berger, J. (1980), *Statistical Decision Theory, Foundations, Concepts, and Methods*, Springer-Verlag, New York.

4. Bernardo, J. M., and Smith, A.F.M. (1994), *Bayesian Theory*, John Wiley & Sons, New York

5. Bolstad, W. M. (2007), *Introduction to Bayesian Statistics: Second Edition*, John Wiley & Sons, New York

6. Bowerman, B. L., and O'Connell, R. T. (1990), *Linear Statistical Models: an Applied Approach*, PWS-Kent, Boston.

7. Carlin, B., and Chib, S. (1995), Bayesian model choice via Markov chain Monte Carlo methods, *Journal of the Royal Statistical Society, Series B*, Vol. 57, pp. 473-484.

8. Carlin, B., and Gelfand, A. (1990), Approaches to empirical Bayes confidence intervals, *Journal of the American Statistical Association*, Vol. 85, No. 409, pp. 105-114.

9. Carlin, B. P., and Louis, T. A. (2000), *Bayes and Empirical Bayes Methods for Data Analysis: Second Edition*, Chapman & Hall, London.

10. Cox, D. R.,and Hinkley, D. V. (1974), *Theoretical Statistics*, Chapman and Hall, London.

11. DeGroot, M. H. (2004), *Optimal Statistical Decisions*, Wiley Classics Library, John Wiley & Sons, New York.

12. Dey, D., Muller, P., and Sinha, D. (1998), *Practical Nonparametric and Semiparametric Bayesian Statistics*, Springer, New York.

13. Efron, B. (1986), Why isn't everyone a Bayesian, *The American Statistician*, Vol. 40, No. 1 pp. 1-11.

14. Efron, B. (1996), Empirical Bayes methods for combining likelihoods, *Journal of the American Statistical Association*, Vol. 91, No. 433, pp. 538-550.

15. Feller, W. (1968), *An Introduction to Probability Theory and Its Applications, Volume 1, Third Edition*, John Wiley & Sons, New York.

16. Fergusson, D. M., Boden, J. M., and Horwood, L. J. (2006), Circumcision status and risk of sexually transmitted infection in young adult males: an analysis of a longitudinal birth cohort, *Pediatrics* 118, pp. 19711976.

17. Fisher, R. A. (1922), On the mathematical foundations of theoretical statistics, *Phil. Trans. Roy. Soc. London A*, 222, pp. 309-368. Reprinted in *Breakthroughs in Statistics 1* (1991), (S. Kotz, and N. L. Johnson, Editors), Springer, Berlin, pp. 11-44.

18. Gamerman, D. (1997), *Markov Chain Monte Carlo*, Chapman & Hall, London

19. Gelfand, A. (1996), Empirical Bayes methods for combining likelihood: comment, *Journal of the American Statistical Association*, Vol. 91, No. 433, pp. 551-552.

20. Gelfand, A., and Smith, A.F.M. (1990), Sampling-based approaches to calculating marginal densities, *Journal of the American Statistical Association*, Vol. 85 No. 410, pp. 398-409.

21. Gelman, A., Carlin, J. B., Stern, H. S., and Rubin, D. B. (2004), *Bayesian Data Analysis: Second Edition* , Chapman & Hall/CRC, London, pp. 182184.

22. Gelman, A and Rubin, D. B. (1992), Inference from iterative simulation using multiple sequences, *Statistical Science*, 7, pp. 457-511.

23. Geman, S., and Geman, D. (1984), Stochastic relaxation, Gibbs distributions, and the Bayesian restoration of images, *IEEE Transactions on Pattern Analysis and Machine Intelligence*, Vol. PAMI-6 No. 6, pp. 721-740.

24. Gilks, W. R., Richardson, S., and Spiegelhalter, D. J. (1996), *Markov Chain Monte Carlo in Practice*, Chapman & Hall, London

25. Gilks, W.R. and Wild, P. (1992), Adaptive rejection sampling for Gibbs sampling, *Applied Statistics*, Vol. 41, No. 2, pp.

26. Gill, J., *Bayesian Methods: A Social and Behavioral Sciences Approach, Second Edition*, Chapman & Hall, London

27. Green, P. (1995), Reversible jump Markov chain Monte Carlo computation and Bayesian model determination, *Biometrika* 82, pp. 711 732.

28. Hastings, W.K. (1970), Monte Carlo sampling methods using Markov chains and their applications, *Biometrika* 57, pp. 97-109.

29. Hobert, J. P., and Casella, G. (1996), The effect of improper priors on Gibbs sampling in hierarchical linear mixed models, *Journal of the American Statistical Association*, Vol. 91, No. 452, pp. 1312-1316.

30. Hosmer, D. W. and Lemeshow, S., (1989), *Applied Logistic Regression*, John Wiley & Sons, NY

31. Hosmer, D. W. and Lemeshow, S., (1998), *Applied Survival Analysis*, John Wiley & Sons, NY

32. James, W., and Stein, C. (1961), Estimation with quadratic loss, *Proc. Fourth Berkeley Symp. Math. Statist. Prob.*, Vol. 1, pp. 311-319

33. Jaynes, E. T. (Author) and Bretthorst G. L. (Editor) (2000), *Probability Theory: The Logic of Science*, Cambridge University Press, Cambridge.

34. Jennrich, R.I. (1995), *An Introduction to Computational Statistics: Regression Analysis*, Prentice Hall, Englewood Cliffs, NJ.

35. Klein, J. P., and Moeschberger, M. L. (2003), *Survival Analysis: Techniques for Censored and Truncated Data, Second Edition*, Springer, New York.

36. Lawless, J. F. (1982), *Statistical Models for Lifetime Data*, John Wiley & Sons, New York.

37. Lee, P. M. (2004), *Bayesian Statistics: An Introduction, Third Edition*, Hodder Arnold, London.

38. Lunn, D.J., Thomas, A., Best, N., and Spiegelhalter, D. (2000), WinBUGS – a Bayesian modelling framework: concepts, structure, and extensibility, *Statistics and Computing*, Vol. 10, pp. 325-337.

39. McCullagh, P. and Nelder, J. A. (1989), *Generalized Linear Models, Second Edition*, Chapman & Hall/CRC, London.

40. Medhi, J. (1994), *Stochastic Processes, Second Edition*, John Wiley & Sons, New York.

41. Metropolis, N.A., Rosenbluth, A.W., Rosenbluth, M.N., Teller, A.H., and Teller, E. (1953), Equations of state calculations by fast computing machines, *Journal of Chemical Physics*, Vol. 21, pp. 1087-1092.

42. Mood, A. M., Graybill, F. A., and Boes, D. C. (1974), *Introduction to the Theory of Statistics*, McGraw-Hill

43. Myers, R.H., Montgomery, D. C., and Vining, G.G. (2002), *Generalized Linear Models: With Applications in Engineering and the Sciences*, John Wiley & Sons, New York.

44. O'Hagan, A. and Forster, J. (2004), *The Advanced Theory of Statistics: Vol. 2B: Bayesian Inference*, Hodder Arnold, London

45. Nelder, J. A. and Wedderburn, R. W. M. (1972), Generalized linear models, *Journal of the Royal Statistical Society, Series A*, Vol. 135, pp. 370-384.

46. Neter, J., Wasserman, W., and Kutner, M. H. (1990), *Applied Linear Statistical Models, Third Edition*, IRWIN, Homewood IL.

47. Pawitan, Y. (2001), *In All Likelihood: Statistical Modelling and Inference Using Likelihood*, Clarenden Press, Oxford University Press.

48. Propp, J. G., and Wilson, D. B. (1996), Exact sampling with coupled Markov chains and applications to statistical mechanics, *Proceedings of Seventh International Conference on Random Structures and Algorithms, Random Structures & Algorithms* 9, pp. 223-252.

49. Propp, J. G., and Wilson, D. B. (1998), Coupling from the past: a user's guide, microsurveys in discrete probability, *DIMACS Series Discrete Mathematics and Theoretical Computer Science*, American Mathematical Society, pp. 181-192.

50. Ross, Sheldon M. (1996), *Stochastic Processes, Second Edition*, John Wiley & Sons, New York.

51. Savage, L. J. (1976), On Rereading R. A. Fisher, *The Annals of Statistics*, Vol 4., No. 3, pp. 441-500.

52. Smith, A.F.M., and Gelfand, A.E. (1992), Bayesian statistics without tears: a sampling-resampling perspective, *The American Statistician*, Vol. 46, No. 2, pp. 84-88.

53. Smith, A.F.M., and Roberts, G.O. (1992), Bayesian computation via the Gibbs sampler and related Markov chain Monte Carlo methods, *Journal of the Royal Statistical Society Series B*, Vol. 55, No. 1, pp. 3-23.

54. Spiegelhalter, D. J., Best, N. G., Carlin, B. P., and van der Linde, A. (2002), Bayesian measures of model complexity and fit (with discussion), *Journal of the Royal Statistical Society, Series B*, Vol. 64, No. 4, pp. 583-639.

55. Stein, C. (1956), Inadmissibility of the usual estimator for the mean of a multivariate distribution, *Proc. Third Berkeley Symp. Math. Statist. Prob.*, Vol. 1, pp. 197-206

56. Swanson, N., Devlin, G., Holmes, S., and Nunn, C. (2007), Very long-term mortality after primary angioplasty for myocardial infarction (PAMI): effects of changing interventional practice, *Heart, Lung, and Circulation*, Vol. 16, pp S1-S201, Elsevier

57. Tierney, L. (1991), Exploring posterior distributions using Markov chains, *Computer Science and Statistics 23rd Symposium at the Interface* Edited by Keramidas, E.M., pp. 563-570

Index

WILEY SERIES IN COMPUTATIONAL STATISTICS

Billard and Diday · Symbolic Data Analysis: Conceptual Statistics and Data Mining
Dunne · A Statistical Approach to Neural Networks for Pattern Recognition
Ntzoufras · Bayesian Modeling Using WinBUGS
Bolstad · Understanding Computational Bayesian Statistics